Scotland

MAPPING THE ISLANDS

PAPA STOUR
or
GREAT PAPA
Papa Sond
Kamna Voe
Deip daill
ZEETLAND
ofte
EASTING
SHEATLAND
By de Hollanders
HITLAND
VELA
Vela Sond
WALS
Norbie
Melbie
Hogsetter
Busta Voe
Bura Voe
Onifirth
SANSTING
Rewyk
Celi Voe
Gilt Rumple
of Goud Start
Bixsetter Voe
Weis dail
TINWAL
Scallowa
Casteel
Kerk
Gwelsta Voe
Laxfort Voe
GULBER WYK
Deals Voe
Scotland
Kabesta
Rode head
de brothers
Grinholm
de Soldins
Scorhead
WEST BURRA
East Burra
Clist Sound
Trondri
S = BURGH
West Quaerd
Fladebuster
Bremer
Helli Nes
Sond
Bresay
De Baert van Bresay
BRESAY
Kirk
Eath Voe
Eath
Gulbins berg
Nos
Nos-head of Hangl p
NESTING
Catfoord Voe
N.Laxfort
Bulsten
Moul
Houstal
Vodders
Hog Neip
of het Hoogh
South Capa
North Capa
Rumll
WHALSAY
Groene Eyland genaemt
Linge
is in a 15 Vadem Water
Luna
Holm
LUNASTING
Colle firth Voe
Swibing Voe
Lax Voe
Snelbeu
Levenp
Lonning
Head
Nes
DELTING
Deal Voe
Scats ta
Gairth
Wahlerholm
Linga
Fisch holm
Rumbe
Glus Eyland
Hail holm
Het groote Voe van Bree
NORD MAV
Roone
Meris Grind
Nibon
Busta
Mangster Voe
Hamer
Booth une
Firth
Kirk
Kirk hous
Hooge lan
Drong
Door holm
Stennes
Eyland
Ur
Vementri
Great
Rooe
Swarbaksmin
Papa
Lith
Chines
Papa
Grinholm
Hilesa
Hillsa

Scotland

MAPPING THE ISLANDS

Christopher Fleet, Margaret Wilkes
and Charles W.J. Withers

in association with the National Library of Scotland

First published in Great Britain in 2016 by
Birlinn Ltd
West Newington House
10 Newington Road
Edinburgh
EH9 1QS

www.birlinn.co.uk

in association with
The National Library of Scotland

www.nls.ac.uk

ISBN: 978 1 78027 351 8

British Library Cataloguing-in-Publication Data
A catalogue record for this book is available on
request from the British Library

Typeset and designed by Mark Blackadder

PREVIOUS PAGE.
Josua Ottens/Reinier Ottens, *Nieuwe paskaard van Hitland met de daar omleggende eylanden…* (1745).

Printed and bound by Livonia Print, Latvia

CONTENTS

POMONA or THE MAIN LAND
HOY
WALLS
SOUTH WALLS
SOUTH RONALDSHA
SHAPINSHA
SCAPA FLOW
the Stream of Tide scarce sensible
HOLM SOUND
WATER SOUND
Deer Sound
Plenty of Cod and Ling along this Coast
The Tide here runs South the first Two Hours of Flood then North Ten Hours viz. till Low Water
Marwick Head
Marwick
Birsa
Costa
Burgar
Eva
Woodwick
Crook
Hall
Gorsness
Apitown
Wald
Burness
Manse
the Loch of Stromness
Stenhouse
Bridge
Yestnaby
Skeel
Holorow
Black Crag
Strömness
Bracknefs Point
Velocity
Bue
Holm
Bow
Gremsa
Clestron
Cubister
Houton
Orfer
Swanbuster
Walk Mill
Wart Hill of Orfer
Wideford Hill
Firth
Cursater
Holm of Grimbuster
Damsa
Severock
Grain
Mount Wyland
Kirkwall
Head
Hill
Scapa
Gatnip
Inganefs Bay
Inganefs Head
Yensta
Tankerness
Saba
Campston
Stembuster
Gremshall
Paplay
Lamaness
Rose Nefs
Brebuster
Sands
Rerwick
Shells
Westshore
Hoyquoy Head
Glimsholm
Hunda
Burra
Ness
Cara
Hope
Widewall
Clets
Newark
Grimnefs Head
Keam of Hoy
Hoy Hill
Rackwick
Bring
Lyrva Burn
Pegal Head
Pegal Burn
Snuck
Mill Bay
Ore Hope
Ore
Air
the Berry
Red Head
Melseter
Long Hope
Fea
Kirk Hope
Cava
Rifa
Fara
Flota
Pan Hope
Rona
Stangar Head
Switha
Cod and Ling Here
Newaholm
Grinds
Crow Taing
Hoxa Head
the Barrel of Butter
Rocky Ground
Wire
Wire Sound
West Ness
Howa Sound
Swine Holm
Gairsa
Galt Skerry
Galt
Viantro
Bora
Tainga
Grass
Vasa
Langa
Strombery
Thieves Holm
Eller Holm
Wilands
Holland
Kirbuster
Sound
Sandstown
Foot
Hacks Ness
the Greenholm
Warness
Eda Groina
Rocky

FOREWORD

On the island of Orkney, where I was born, history is written in the maps drawn up by its many invaders. Maps showing its ancient stone circles, maps showing its Viking and Pictish remains, early Christian maps, and then the painstaking maps of the eighteenth century, which are the only record we have of places and buildings which are today only ruins. In my 1946 *Inventory of Orkney*, one of a three-volume set produced by the Royal Commission on the Ancient and Historical Monuments of Scotland, there is a map from 1750, of the Parish of Birsay, close to where I grew up. It shows a coastline, now almost unrecognisable because of erosion, the Earl's Palace, with its gardens, the church and the old manse, a bridge, the broch of Birsay, and a ruined building with a nine-foot-long grave marked beside it. Not much of this remains today. But it is there on the map. It existed, and it tells us about our past.

Scotland's islands were made for maps. No modern-day sailor would attempt to thread his way through the channels, skerries, tideways or shoals of the Hebrides or the Northern Isles without the most detailed charts; and so it was for their predecessors. Maps were not just a source of information, they were a way of avoiding a watery grave. When, sometime in the 1960s, my father and brother brought a skiff up from Tarbert, Loch Fyne, meaning to pass through the Caledonian Canal to Inverness, they were guided only by an AA road atlas, which, not surprisingly, failed to include the Gulf of Corryvreckan, north of Jura. The gods were kind to them, and they survived. What they needed was a decent sea map. Their nautical predecessors knew this better than they did.

The pages of this wonderful guide to the historical and geographical nature of Scotland's islands tell us as much about our country as many an academic treatise. We can see how place names stretch back to the earliest times, how our ancestors imagined the shape of the islands long before they were able to define them, and, as maps evolved, we understand how the art and science of cartography developed. Names slide back into ancient times: Glasco, Sterling, Paslei, Dombroton, Gallovidia, Donfermilg, Lacus Levinus, Aberdonia – these we can decipher. But whatever happened to Moravia, Maria, Buthania, Arana or Argadia? – names that have evolved so far from their ancient roots that they are unrecognisable today.

Maps were produced not just by navigators but by those for whom property was important, boundaries commercially

Opposite. R. Sayer & J. Bennett, *A new chart of the north coast of Scotland with… the Orkney Islands* (1781).

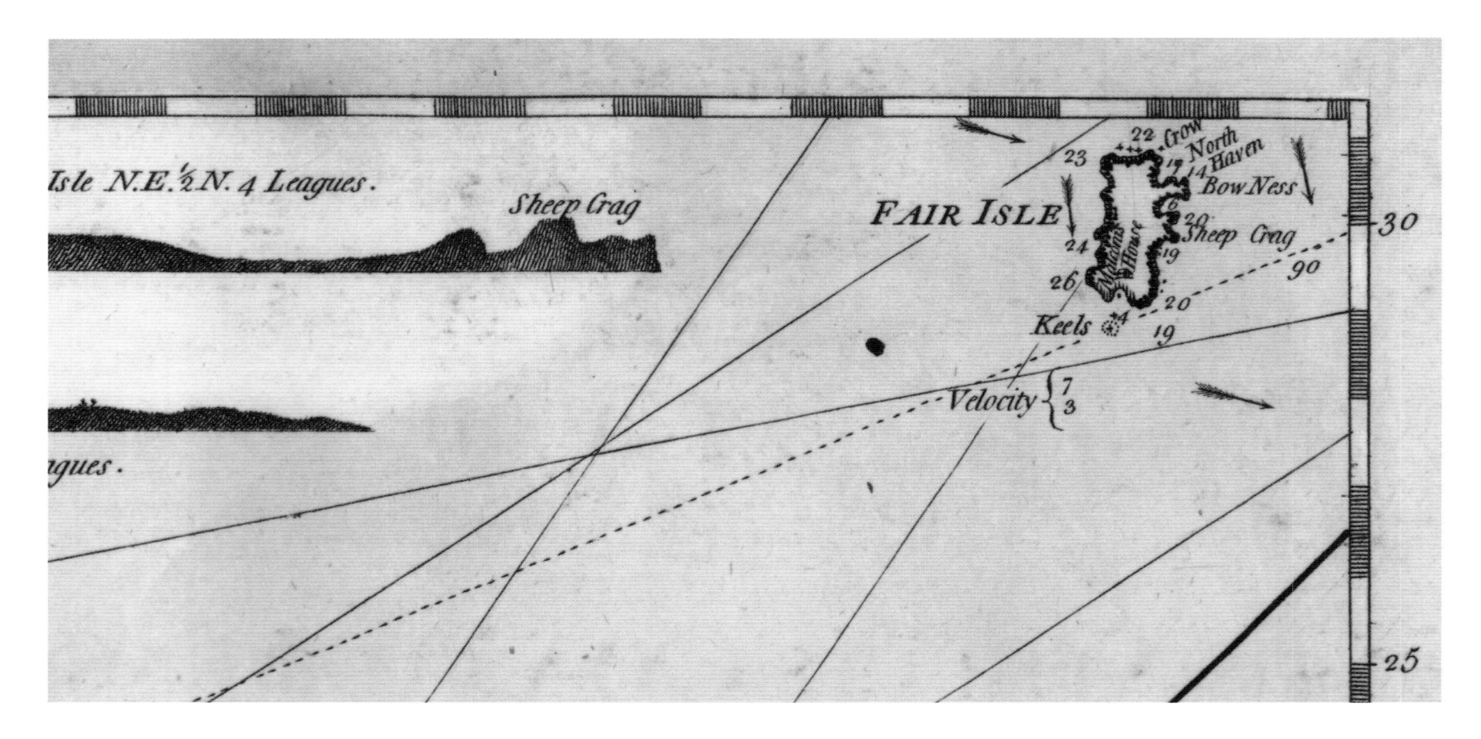

sensitive, and the limits of kingdoms or regions a vital part of the exercise of power. The Lords of the Isles, who ruled most of the Hebrides in the fifteenth century and owned more land than the kings of Scotland, needed to know precisely where their territory began and ended. In the eighteenth century, some of the most beautiful maps of Scottish land were drawn up by estate owners, not just to boast of the extent of their holdings but to pin down who owed them rent and how far their sheep should be allowed to roam.

When, in 2014, the Saltire Society named *Scotland: Mapping the Nation* its Scottish Research Book of the Year, it did so in the knowledge that that book was breaking new ground in charting the emerging shape and nature of the country. Now its authors have turned their attention to the islands that lie off Scotland's coast. What emerges is a fascinating exercise, not just in understanding Scotland's relationship to its remoter parts but in reminding us of the romance of the past: Columba landing on Iona, Bonnie Prince Charlie escaping to Skye, the crofters of the eighteenth century leaving for America, the final abandonment of St Kilda. 'There are people who find islands somehow irresistible', wrote Lawrence Durrell. 'The mere knowledge that they are on an island, a little world surrounded by the sea, fills them with an indescribable intoxication . . . they are places where different destinies can meet and intersect in the full isolation of time.'

Here, in these pages, are contained not only the isolation of islands, their magic and their mystery, but also the way that map-makers, from the earliest times, have striven to link them to the mainland, to make them part of our nation's history and to define its present shape.

Magnus Linklater
Edinburgh
August 2016

PREFACE AND ACKNOWLEDGEMENTS

Maps fascinate. That they do and do so in ways beyond their seemingly obvious utilitarian function – 'this is how I get from here to there', 'so this is what this place looks like' – is clear. Yet it is less clear quite why. Most maps are visually delightful. It takes real skill to represent parts of the world in reduced form, to show the shape of the land and, using a language of symbols, lines and colours, to turn map readers into map users. The paper landscape revealed to us is a work of art. But all maps are also visually deceitful. The skill in showing the world or parts of it in stylised form is a skill of selection, a matter of omission, even of misrepresentation, for how can one show all that is there and in the correct proportions? The paper landscape revealed to us is thus also always a work of artifice. That is not to say that maps are not 'accurate', that they are not truthful documents. We trust in maps to be faithful to the world. It is to say that, rather than taking maps to be things which simply reflect our world, it is more appropriate to understand maps as documents and devices which help produce our understanding of the world. From that understanding, actions follow: prompted by maps, we engage with the world in ways which may require new and different maps.

Maps are works of and about geography. They show, in a variety of ways, parts of the world. Maps are also works of history and objects which themselves have a history. Today, we can go into bookshops and expect to find a range of maps for sale: guidebooks to cities elsewhere, in the form of atlases, as bus or tram maps, as more detailed series-based maps for hill walking, and so on. Maps are no longer just paper things. They appear on mobile electronic devices and portable navigation systems, the key locations symbolised and even 'clickable', the physical landscape reduced to colour shades without a contour line in sight. In the past, too, maps were not just paper things: cloth maps hung on walls, 'dissected maps' – jigsaws in modern parlance – taught eighteenth-century children Britain's place in the world. But where we moderns encounter maps in one form or another often, that was not so of our forebears. Maps were relatively few in number, were seen more by kings and courtiers than by the common folk, and some maps were documents of state power and political dominion.

Islands fascinate too. This book looks at the history and geography behind the many maps of Scotland's islands. Our focus is not just in understanding the history and the geography of the islands themselves but also in showing how that history and geography has been realised in and produced through the art, the artifice and the authority of maps. It is the first book to take the maps of Scotland's islands as its central focus. As importantly, our concern is with the mapping behind the maps – that is, with the several technical, political, institutional and artistic processes by which Scotland's islands have been at various times and in various ways made real in map form. Knowing where things are in geographical space is a fundamental need. Knowing what relationships in space one feature has with another is no less important. These issues are perhaps even more important for maps of islands given that, by definition, islands are land surrounded by water. Maps of islands, we might presume, should have always the compelling requirements of accuracy of location and dimensions if not of their interior content. As we show, however, Scotland's islands did not so readily come into view. Accuracy was always something striven for, seldom achieved. Some islands refused to be easily positioned. Others were positioned but not given their proper dimensions. For yet other islands, theirs is a geography of transience, a twice-daily emergence in certain tidal conditions. How does one show on a map what is there but not always visible? Quite what map makers took the purpose of their map to be, and so how they showed islands, will be shown to differ greatly. In a very profound sense, Scotland's islands have been powerfully shaped – on paper and in the imagination – by being produced in map form. And, in turn, knowing how to show islands has been a significant element in Scotland's map and mapping history, notably of charts and hydrographic mapping. Distinguishing sea from land, one island from another, what is an island and what not, is vital for map makers and users alike.

It is a pleasure to record our intellectual debts to the many people who have assisted in the writing and the production of this book. Hugh Andrew of Birlinn encouraged us in the wake of our *Scotland: Mapping the Nation* to consider an islands-based book and never lost faith even as other demands upon our time and our own doubts seemed at times to make this project unrealisable. We trust that his belief in us has not been diminished by what we have produced. Andrew Simmons, Mairi Sutherland and the staff at Birlinn have been patient and good-humoured and it is a pleasure to again benefit from the skills of Mark Blackadder for the book's design. We are grateful too for the splendid Foreword by Magnus Linklater. The argument of the book is heavily dependent upon its many map images. Finlay MacLeod kindly read our 'Naming' chapter in draft and we are grateful to him. John Scally, National Librarian and Chief Executive of the National Library of Scotland, has been supportive and encouraging throughout. We hope that as a joint production the book will serve not only to strengthen the link between the National Library and Birlinn, but also to promote the Library's strategic commitment to research in works which will make further use of its rich and important collections to engage audiences in Scotland and beyond. Above all else, it is our hope that readers will come to regard Scotland's islands in a new light and, more, that they will appreciate Scotland's map histories as a way of understanding its islands' geographies.

Image credits

Unless stated otherwise, all images are reproduced courtesy of the National Library of Scotland. We gratefully acknowledge all other sources for non-NLS map images in the captions for the respective individual figures.

Place names

A note on place names is necessary. In the last few years, the Ordnance Survey's Gaelic Names Policy has emerged as part of its role as Britain's mapping agency and through the Ainmean-Àite na h-Alba (AÀA) partnership (Gaelic Place-Names of Scotland). This policy both supports and is part of the greater presence and role of the Gaelic language in Scotland evident in the Gaelic Language (Scotland) Act of 2005, the National Gaelic Language Plan 2012–2017, and, of course, in the use of Gaelic by the 87,506 persons in Scotland who declared a competence in understanding the language in the 2011 Census (57,375 to speak Gaelic; 38,409 to speak and read it; 32,191 to speak, read and write the language). Although Gaelic speakers are found throughout Scotland, the heartland of the language remains the Highlands

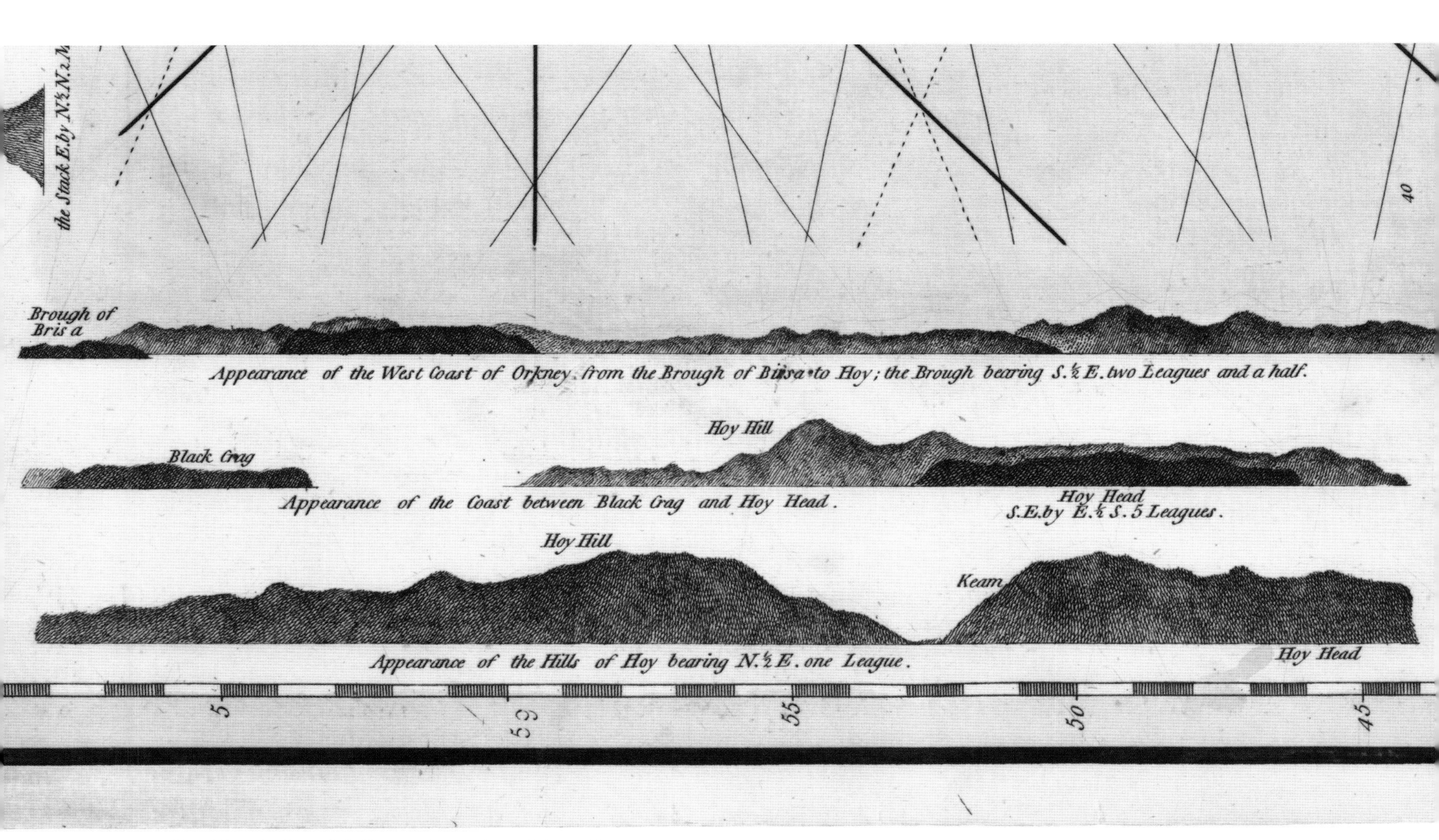

Detail from R. Sayer & J. Bennett, *A new chart of the north coast of Scotland with… the Orkney Islands* (1781).

and, particularly, the western islands. In its maps of the Western Isles (Na h-Eileanan an Iar), Ordnance Survey has put the policy into practice by producing Gaelic-language maps with Gaelic and/or English being used for names in the natural environment, the man-made environment, and, where appropriate, for administrative divisions. We follow this official policy by giving the English and the Gaelic name for all those features of the natural, man-made, or administrative environments located in the Western Isles which are the subject of discussion here (thus Western Isles/Na h-Eileanan an Iar, Lewis/Leòdhas, and so on). Where reference is made to Gaelic place names and name elements in other parts of Scotland, we note only the use of the language and the meaning of the named feature in question but do not give the twin name form as above.

Strauer
Dorno
Lindorna f.
Nardus f.
Canoria
Sinus salutis
Hispana f.
Nessa f.
Emuernes
Elgen
MORAVIA
ROSSIA
Banf
BUTHANIA
Buzhain
Slanis
IONA
Rozhmai
Spea f.
Stermaggi
Donas f.
MARIA
Aberdonia
Dea f.
ILA
BRA
Couc
Dononer
LOQUABRIA
GRAMPIUS
IAI BAC
MARNIA
Loreston
Dunkel
Taus f.
Eska f.
Brechin
ARGADIA
Erna f.
S. Ioanes
ANGUSIA
Dodesia
Lacus leuinus
Dumblain
Lacus lomunde
ALBANIA
Porthea
Donfermilg
FIFA
Kinghorn
Monros
S. Andreas Metropolis
Avana
Lacus
Dombroton
Sterling
Glasco
Lizhco
Bas
Air
Edingburg
Hamelton
Paslei
GALLOVIDIA
Dumbar
Dus
MARCIA
Kelson
Solvus f.
Tueda f.
Beruicum
Creus f.
Galcis
Ken f.
Nus f.
Sinus
S. Ninianus
Wigton
Kirkobro
Dunfres
Norham
Bambrog
Insula Sacra
Anwik

CHAPTER ONE

INTRODUCTION

> I have noticed from the study of maps
> The more outlying the island –
> The further out it is in the remote ocean –
> The stronger the force that pulls us toward it.
>
> David Greig

David Greig's play *Outlying Islands* captures wonderfully what we know and feel variously as the 'lure of islands'. As Robert, one of the lead characters, puts it in the play's opening lines, 'I have noticed that something draws us towards outlying islands. Some force pulls.' The allure of islands is in part because they seem 'far off' – not necessarily in a geographical sense (though that may be so) but that they are somehow 'other', separate from our point of reference certainly, but also not of it by type, character, in their essential features. And when islanders appear, they seem also somehow doubly resourceful – if they are not being depicted as stereotypically exotic or unknowing of the 'real world' – their island homes at once a refuge and a responsibility only to themselves. Island life, wherever and whatever it is, seems altogether less urgent, less immediate.

Although they exert this attractive pull, islands may not always be easily distinguished, even when they are solid and inhabited places – and not all are, of course. On summer days, distant islands seem to move in the air, to float free of surrounding seas, to shimmer in the heat. In poor weather, as driving rain and ocean spray make solid and liquid hard to tell apart, islands can appear alarmingly indeterminate. Yet knowing precisely what is an island and what is sea or water matters, for sailors, for islanders and for others. The making of maps of islands, and the careful study and use of them, can be a matter of life and death.

Islands pull more people than just playwrights, of course, and do so in diverse ways. Think of desert islands, of islands as a site of danger, intellectual retreat, solitary confinement, laboratory experimentation, of the island as paradise – each a stock resource in cartoons, literature, politics, science fiction and science fact, and tourist brochures. All these images, to generalise, are 'outsiders' views'. To insiders, islands are home: livings have to be made there even if that involves trips to 'the mainland' or further afield. Island life is simply what it is, neither slower nor more exotic than anywhere else. To those

Opposite. Detail from Paolo Forlani/George Lily, *Scotia: Regno di Scotia* (*c*.1561).

who live on them, islands are much more than the objects of others' representations.

Maps similarly have powerful allure. The attraction of maps rests, in part, in that they 'collapse' time – here, we are invited to consider, is what this place, landscape, or island looked like at some point in its past. That is in one sense true: the discipline of map history to an extent rests on the study of maps as documents recording 'lost' and present geographies. Past geographies are seldom wholly lost, of course. Relict features from earlier times survive today or in whichever historic present is signalled by the date of the map's making or its publication. In another sense, the lure of maps as records of the past is just that – a lure. Vitally important as they are in so many ways, maps are much more than the objects of others' representations.

Maps are not mirrors to the world, but stylised versions of it. They symbolise geographical complexity: think of contours, spot heights, dots of different sizes to depict cities, towns and villages, and so on. Symbolisation is used because it is impossible to do otherwise. Maps cannot easily be produced at the same dimensions as the objects they purport to represent, so they use a language of scales and symbols to reduce and simplify, to omit, to misrepresent even. Maps may wonderfully convey the lie of the land, but they may also lie about the land, selectively depicting it to suit the map maker's purpose. Maps are, in a twin sense, power-full documents. Because they do not show the world in complete form, maps reflect particular intentions in what they show 'within' themselves (even if they do not advertise how or why those features and not others have been shown). And because maps prompt action, they always produce an effect 'beyond' themselves – this symbol we take to mean 'steer clear of shallow waters', these to be read as rocks or hills or marking safe passage, and so on.

Useful as they are as tools for finding our way, for locating places, and as geographical and historical documents, maps also have a history. Map history recognises that what may appear as straightforward questions – 'What is a map?' 'What does a map do?'– may not so readily produce straightforward answers. This is not only because maps come in different sorts and are produced in different ways and with different intentions. What were once engraved, printed and published as paper documents, either as single sheets or in atlases, are now commonly available digitally on portable electronic devices and satellite navigation systems. Maps have a cultural currency today – appearing in a variety of forms (fig. 1.1 for example) – that they did not so commonly have in the past. It is also because the purposes behind maps are so varied, and because those different purposes need to be understood in terms of the culture in which the map was made – in other words, its originating context, geographically, historically, intellectually, politically.

For these reasons, it is helpful for the more detailed discussion and illustrations which follow to present, in outline, a map history of Scotland's islands and to consider the different ways in which islands have been mapped.

Towards a map history of Scotland's islands

Both Scotland, and the map of Scotland, are relatively modern inventions. Scotland the Country only becomes so in geographical and historical terms following the incorporation of the Northern Isles into Scottish administration from about 1470, Orkney and Shetland having been under Norse rule before then as were the Western Isles/Na h-Eileanan an Iar until the late thirteenth century. Scotland the Nation, or, perhaps more properly, Scotland the Kingdom – that is, in outline, what contemporaries understood Scotland as a realm to be – first appears in the mid sixteenth century. When Scotland does appear as the central feature in maps, its islands in particular are, in the main, geographically imprecise in the eyes of their map makers (fig. 1.2). When Scotland is mapped in 1595 by the Flanders-based map maker Gerard Mercator, Europe's if not also the world's leading map maker at that time, its islands are more recognisably shaped and positioned to the modern eye: several have settlements and other locational features (fig. 4.1a). Mercator's map of 1595 was in part based on a map of 1583 by the French cosmographer and map maker Nicolas de Nicolay (fig. 1.3). The nation is 'framed' in these works, its bounds delineated. But they are not complete works or, in our terms, 'accurate' maps. Nor should they be read as such: modern assumptions about planimetric accuracy and completeness of view are not appropriate in understanding the maps of this period.

Several points may be made of this evidence. Scotland's islands appear on maps by the later 1500s as part of the shaping of the nation as a whole, rather than as the object of particular focus. Scotland is mapped – as a geographical entity, a kingdom and a nation – because of the growing importance of maps as political and symbolic documents in association with the birth of the then 'modern' European nation state: territorially defined, ruled by a monarch, administered by a developing civil service and protected by a standing army. These maps may have been how Scots and others saw themselves, geographically speaking, although few persons outside the court and parliament would have had sight of them, but they were not the work of Scots themselves. This sense of maps providing an outsider's view, looking at the shape of Scotland's islands but not in detail at their content, is apparent in an early map of Lewis/Leòdhas and Harris/Na Hearadh whose origins are obscure but which may have to do with state relations and island politics following the birth of Britain as a united kingdom in 1603 (figs 1.4 and 1.5).

At the same time, some maps – more properly, sea charts – did feature islands as part of their central focus. The *isolario*, a map or collection of maps designed to show islands often in pictorial, even stylised ways, has its origins in fifteenth- and sixteenth-century Italian map making as a kind of illustrated island guide initially for the use of sailors and merchants in the Mediterranean. In intention, although not always in physical form, it is closely related to the portolan chart, derived from the Italian *portolano*, a form of chart for seafarers. In portolan charts, coasts were shown with the main useful features such as ports and harbours labelled, often at right angles to the coast in order to give clarity to the feature or its location.

FIGURE 1.1
The association between the cultural 'power' of the map and island sources of quality dairy produce is clear from this milk carton showing, in profile, a dairy cow with the Inner Hebridean islands of Mull, Tiree and Coll in outline form on its flanks. Although this is, strictly, an item of 'cartographic ephemera' ('cowtography'?), the commonly recognised shape of the islands is presumed sufficient to carry the marketing message.
Source: One Pint Milk Carton (*c.*1992). Reproduced with the permission of J. and C. Reade & Sons, Sgriob-ruadh Farm Dairy, Tobermory, Isle of Mull.

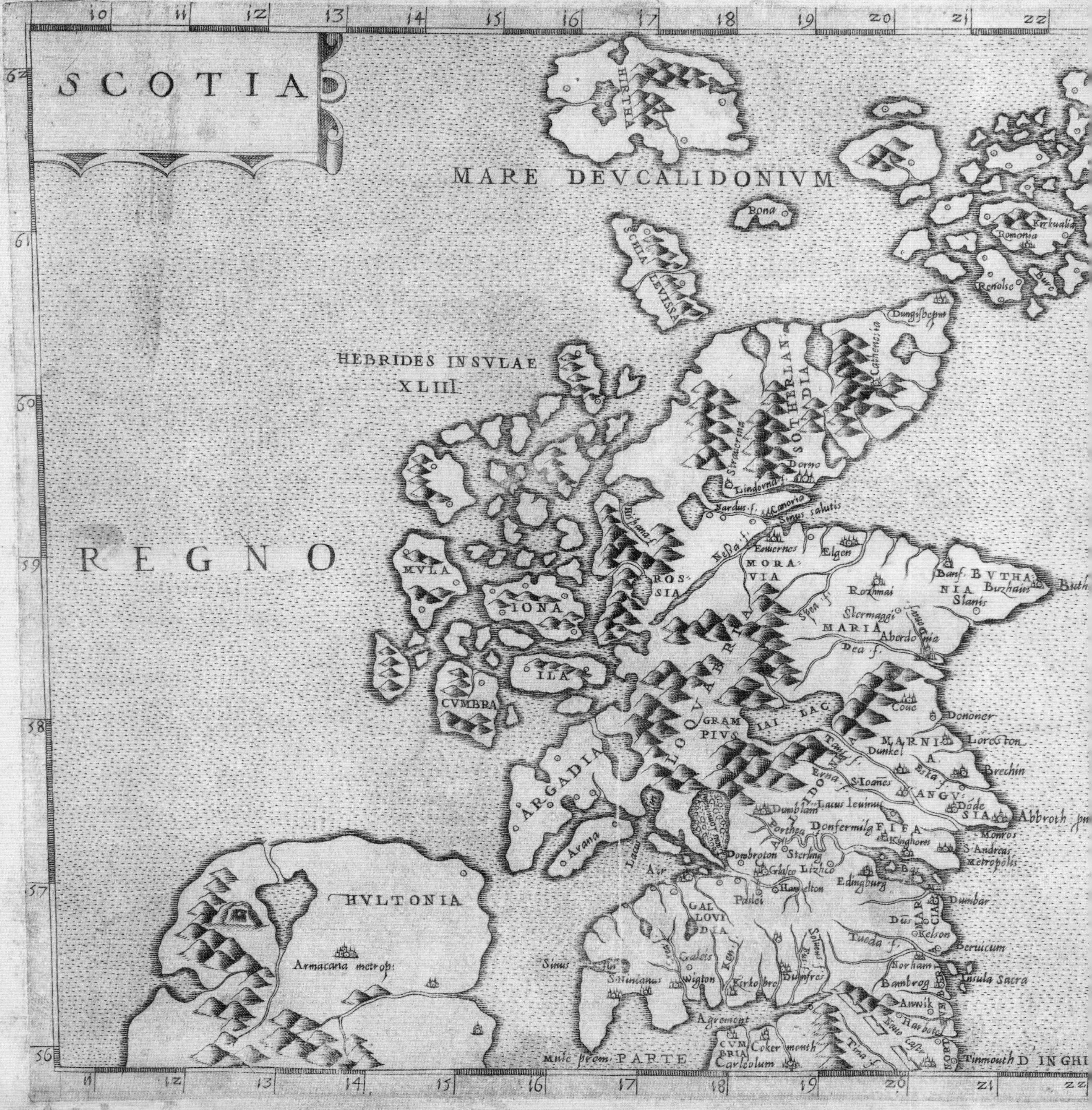

SCOTIA

MARE DEVCALIDONIVM

HEBRIDES INSVLAE XLIII

REGNO

HIRTHA

Bona

SCHIA

LEVISSA

Kirkualia

Romonia

Renolse

Dungisbepnt

SOTHERLAN DIA

Cathenesia

Strathorma

Dorno

Lindorna f.

Nardus f.

Canovia

Sinus salutis

Bishana f.

Nessa f.

Enuernes

Elgen

MORA VIA

ROS SIA

Banf.

BVTHA NIA

Buzhaus

Slanis

Buth

Rozhmai

Spea f.

Skermaggi

MARIA

Aberdonia

Dea f.

MVLA

IONA

ILA

CVMBRA

Coue

Dononer

GRAM PIVS

LOQVABRIA

IAI BAG

MARNI A.

Loreston

Dunkel

Erka f.

Brechin

Tauus f.

Erna f.

S. Ioanes

ANGV SIA

Dodo

Abbroth

ARGADIA

Avana

Lacus leuinus

Dumblam

Porthea

Donfermilg

FIFA

Kinghorn

Monros

S. Andreas metropolis

Dombroton

Sterling

Glasco

Lizhco

Air

Hamelton

Edingburg

Paslei

HVLTONIA

Armacana metrop:

GAL LOVI DIA

Dumbar

Dus

MARCIA

Kelson

Tueda f.

Beruicum

Sinus

Galeis

S. Ninianus

Wigton

Kirkobro

Dumfres

Norham

Bambrog

Insula Sacra

Anwik

Harbote

Agremont

CVM BRIA

Coker month

Carleolum

Tina f.

Tinmouth

Mule prom.

PARTE

D'INGHI

FIGURE 1.2
(*Left*) *Scotia: Regno di Scotia*, which dates from about 1566 and possibly from 1561, is thought to be the work of Paolo Forlani, a Veronese map maker working in Venice. It is also based in part on a map of 1546 by an English cleric, George Lily, who may have had access to continental European geographical sources during his time as a Catholic exile in Rome in the wake of the Reformation. There has been some careless copying, probably by Forlani: Arbroath is north of Montrose, for example, and several names which appear on Lily's map are omitted here. The depiction of the islands repays close attention, not least for the relative positions and sizes of the principal islands. Shetland is not shown. The Orkneys are shown, and, although they are not accurately delineated, are more correctly positioned in relation to the mainland than either the Outer Hebrides/Innse Gall or, especially, Skye and Lewis ('*Schia. Levissa*') off the north-west mainland. Iona is out of proportion to Mull (as it is to all other islands) and shown at the east, rather than the west, of that island. Cumbra – the Cumbraes in modern context – is out of proportion and wrongly positioned. Most errant of all is 'Hirta', the main island in the St Kilda group, here shown to the top of the map as Scotland's largest single island. Lily's reliance upon written sources, with their tendency to emphasise certain places and features, particularly distant islands or religious centres, may partly explain some of these obvious 'distortions' in the map to modern eyes.
Source: Paolo Forlani/George Lily, *Scotia: Regno di Scotia* (*c.*1561).

FIGURE 1.3
(*Overleaf*) Nicolas de Nicolay was a widely travelled French geographer and army officer of the mid sixteenth century and author of several regional descriptions of France as well as a treatise on navigation. His chart of 1583 was based on a circumnavigation of Scotland in the 1540s (see also fig. 6.14) undertaken at the behest of King James V of Scotland and under the command of Alexander Lindsay (sometimes 'Lyndsay'). Nicolay's chart was reproduced almost two centuries later when John Adair copied it closely as part of his own endeavours to chart the coast and islands of Scotland, a story explored in more detail in chapter 4. It is Adair's copy that is reproduced here.
Source: John Adair, *A True and Exact Hydrographical Description of the Sea Coast and Isles of Scotland* (Edinburgh, 1688).

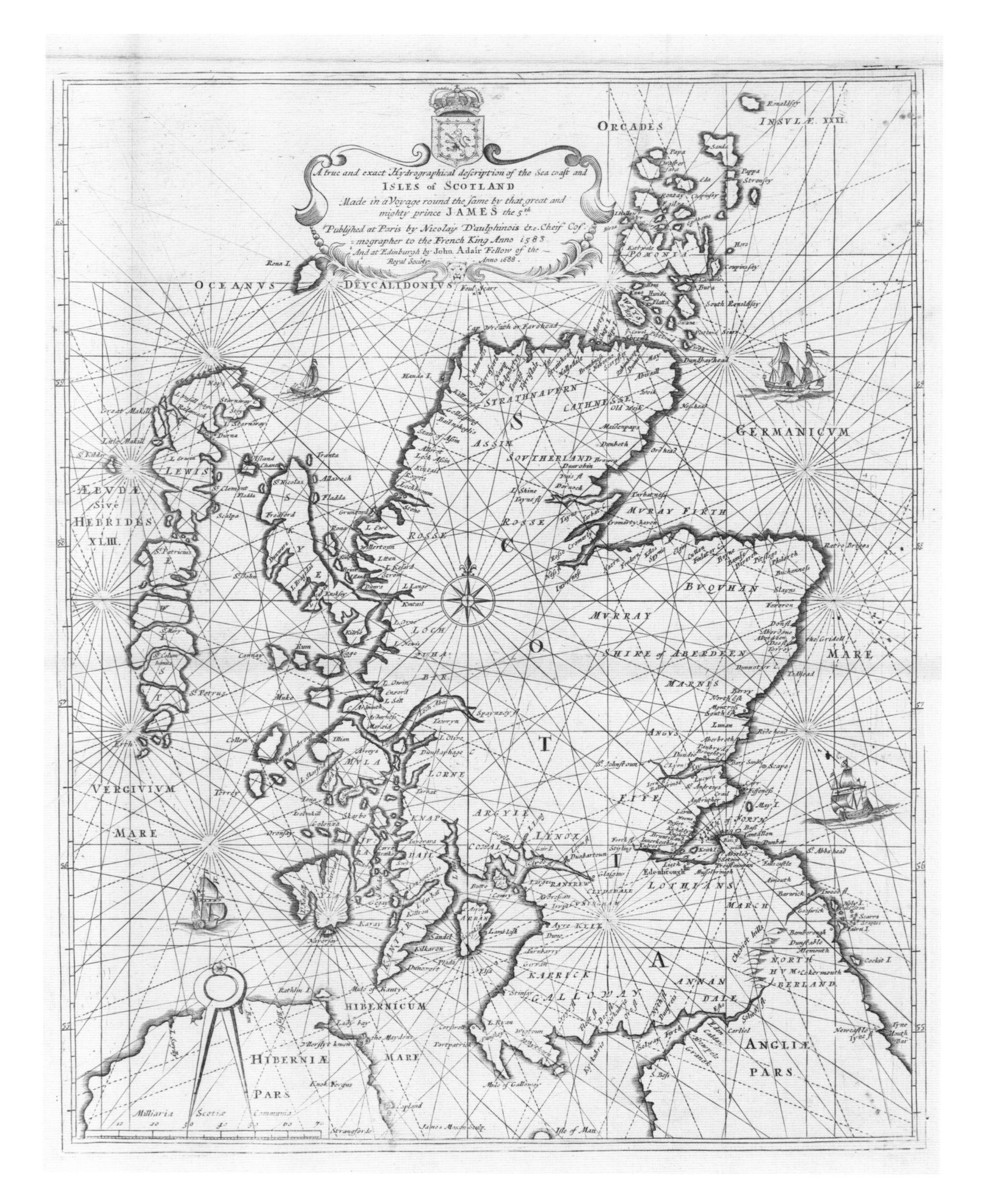
A true and exact Hydrographical description of the Sea coast and
ISLES of SCOTLAND
Made in a Voyage round the same by that great and mighty prince JAMES the 5th
Published at Paris by Nicolay D'aulphinois &c. Cheif Cosmographer to the French King Anno 1583
And at Edinburgh by John Adair Fellow of the Royal Society Anno 1688.
ORCADES
INSULÆ XXXI
POMONIA
OCEANUS DEUCALIDONIUS
Rona I.
Foul Scarr
ÆBUDÆ Sive HEBRIDES XLIII.
LEWIS
STRATHNAVERN
CATHNESSE
ASSIN
SOUTHERLAND
ROSSE
MURAY FIRTH
BUQUHAN
MURRAY
SHIRE of ABERDEEN
MARNIS
ANGUS
FIFE
LORNE
ARGYLE
LYNOX
KNAP
CONAL
CLYDSDALE
LOTHIANS
MARCH
KYLE
KARRICK
GALLOWAY
ANNAN DALE
NORTH HUMBERLAND
GERMANICUM
MARE
VERGIVIUM MARE
HIBERNICUM
HIBERNIÆ PARS
HIBERNIÆ MARE
ANGLIÆ PARS
S C O T I A
Edenbourgh
Glasgow
Dunbar
Berwick
Isle of Man
Mule of Galloway
Milliaria Scotiæ Communia

FIGURE 1.4
This anonymous and untitled map, drawn in ink on paper, is thought to date from the early seventeenth century, possibly from *c.*1610 and maybe as late as 1630. Although it is unsigned, there are clues as to its origin and purpose. The fact that there is considerable coastal detail and little shown of the interior suggests it was made by a seaman, possibly an Irish sea captain or soldier. Above the scale bar a sea creature, perhaps a sea otter or, it has been speculated, the *onchú* (in Irish, a heraldic device or animal associated with Ireland, as were the lion and the unicorn for England and Scotland) holds aloft a flag emblazoned with a harp, another Irish symbol. Irish forces opposed to Elizabethan rule in Ireland received support from Hebridean island clans in the late sixteenth century, and a force loyal to the Queen was sent from Ireland in 1615 to oppose the interests of Angus Og MacDonald on Islay (who was that year executed for treason). It is also possible that the map was associated with attempts at this time to establish a more commercial base for fishing in the Western Isles/Na h-Eileanan an Iar. The principal features of the coastline are identifiable, some less than others: 'The Piggmes of Ness', for example, is Eilean Luchraban, a small island near the Butt of Lewis. Inland, the only features marked apart from 'The Eille forrist' and 'The Great Fforrist', are churches, four of the six identified being named. It is interesting to compare these named places and the map outline with contemporary printed maps (cf. figs 1.5 and 3.2).
Source: Anon., Map of Lewis and Harris (*c.*1610–30?). Reproduced with the permission of the National Library of Ireland.

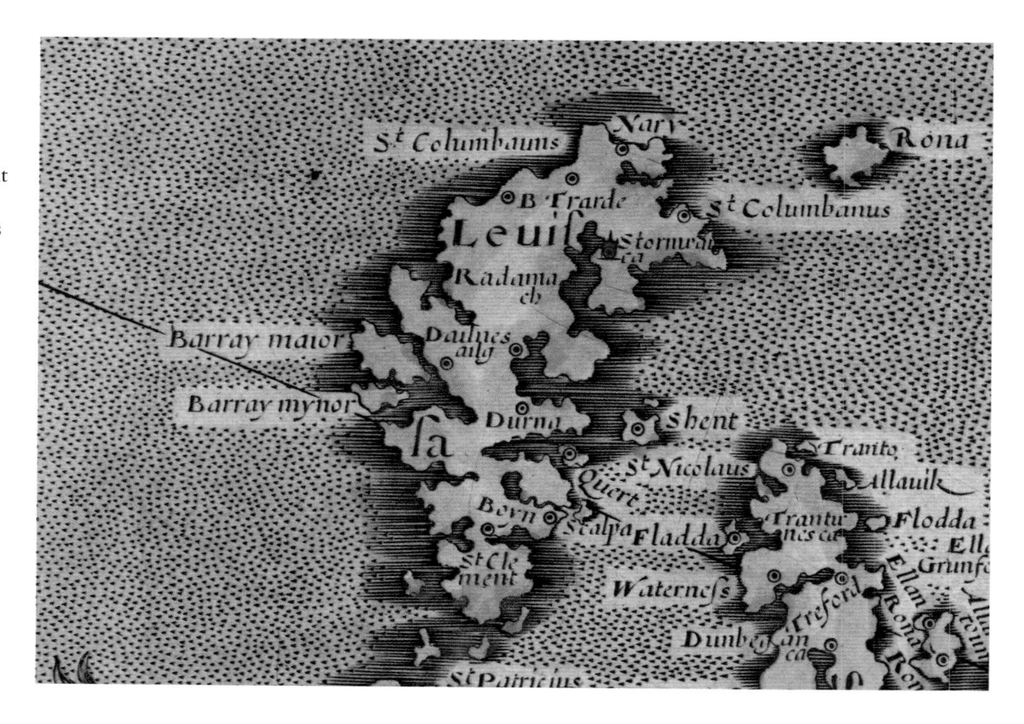

FIGURE 1.5
This detail is from the first map of Scotland to be included in William Camden's *Britannia*, the major topographical and historical work of this period. The map was engraved by William Hole, a versatile portrait engraver based in London, who copied the detail and outlines primarily from Mercator's 1595 map of Scotland (fig. 4.1a).
Source: William Hole, *Scotia Regnum* (London: Bishop & Norton, 1607).

Navigational lines or lines of constant compass bearing would also be included – rhumb lines or loxodromes as they are called – as an aid to practical navigation. Hazards to navigation would commonly be marked as dots or crosses. Scotland appears in several portolan charts, splendidly if confusingly in one example dating from about 1560 by a Cretan ship's master (fig. 1.6 and pp. 32–3).

By the late sixteenth century, Scots were mapping their nation and kingdom and islands for themselves. The surviving map work and geographical descriptions produced between about 1583 and 1611 by the Scottish chorographer and clergyman Timothy Pont have left us a uniquely important picture of Scotland in seventy-eight manuscript maps on thirty-eight sheets. Yet his coverage is incomplete. It is not known whether he travelled throughout the Western and Northern Isles. Only one map in detail of an island survives, that of South Uist/Uibhist a Deas (fig. 1.7). Pont's work is significant for several reasons and features in the following chapters in different contexts. It is important to recognise, however, that his work was not published in his lifetime. When his maps and texts did appear, in volume V of Joan Blaeu's *Theatrum orbis Terrarum sive Atlas Novus* (1654), several had been supplemented with the addition of work from Blaeu and from the geographer-map maker Robert Gordon, and his son James Gordon, minister of Rothiemay. The publication by Blaeu of this work reflected the leading role occupied by Low Country map makers in cartographic production. Scottish islands feature in the work in different ways (fig. 1.8). The fact that the work was published first in Latin – later editions appeared in Dutch, French, Spanish and German (but in English not until 2006) – indicates that this work was the preserve of a literate few rather than being in the public domain, as maps more commonly are now.

By the end of the seventeenth century, maps were more widely available in the public sphere and more widely understood, in Scotland and elsewhere. There are several reasons for this. Maps were being produced for a range of more specialist needs. By this period, three interrelated 'modes' of

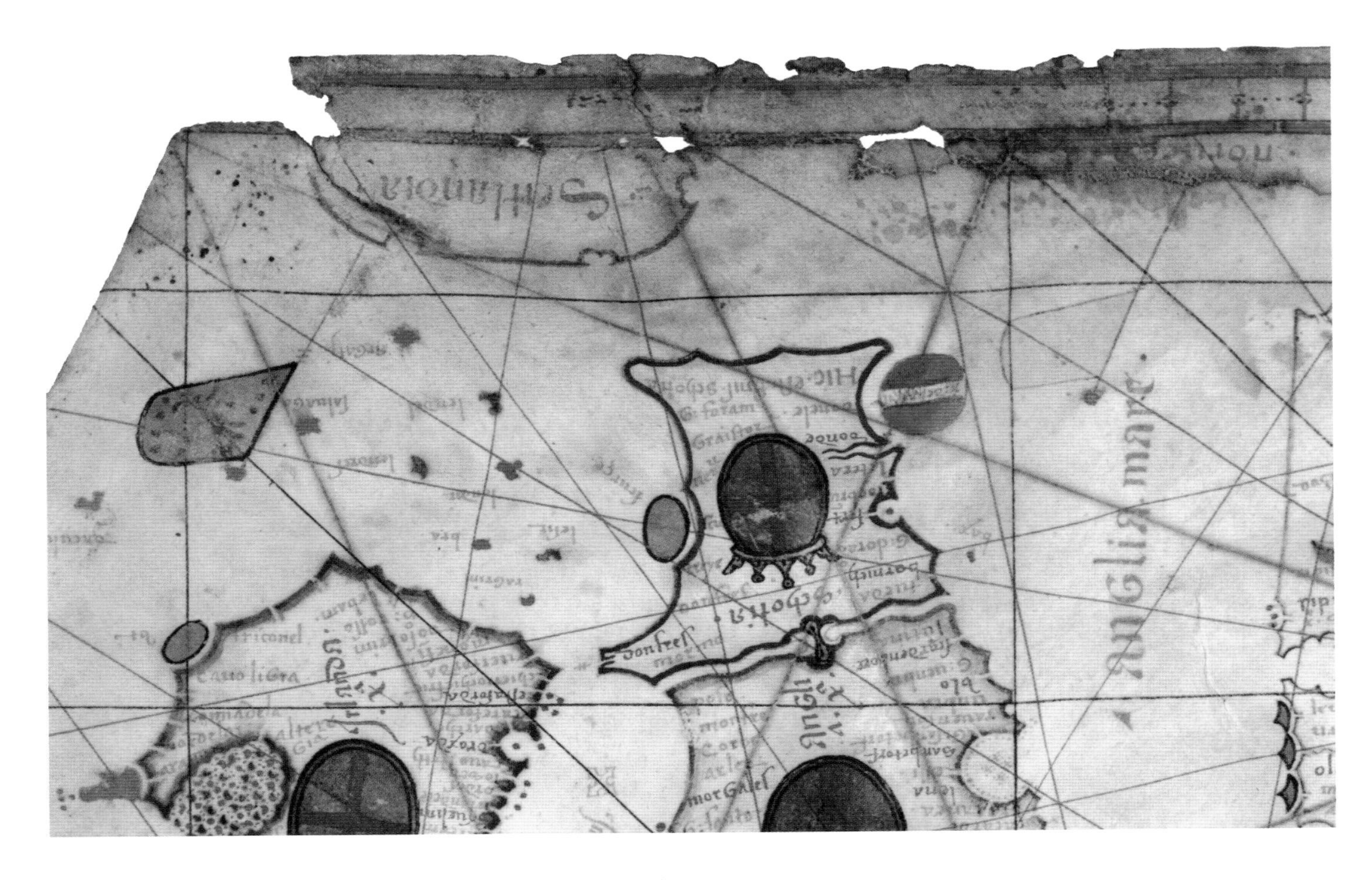

FIGURE 1.6
This detail of Scotland and its outlying islands is taken from a portolan chart of the Mediterranean drafted by Georgio Sideri around 1560. Sideri, often called 'Calapoda' or 'Callapoda' after the name of the town on Crete where he was born, shows Scotland connected to England by a sort of bridge or 'clasp' between the two nations. At Scotland's north-east tip is the mythical island of 'till' marked like a 'No Entry' sign. Islands off the west coast of the Scottish mainland are marked but with no pretensions of accuracy as to their size, position relative one to another, or name. It is likely that the feature labelled 'Schlandia' to the top of the map frame is Shetland but we are given no clue as to the identity of the un-named medallion-like island to the west of the mainland or of the near-rhomboidal island further west. On the mainland, other features are given: the rivers 'Tueda' (Tweed), 'fert' (Forth), and 'latara' (Tay) are shown as well as the ports of 'bernith' (Berwick), 'donde' (Dundee) and 'donfres' (Dumfries). *Source*: Georgio Sideri (Calapoda/Callapoda), Portolan Chart of Europe (*c.*1560?).

map making may be discerned: maritime charting; small-scale mapping of the world or of its various regions and countries (a form of chorography, itself part of general geographical description); and large-scale terrestrial survey. Large-scale maps were increasingly used in urban planning and in estate management. Small-scale maps of the world began to influence public and scientific understanding of the dimensions of nations, continents and outlying islands. Explorers, merchants, travel writers, naval officers and government officials alike all had recourse to maps as a means of directing, and reflecting, the endeavours of navigators and cartographers in putting the world to shape. People became more map literate. Increasingly, maps were seen less as symbolic, even artistic, ways of representing the world and more as realistic depictions of it (maps

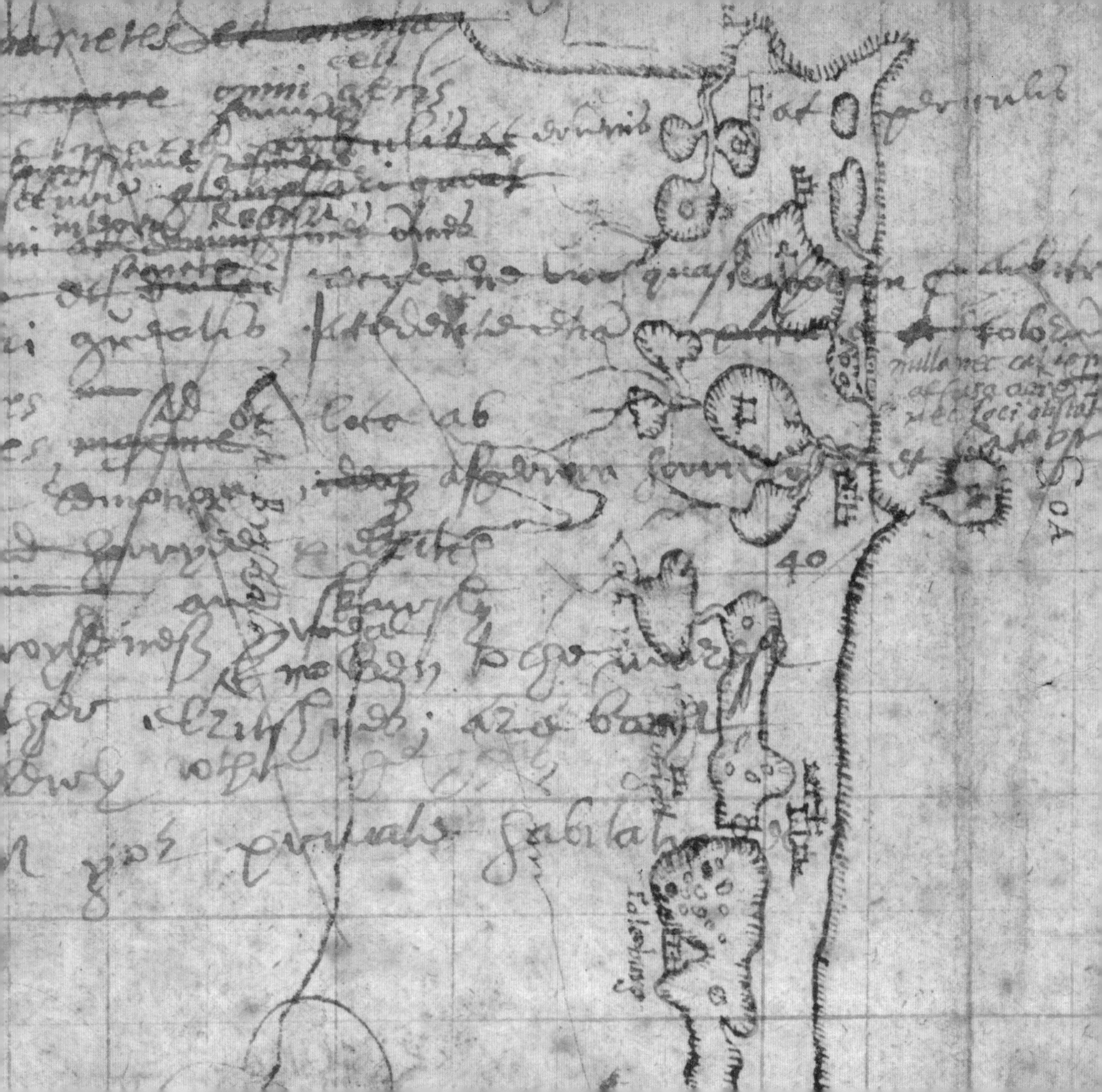

Frobost
Loch Eynort

FIGURE 1.7
(*Previous spread*) Timothy Pont's map of South Uist/Uibhist a Deas is typical of his stylistic mix of textual description and mapped outline, in sketch plan view, of geographical features. This combination of map and text was a distinctive feature of chorography. Some place names are given but the coverage is uneven – the west coast being more inscribed than the east coast of the island. We do not know when Pont undertook this work, whether he was accompanied by a Gaelic-speaking native to facilitate it, or, in relation to the many missing islands, why his survey of island Scotland is as incomplete as it is. The engraved and printed version of this map, which was part of Joan Blaeu's *Atlas Novus* (1654), appears here as fig. 5.2.
Source: Timothy Pont, South Uist, Pont [36] (*c.*1583–1614).

FIGURE 1.8
(*Left*) In this map of Arran, Blaeu uses a commonplace style, the plan view, to show the island's shape. By modern conventions, this representation is neither accurate in shape nor correctly aligned or positioned in terms of latitude and longitude. Blaeu uses small 'mole-hill' features to symbolise topography and suggests, by their relative size, that the topography is more mountainous in the west of the island (in reality Arran is hillier to the north around Goat Fell). In his depiction of Ailsa Craig, Blaeu retains this pictorial symbolism rather than use a plan view: he did not know the feature from first-hand experience and is only figuratively indicating the location of the smaller island.
Source: Timothy Pont/Joan Blaeu, *Arania Insula in Aestuario Glottae [vulgo], The Yle of Arren in the Fyrth of Clyd* (Amsterdam: Blaeu, 1654).

are always a blend of art, science and technical competence in the way they depict things). From the early eighteenth century, throughout the Enlightenment and the centuries that followed, maps were taken to be utilitarian documents to be used in delimiting territorial jurisdiction, for example, or in outlining the 'true' course of rivers as determined by 'accurate survey' or to give the correct shape and position of geographical features – such as islands (issues we discuss in chapter 6, for example).

In Scotland, the mapping of islands as the central object of that mapping dates from the 1680s and the work of the mathematician and map maker John Adair. Adair, supported in his endeavours by the Scottish Parliament, was both maritime map maker and regional map maker in terms of the 'modes' of map work identified above. For reasons which are explored more fully below (see chapter 4), his work was never completed but it marks nevertheless a distinct chapter and new beginning in Scotland's map history. Different types of map and practices of mapping emerged in relation to these distinct but related genres of map making. *Isolarii* and *portolani* were no longer in common use and had in any case served more as decorative cultural objects than as directly functional maps. They were replaced by charts, maps of the sea which included islands, and by hydrography or hydrographic mapping, maps of seas and waters and islands produced using mathematics and a distinctive cartographic language suited to that purpose.

In his *Treatise of Maritim [sic] Surveying* (1774), for example, the Scottish hydrographer Murdoch Mackenzie (whose work is considered below) went to some lengths to document the possibilities open to him. 'A General, or Nautical, Chart', wrote Mackenzie, 'is commonly constructed and drawn by some experienced Seaman from his own Observations and Journals of the Courses and Distances from Headland to Headland, and Point to Point; with the intermediate Spaces filled up from such Charts or Maps as are at hand, or are in most Repute'. 'Abbreviated Collections of Draughts' were designed to show 'a great deal of Coast in a small Compass [scale]'. A 'Memorial Sketch' and an 'Eye-Sketch' need little explanation, the one done from memory, the other without any measurement of distances (figure 1.4 might fit this description). An 'Ambulatory Draught' combined something of the two: '[this] is made by walking along the Shore, taking the Bearings from Point to Point with a Compass, estimating their Distances by the Eye, and sketching the Figure of the Coast between them'. A 'Disjunct Survey' is 'when the Harbours, Bays, or Islands in any Country, are surveyed separately in a geometrical manner . . . all the Rocks, Shoals and Channels are supposed to be carefully examined'. Warming to his task of classification and self-justification, Mackenzie further distinguished the 'Lineangular Survey', when 'the Coast is measured all along with a Chain, or Wheel, and the Angles taken at each Point and Turn of the Land with a Theodolite, or magnetic Needle', and the 'Trigonocatenary Survey', in which, from 'one long Baseline', a connected 'Series of Triangles' was taken along the several 'Heads, Points and Flexures of the Shore'. 'Orometric Survey' involved a baseline to measure summits and mountains. 'Stasimetric Survey' is 'when the Mutual Distances of three, or more, proper Objects are carefully measured; and by Means of these Objects, the Position and Distance of all Stations along the Coast determined trigonometrically, each with its respective Station alone, independent of one another'.

These distinctions were never apparent in Mackenzie's own work, nor were they widely employed by his peers. They represent differences in emphasis rather than strictly different matters of practice. Even so, the language is instructive. Making maps, Mackenzie is saying, especially those of islands and sea coasts, involved mathematics, first-hand observation, instrument use, trust in the 'repute' of other map makers, trust in the instruments and the angles they produced, careful measurement and delineation, and sketching what had not been directly measured or could be recalled from memory. This could be the work of individual surveyors and chart makers – and for many practitioners at the end of the eighteenth century, map making was something done alongside geographical teaching, engraving and printing, and book and journal publishing. By this period, maps of islands were a distinctive genus with numerous species and types in a complex ecology of print and visual culture.

During the eighteenth century, map making also became more widely and more formally the work of government institutions and was seen as a state-making science undertaken by military engineers and trained surveyors. Many European nations undertook nation-wide terrestrial surveys, with France leading the way. In Scotland, the work of William Roy and others in coordinating the Military Survey between 1747 and 1755 reflected the needs of the Hanoverian ascendancy to know what and who was where in 'North Britain' in the wake of Jacobite dissent (see chapter 5). The demands for speed of completion, rather than completeness of coverage, meant that only a few of Scotland's islands are shown in Roy's 'Great Map' as it was known (see fig. 6.14). The more northerly and western islands do not feature.

In Britain, the systematic and state organised mapping of islands and the nation's coasts followed the foundation in 1795 of the Hydrographic Office of the Admiralty. At first, however, progress in hydrography and in maritime charting was hindered by too few resources, the distractions of the Napoleonic Wars and disagreements over the procedures and standards to be followed. The Hydrographic Office released its first officially published Admiralty chart in 1800, but the coordinated surveying and mapping of Scotland did not begin until 1815, with work on the east coast. The Northern Isles were surveyed from the 1820s. From the 1830s onwards, working slowly north and north-westwards, marine surveyors charted Scotland's islands, often doing so before their terrestrial counterparts in Ordnance Survey.

The resultant Admiralty charts represent a distinctive and important map type (fig. 1.9). Coastlines are shown with instrumentally derived accuracy, both high and low water marks being given. A range of features important to mariners was included: the position of rocks and sandbanks and other navigational hazards; depths in fathoms; tidal directions, and the like. Charts also depicted topographic features if they were prominent and therefore useful to navigation, and, commonly, a view of the island or safe anchorages was shown as if from sea level, the viewpoint of the approaching sailor. Admiralty charts were continuously updated through new survey work and on the basis of correspondence and the word of sailors. Obsolete or incorrect information could, after all, be fatal. Admiralty charts feature throughout this book. Produced by government officials working in standard ways with increasingly accurate instruments, they are documents for determining (and continually revising) the dimensions and positions of Scotland's islands. Less obviously, though no less importantly, they also illustrate users' faith in maps: because of its accuracy and currency, information 'within' the chart can be relied upon when planning action 'beyond' the chart.

By the mid nineteenth century, maps were a commonplace feature in books and even in newspapers, as well as a common form of graphic depiction in the sciences. Their greater currency and diversity of form was a consequence of different purposes and of different audiences' needs. Admiralty charts would not have been common in the Victorian household; the maps of James Wyld, map maker and geographical entrepreneur, certainly were (fig. 1.10). In the mid and later nineteenth century, Scottish map-making companies such as W. & A. K. Johnston and the Bartholomew firm established world-wide reputations for their quality and accuracy. Maps became ever more popular and powerful forms of visual representation. Advances in technology meant that mapping became capable of finer detail, could use colour washes, and could be undertaken as planimetric, topographic, or as thematic images where all other geographical content is subordinated to particular needs (fig. 1.11 for example).

There is a tendency to associate 'modernity' with betterment, to assume that things now – maps included – are an improvement upon things then. Certainly, maps are today more common, more used, read and understood, and in many ways more accurate, than was the case in the past. In map history, however, such an interpretation is not appropriate: terms such as 'accuracy', 'map reading', even 'map', had different meanings in the past. The example of Jura, which was cartographically 'lost' by Google for several weeks in 2013, stands as a splendid example to prove this point (fig. 1.12): in maps,

Fishing Bank
Slack tide ¾ of an hour after H. & L. water at Tobermory.
Ardmore Pt.
Bloody Bay
New Rocks
Mc. Lean's Nose
LOCH
Runa Gal
Tobermory
Poor House
Calve I.
Stot Hill
Loch na Meal
Speine More
Speine Beg
Cnoc Caorach
Aros Water
M
U
Bein Craigach
Cairn mor
Dervaig
Church
Calgary Bay
Cruachan Sheila
Bein Bhuidle
Beinnanairich
Cruachan
Burgh Bay
Bogha mhor
L O C H
T U A D H
Ballygown
Cnoc Buidhe
Runa Gal or Sound of Mull Lighthouse
from the Westward
Rum Island
Ardnamurchan Lighthouse
Ardnamurchan Point
General appearance to the North from off the West entrance of the Sound of Mull.
Ardnamurchan Lighthouse
from the South
Ardnamurchan Point
Directions
A Left shoulder of Eorsa on with Cruach.
B Knoll open of Airdcailleach.
C Left end Eiln. Dubh and Right end Giasgil Rocks.
D Upper cliffs of Cairn a Burgh just open.

FIGURE 1.9
(*Left*) This Admiralty chart of part of Mull, published in 1866, shows a number of the features common to this map type. This map is also distinctive, however, in that the inland and largely upland topography of the island was not a central concern of the maritime map makers and so has been omitted, apart from a coastal 'strip'. The space vacated has been filled with sketches of the coast as sailors would see it. Note the compass rose, another standard feature of Admiralty charts, and, in the Sound especially, the wealth of detail about sea depth and the characteristics of the sea floor. The surveying for this map, in 1863–64, was led by Captain William Henry C. Otter, a major figure in the hydrographic mapping of west Scotland in the mid nineteenth century.
Source: Hydrographic Office, *Sound and North-West Coast of Mull*. Admiralty Chart 2155 (1866).

FIGURE 1.10
(*Right*) James Wyld (the younger) was appointed Geographer to Queen Victoria in 1839 (a position held by his father before him). Wyld was widely known in mid Victorian Britain as something of a geographical showman although he was also Liberal MP for Bodmin for over twenty years from 1847. His most spectacular geographical enterprise was his 60-foot-high 'Great Globe', which was erected in London's Leicester Square in 1851 to coincide with the Great Exhibition. At once ridiculed yet very popular, the Great Globe was used by Wyld as geographical advertising space for his map and travel publications and as a platform for him to launch impassioned speeches about the inaccuracies of Ordnance Survey in comparison with his own work. This map of 1846 was probably aimed at the emerging tourist market and shows newly completed railways (in black) as well as those 'in progress' (in red).
Source: James Wyld, *Scotland with its Islands* (London: Wyld, 1846).

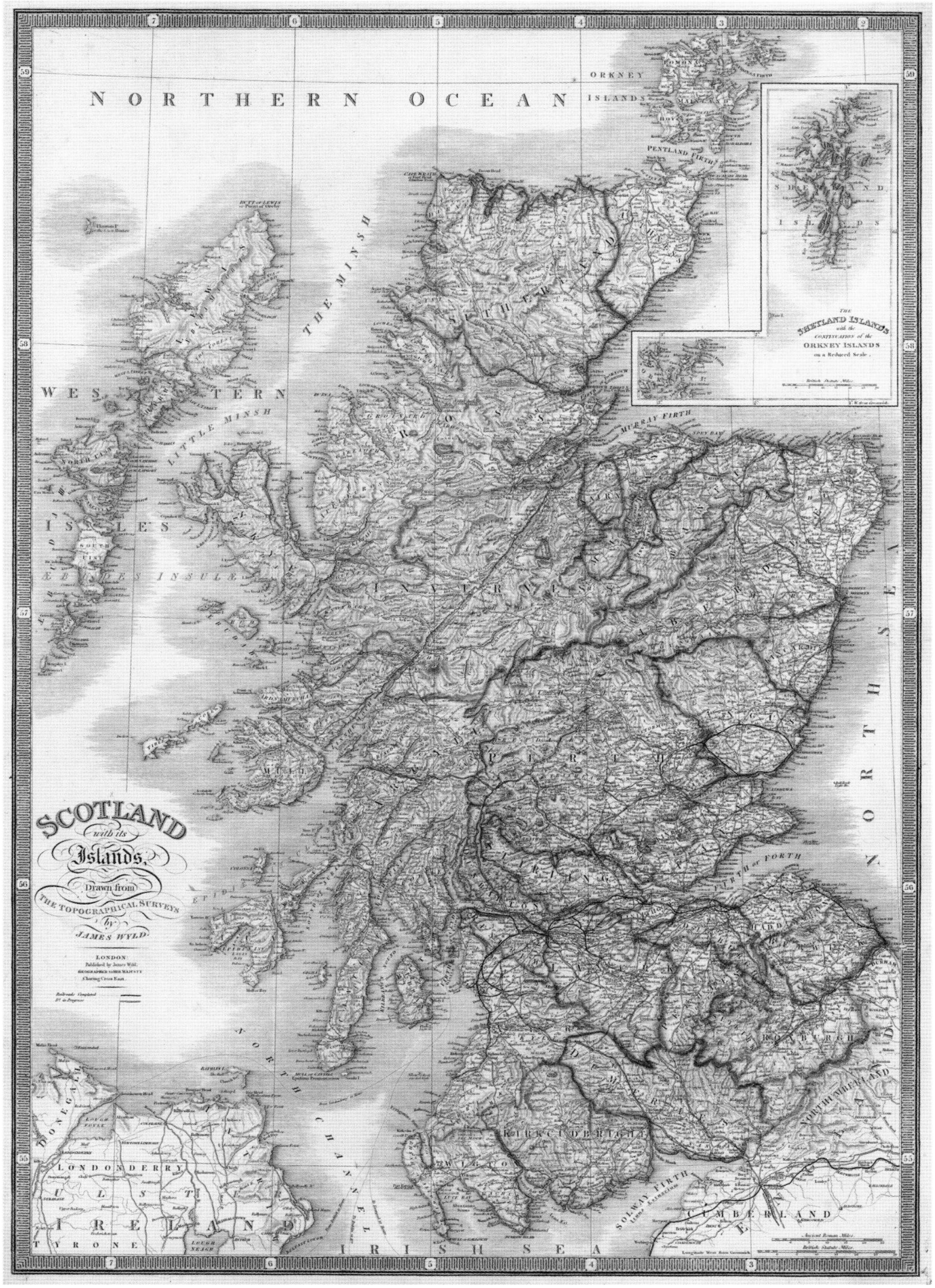

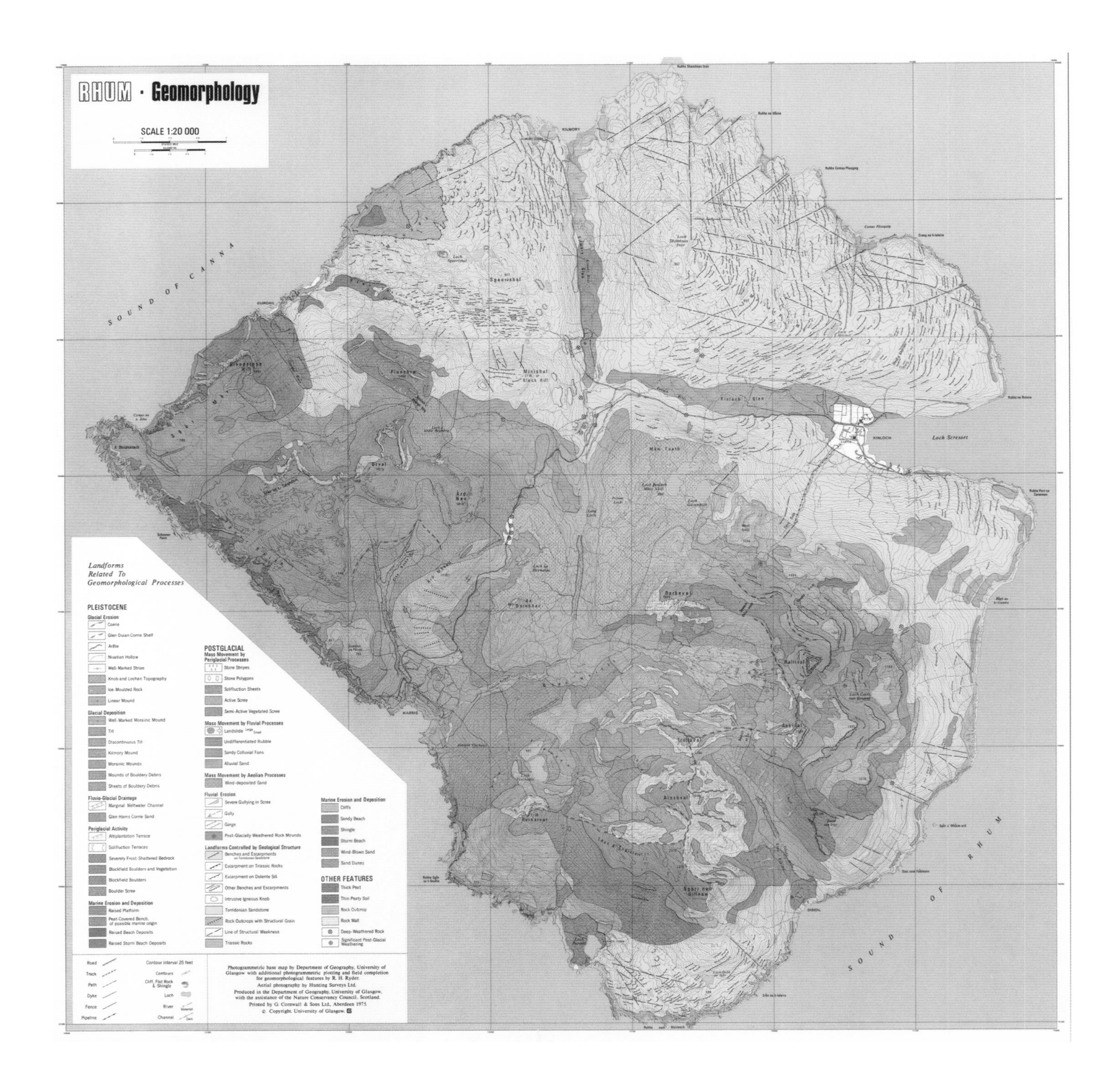
RHUM · Geomorphology
SCALE 1:20 000
SOUND OF CANNA
KILMORY
Kinloch Glen
KINLOCH
Loch Scresort
Barkeval
Hallival
Askival
Trallval
Ainshval
HARRIS
SOUND OF RHUM
Landforms Related To Geomorphological Processes
PLEISTOCENE
Glacial Erosion
Corrie
Glen Duian Corrie Shelf
Arête
Nivation Hollow
Well-Marked Striae
Knob and Lochan Topography
Ice-Moulded Rock
Linear Mound
Glacial Deposition
Well-Marked Morainic Mound
Till
Discontinuous Till
Kilmory Mound
Morainic Mounds
Mounds of Bouldery Debris
Sheets of Bouldery Debris
Fluvio-Glacial Drainage
Marginal Meltwater Channel
Glen Harris Corrie Sand
Periglacial Activity
Altiplanation Terrace
Solifluction Terraces
Severely Frost-Shattered Bedrock
Blockfield Boulders and Vegetation
Blockfield Boulders
Boulder Scree
Marine Erosion and Deposition
Raised Platform
Peat-Covered Bench, of possible marine origin
Raised Beach Deposits
Raised Storm Beach Deposits
POSTGLACIAL
Mass Movement by Periglacial Processes
Stone Stripes
Stone Polygons
Solifluction Sheets
Active Scree
Semi-Active Vegetated Scree
Mass Movement by Fluvial Processes
Landslide Large Small
Undifferentiated Rubble
Sandy Colluvial Fans
Alluvial Sand
Mass Movement by Aeolian Processes
Wind-deposited Sand
Fluvial Erosion
Severe Gullying in Scree
Gully
Gorge
Post-Glacially Weathered Rock Mounds
Landforms Controlled by Geological Structure
Benches and Escarpments on Torridonian Sandstone
Escarpment on Triassic Rocks
Escarpment on Dolerite Sill
Other Benches and Escarpments
Intrusive Igneous Knob
Torridonian Sandstone
Rock Outcrops with Structural Grain
Line of Structural Weakness
Triassic Rocks
Marine Erosion and Deposition
Cliffs
Sandy Beach
Shingle
Storm Beach
Wind-Blown Sand
Sand Dunes
OTHER FEATURES
Thick Peat
Thin Peaty Soil
Rock Outcrop
Rock Wall
Deep-Weathered Rock
Significant Post-Glacial Weathering
Road
Track
Path
Dyke
Fence
Pipeline
Contour interval 25 feet
Contours
Cliff, Flat Rock & Shingle
Loch
River
Channel
Photogrammetric base map by Department of Geography, University of Glasgow with additional photogrammetric plotting and field completion for geomorphological features by R. H. Ryder.
Aerial photography by Hunting Surveys Ltd.
Produced in the Department of Geography, University of Glasgow, with the assistance of the Nature Conservancy Council, Scotland.
Printed by G. Cornwall & Sons Ltd., Aberdeen 1975.
© Copyright, University of Glasgow.

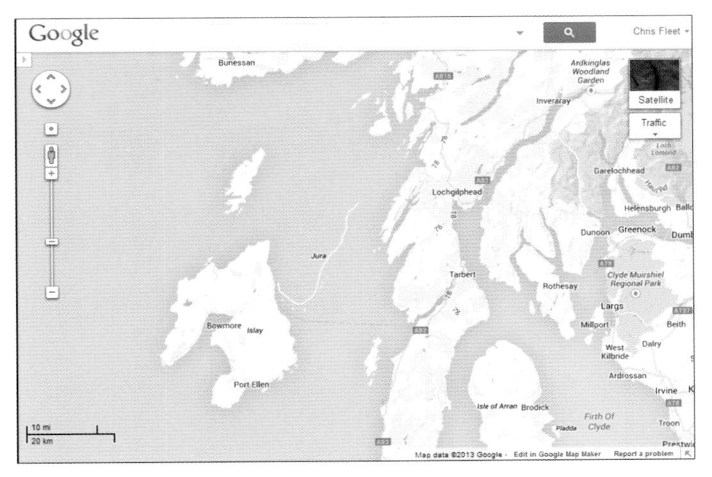

FIGURE 1.11
(*Opposite*) This geomorphological map of Rum in the Small Islands was produced by the Department of Geography at the University of Glasgow in 1975 with the assistance of the Nature Conservancy Council. Produced from a photogrammetric base map with additional evidence added in the field, it is a fine example of a thematic map and indicative of the high-quality maps produced by the Glasgow Department. *Source*: R. H. Ryder/Department of Geography, University of Glasgow, *Rhum – Geomorphology* (Glasgow, 1975). Produced in the School of Geographical & Earth Sciences, University of Glasgow, 1975. © Copyright University of Glasgow.

FIGURE 1.12
(*Above*) Google maps have a definite homogenising quality: everywhere looks rather like everywhere else through the firm's obvious but bland choice of symbols and colours. Their maps are at once familiar, even reassuring, in their sameness of tone and style, but they also erase local difference. Here, they succeeded in erasing a whole island, a fact which was spotted by Jura residents and others in early July 2013. The spectral presence of the main road from Feolin Ferry along the coast to Barnhill in the north of the island hinted at what was missing. By 16 July, after considerable embarrassment and coverage in the national and international media, Google engineers were 'beavering away to fix it' (to 'recover' Jura from its cartographic erasure). By August, Jura was back on the map.
Source: Google Maps, detail of centre-west Scotland, taken July 2013. Map data © 2013 Google.

modernity does not equate to completeness of coverage and accuracy of content. The Jura example also shows that while maps and mapping have a rich history, the mapping of islands has its own distinctive features, not the least of which has been in knowing and showing what is, and what is not, an island.

What is an island?

There is a proper term for the study of islands – 'nissology' – the term derived from the Greek root word for island, *nisos*. Its practitioners (nissologists?) admit to the 'indescribable intoxication' of islands in much the same way as the characters in David Greig's *Outlying Islands*. Yet it is not always clear what an island is, or, rather, we should not assume that what is an island now has always been one or always looked as it does now. Maps are important instruments in these respects. Maps can be used not only to show how islands come into view in different ways, but also to show that what are islands today may not have been so in former times, and vice versa.

The effect and scale of climate change is of enormous political and public concern in the modern era. Our world, our islands and our maps will look different if and when sea levels change significantly. Yet climate change has been a recurrent feature of the geological past. After the last retreat of glacial ice across Scotland, about 20,000 years ago, sea levels were lower than they are at the present day. By about 14,000 years ago, what we know now as the several separate islands making up the Outer Isles/Innse Gall were connected to one another as a single large island. At the same time, the action of the sea can disclose the former extent of islands. Evidence of Bronze Age human habitation revealed in December 2015 by coastal erosion at Tresness on Sanday in Orkney, for example, shows that Scotland's islands had different human geographies in the past (issues addressed in chapter 2). Maps cannot always capture the past patterns of settlement in a place or the rates of change in the landscape but they can help illustrate the results (fig. 1.13a, b).

If maps thus disclose the 'disappearance' of what was once

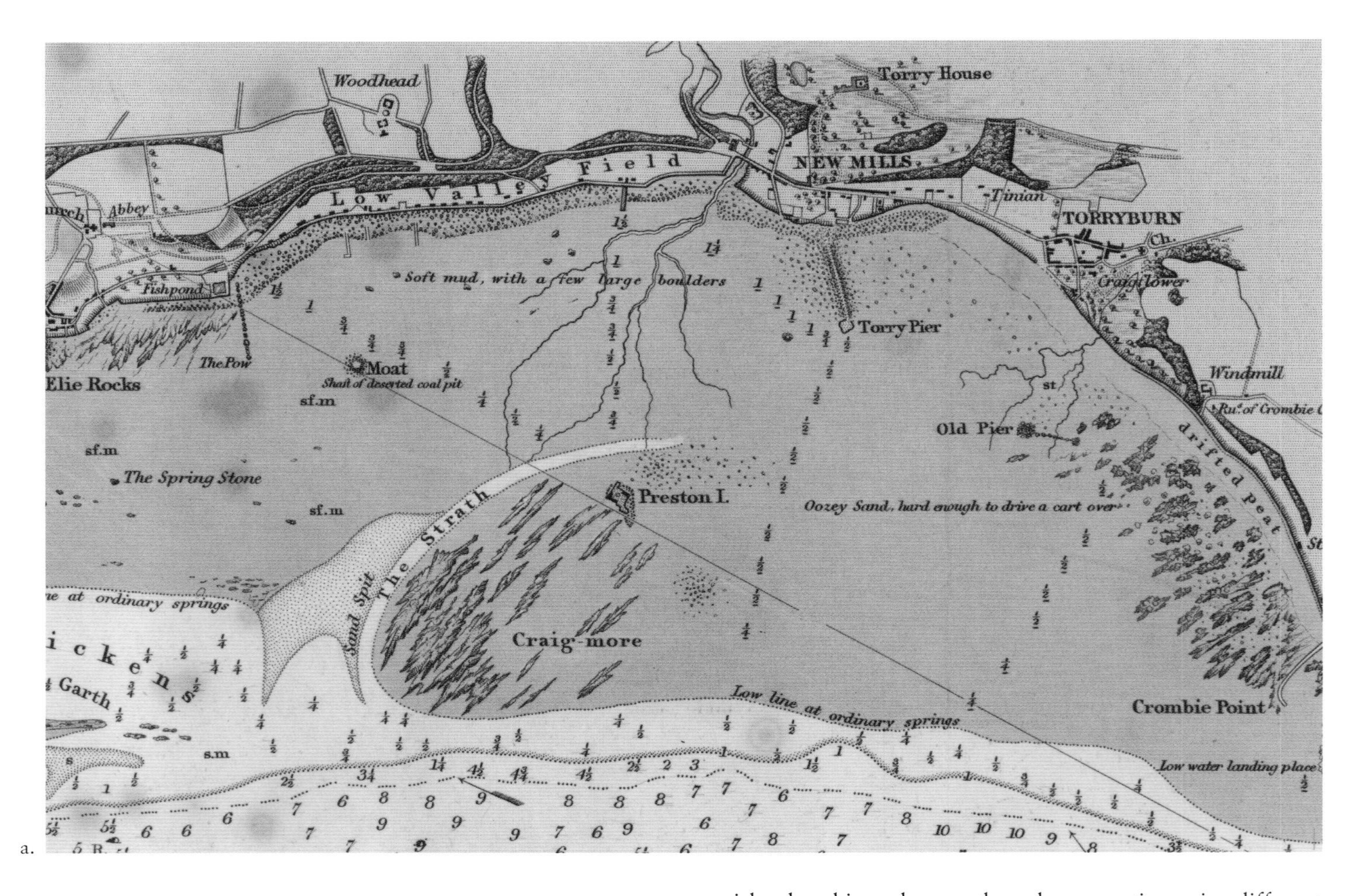

a.

FIGURE 1.13
(*Above and opposite*) In 1860, as this Admiralty chart shows, Preston Island was a small island in the Firth of Forth, about two-thirds of a mile out from Culross (a). Within the last thirty years large-scale reclamation along the coast, principally as the result of the deposition of ash from the Kincardine and Longannet power stations, has made Preston Island no longer an island (b).
Source: (a) Hydrographic Office, *Queensferry to Stirling*. Admiralty Chart 114c (1860). (b) OpenStreetMap, Detail of Firth of Forth near High Valleyfield (2016). © OpenStreetMap contributors.

an island and is no longer, they also, over time, give different shape to islands once identified (see, for example, figs 1.14a, b and 1.15a–c). Maps can help determine that an island is indeed an island. That this is not always obvious may be illustrated by the case of Rockall, some 230 miles west of North Uist/Uibhist a Tuath in the North Atlantic. Known for centuries, Rockall was first taken seriously as a mappable object in the early nineteenth century by the British naval captain and explorer Basil Hall. The reason for his interest was that Rockall did not look like an island at all: in high seas and winds, sea spray obscured the rock, making the island appear, as Hall put it, like 'a vessel under sail'. Even after confirming its place, reported Hall, 'We were deceived by it.'

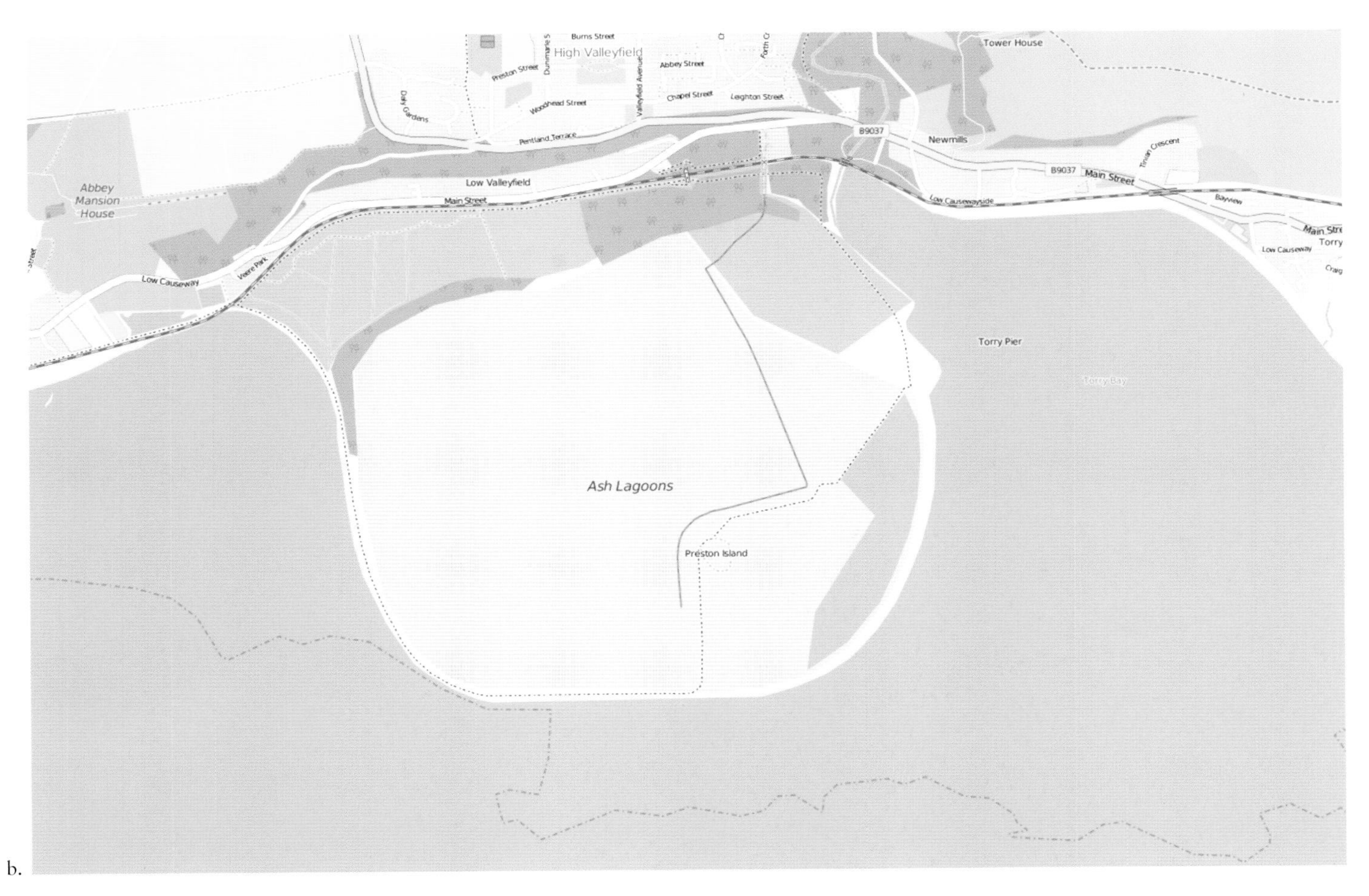

b.

Not to map the island would be to present a danger to shipping. But to map it presented a real difficulty since no cartographer's mark or symbol could be used that did not dramatically over-represent Rockall's size. Naval hydrographers 'solved' the issue by combining two types of representation on one chart: a plan view, exaggerated in scale, and a series of sea-level sketches to show that what could be taken to be a sail at sea was something altogether more solid (fig. 1.16).

It is also the case that one island or island name may be taken for many islands. What is taken to be St Kilda, for example, is, strictly, an archipelago, yet, despite there not being an island of 'St Kilda', the name has stuck in descriptions of Hirta (the main island), Dùn, Soay and Boreray, which together make up this island group in the north Atlantic. The name 'skilder' or 'skilda' is thought to be derived from the Old Norse term 'skilder' meaning 'shields' although the name by which the island is commonly known, 'St Kilda', does not appear in travellers' accounts until the late seventeenth century. The Rev. Kenneth Macaulay, minister of Ardnamurchan, visited St Kilda in 1758. Without the means to undertake a survey in his own right, Macaulay based his map on that of the Gaelic-speaking natural historian and author Martin Martin (fig. 1.17).

Macaulay's map illustrates another sense in which islands and maps work together but not always in terms of accuracy.

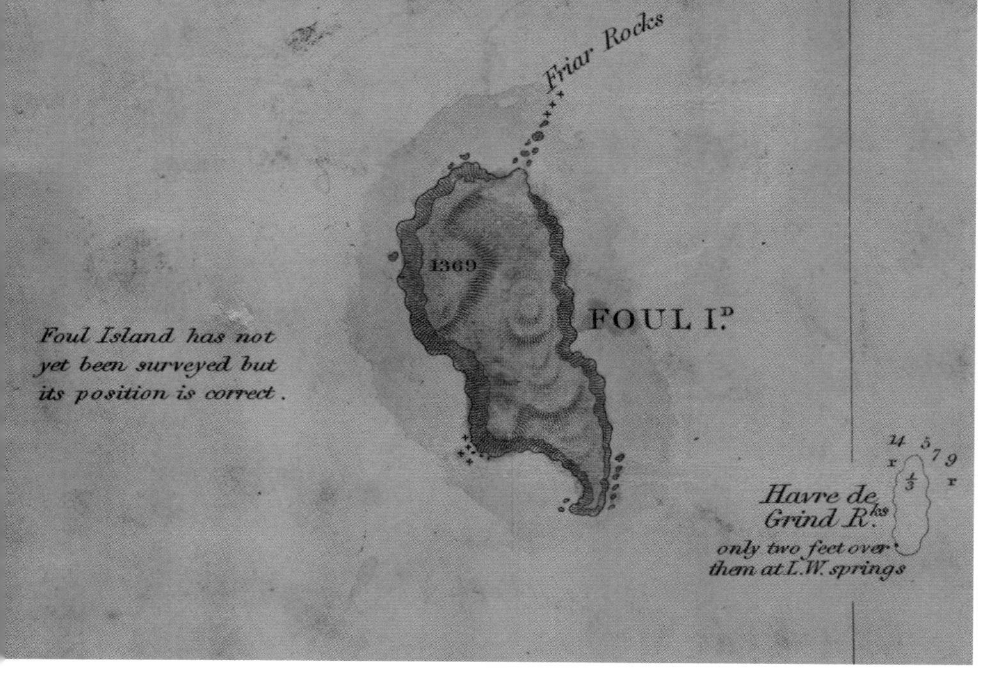

a.

FIGURE 1.14
George Thomas's hydrographic work for the Admiralty in the Northern Isles was still a work in progress when this chart showing the island of 'Foul', Foula in modern place-name terms, was published in 1838 based on survey work from five years earlier. As Thomas noted, the island was correctly positioned, but had not yet been surveyed (a). Nearly forty years later, the island had not only been confirmed in its position but also re-shaped and had its content depicted as a result of map work by Ordnance Survey (b).
Source: (a) Hydrographic Office, *The Shetland Isles*. Admiralty Chart 1118 (1838). (b) Ordnance Survey, One-Inch to the mile, *Scotland, Sheet 125* (surveyed 1877, published 1886).

b.

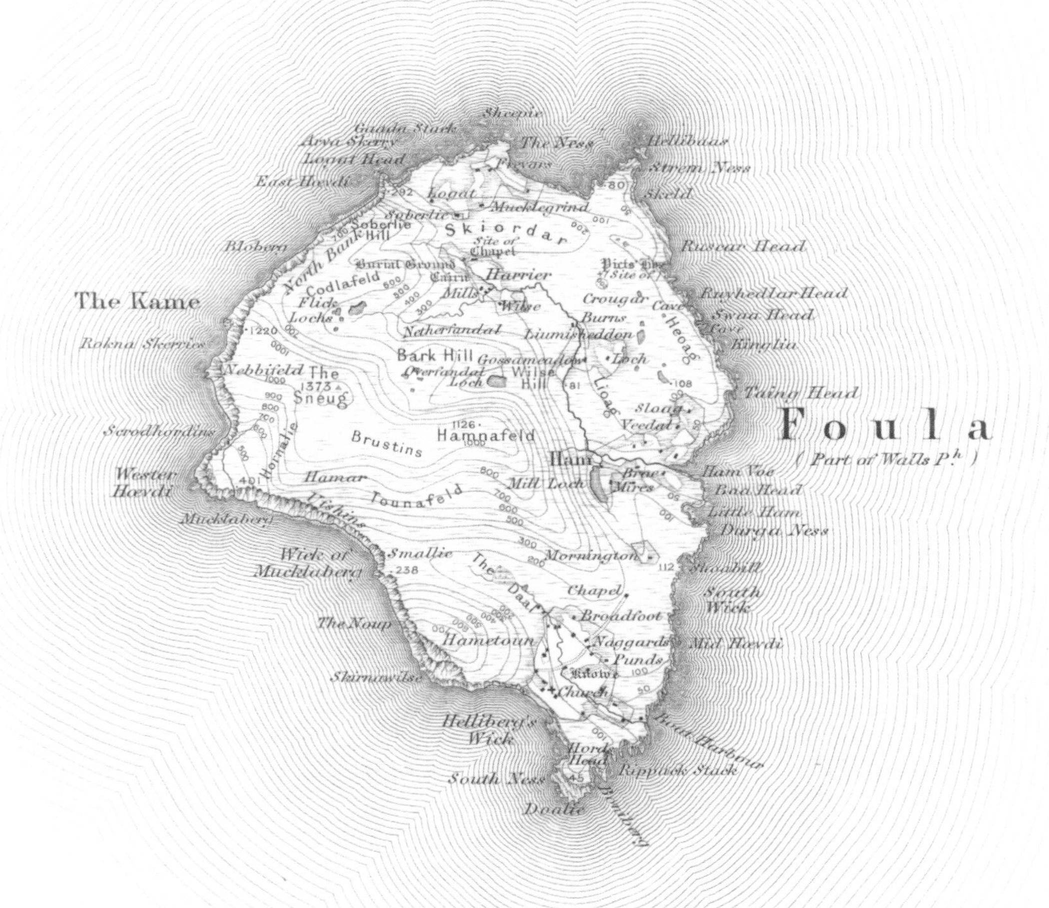

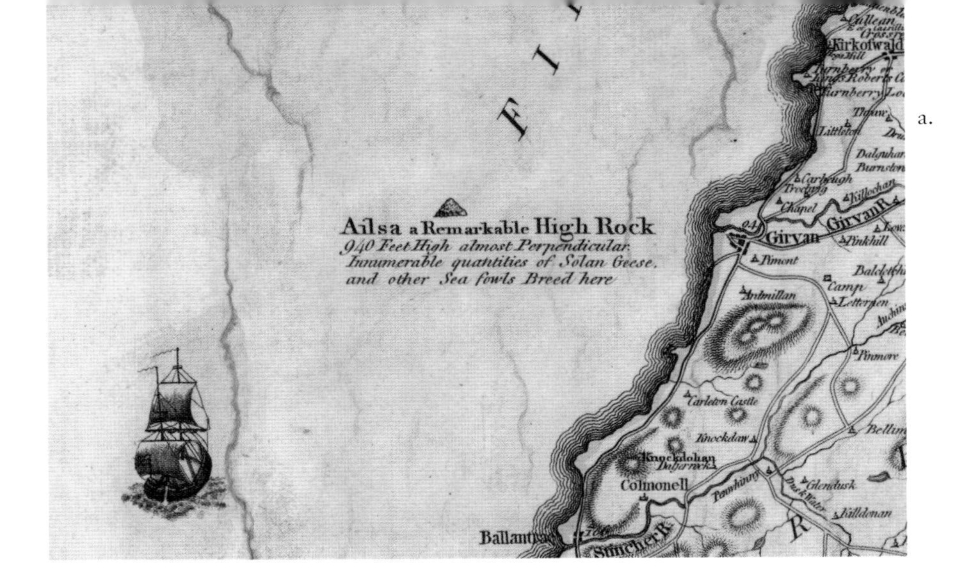

a.

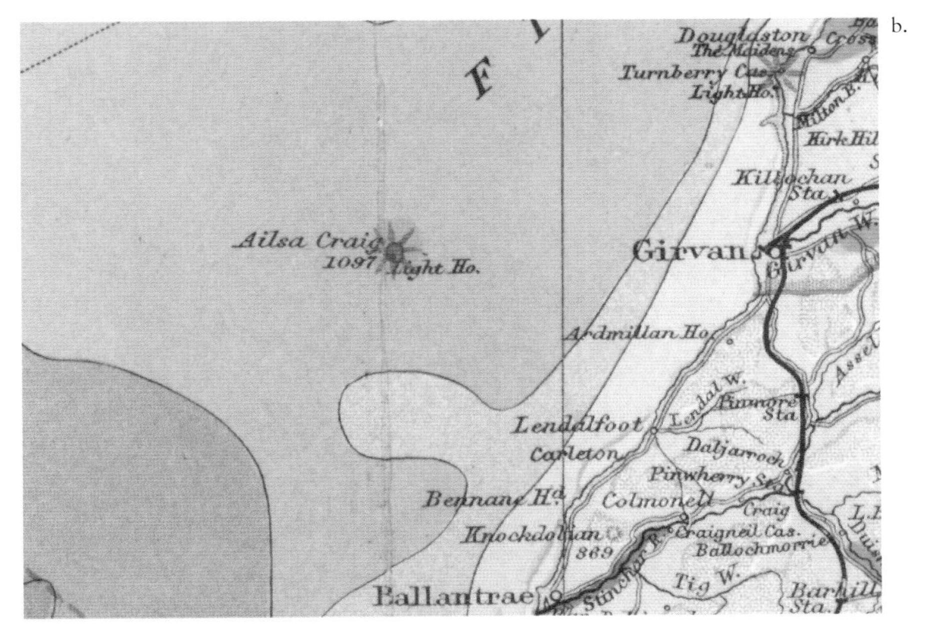

b.

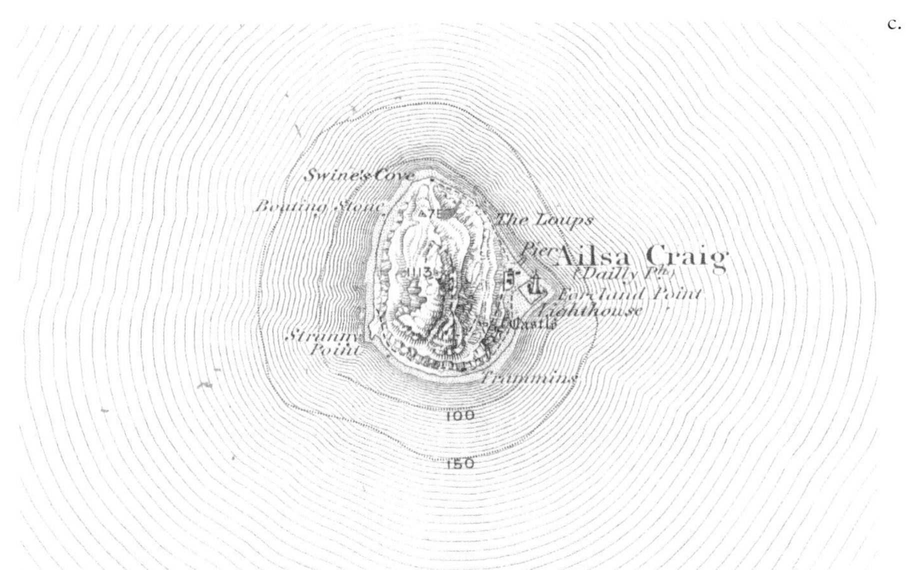

c.

FIGURE 1.15
Ailsa Craig, or 'Paddy's Milestone' as it is popularly known in reference to its role as a way marker on the sea route between Ireland and the western approaches to Scotland, has been differently depicted over time on maps. In the late eighteenth century, and despite the title of his map, which suggests at least a degree of mathematical precision in the undertaking, John Ainslie shows Ailsa Craig in an almost 'pop-up' style and provides a short description of the seabirds associated with the island (a). In the 'Naturalist's Map of Scotland' a little over a century later, the island has been obscured by the representation of its lighthouse, built by Thomas and David Stevenson between 1883 and 1886 (b). Lighthouses were important in navigation – but here John Harvie-Brown and John Bartholomew are signalling the role of island lighthouses as observational points for seabirds, a theme we explore further in chapter 4. By the turn of the twentieth century, a yet different form of cartographic depiction is clear from the Ordnance Survey One-Inch to the mile series (c). Note too that, over time, Ailsa Craig is not only depicted differently but would appear to be rising out of the sea and getting higher: from a 'remarkable High Rock' of 940 feet as Ainslie notes in 1789, to 1,097 feet in 1893, to 1,113 feet by 1905.
Source: (a) John Ainslie, *Scotland Drawn and Engrav'd from a Series of Angles and Astronomical Observations* (1789).
(b) John A. Harvie-Brown and John George Bartholomew, *Naturalist's Map of Scotland* (1893). (c) Ordnance Survey, One-Inch to the mile, *Scotland, Sheet 7* (revised 1902, published 1905).

a.

b.

FIGURE 1.16

In its physical dimensions, Rockall is small – only about 63 feet above high water. In its strategic dimensions, it is much bigger: it was formally annexed by British sailors in 1955 at the height of the Cold War, given fears over its possible use as an observation or listening post by Soviet Bloc forces. In its cartographic dimensions, it is commonly – as here – made to appear larger than it is. The impressive profiles in (a) are taken from Captain Alexander Vidal's survey of Rockall on board HMS *Pike* in 1831, the first to determine the position of Rockall with reasonable precision. In contrast, on the chart itself (b), Rockall is hard to spot among a scattering of depth soundings.
Source: Hydrographic Office, *Rockall and Soundings in its Vicinity*. Admiralty Chart 2870 (1863).

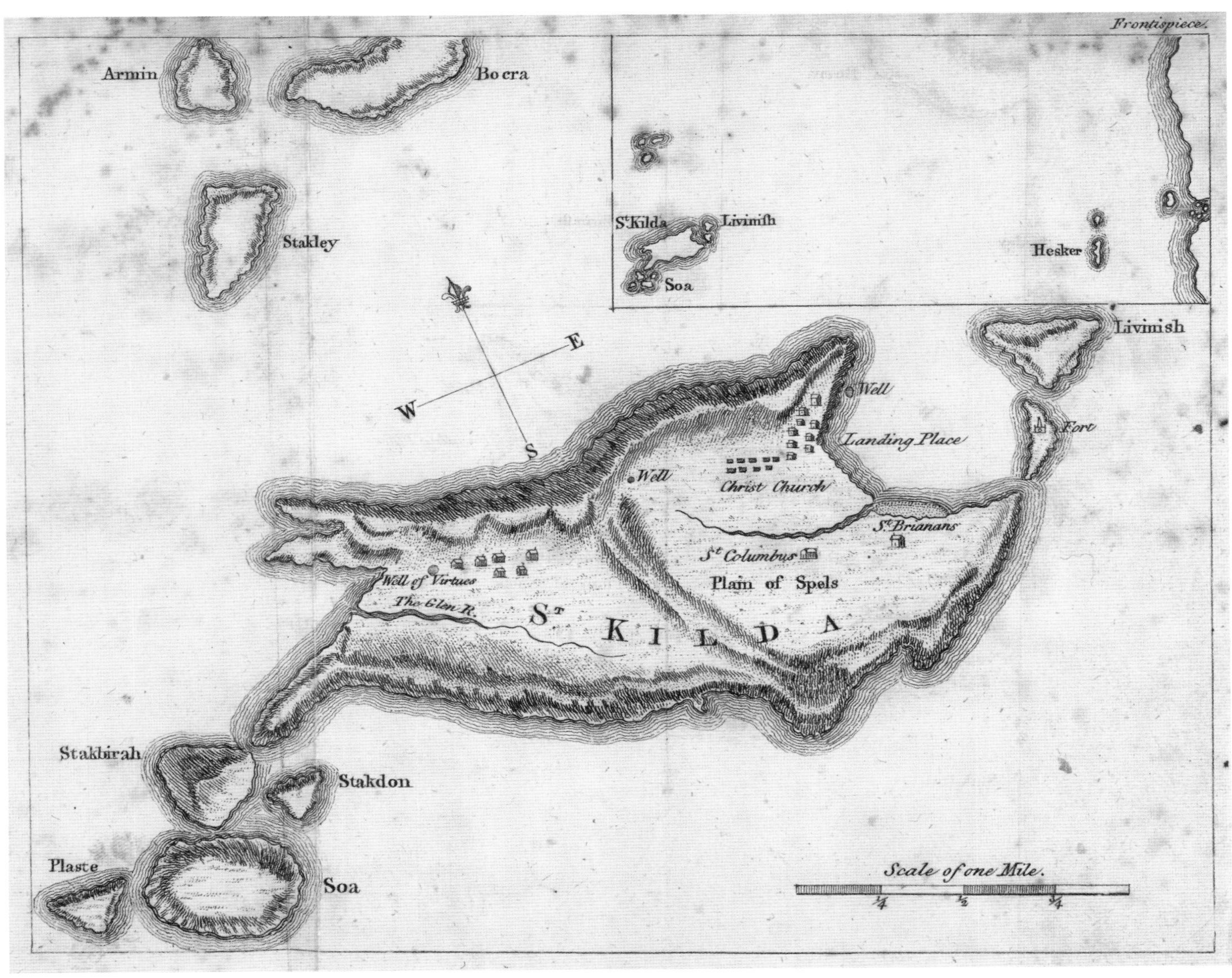

FIGURE 1.17

St Kilda – in reality, the main island of Hirta – as it appears in the Rev. Kenneth Macaulay's map of 1764. In borrowing from the earlier map of the island group as it appears in the work of Martin Martin, Macaulay has made no attempt to show the relative topography of the islands, yet he does show something of the inhabitants' dwellings in stylised diagrammatic form: note, too, the several wells on Hirta. Unclear as to the origins of the islanders, Macaulay offered the view that 'Hirta was first peopled by pyrates, exiles, or malefactors who fled from justice'.
Source: K. Macaulay, Map of St Kilda from *The History of St Kilda* (London: Printed for T. Becket and P. A. de Hondt, in the Strand, 1768).

He has had recourse to a positional 'map frame' or inset. Maps of Scotland the Nation appear as the object of the map maker's art from the late sixteenth century, for the reasons discussed above. Commonly, the Western Isles/Na h-Eileanan an Iar are shown since, with some extension westwards of the map frame, they can easily be included. But this is not always true of the Northern Isles. Because of these islands' locations to the north of the mainland, map makers have had to propose a solution that will be tolerated – and understood – by the maps' users, as well as something that is acceptable in design terms and which makes the best possible use of space on the printed page.

Several possibilities present themselves but they all involve sacrificing accuracy in one way or another. One solution is to leave Shetland out altogether: too far distant from the mainland, it is impossible to include in its true position and, at the same time, represent mainland Scotland at a reasonable scale (fig. 1.18a). Another is to include Orkney on the main map but place Shetland as a separate inset, geographically out of position (fig. 1.18b). Further options involve including Orkney as an inset but leaving out Shetland or including both Orkney and Shetland in their true geographical positions in relation to mainland Scotland. But this often led to an over-abundance of blank sea space and a somewhat ungainly, inartistic and 'over-stretched' appearance. Which is best? (Answers will always reflect the map makers' purpose.) Should relative geographical accuracy prevail, or does one forget about Shetland – and perhaps the Orkneys too – or should one compromise by putting Orkney in its true position but have Shetland as an inset, or treat both Orkney and Shetland as insets and out of true position? All these solutions have been tried (fig. 1.18a–d). This question of the Northern Isles' 'cartographic displacement' is not a problem of the past. To one degree and another, it is evident in the maps which illustrate weather forecasts on television and in the printed media: the Northern Isles (if shown) are depicted as marginal, a mapped afterthought. But such mapped views of islands, far and near, can be subverted more easily now than they could be in the past (fig. 1.19).

FIGURE 1.18
These four maps illustrate the map maker's dilemma when having to depict the Northern Isles. Detail (a), included in the 1607 edition of William Camden's *Britannia*, his chorographical and historical description of Britain, excludes the Shetland Islands, but just manages to include Orkney, squeezed in to the top of the map. The idea of including Shetland in some form of inset when producing a map of Scotland was not common practice before the mid eighteenth century. In his map of 1776 (b), and by diminishing their scale and size in order to allow portions of England and Ireland to be included, the Venetian engraver and publisher Antonio Zatta has had the clever idea of showing the Shetland Islands as though they were 'hanging' from a scroll or roller, thus adding a further artistic element to complement the title cartouche. Since it was common at this time to hang maps on walls, this cartographic conceit and embellishment would have been familiar to his contemporaries. In (c), the 1689 product of a Venetian friar and a Parisian engraver-publisher deals skilfully with the problem of including the Northern Isles by presenting them neatly as two insets, one beneath the other. The map goes further by including the Faroe Islands as the topmost inset. The map makers give their reasons in a note in the inset in question explaining that the Faroes, once part of the Kingdom of Scotland, are now dependencies of Denmark. They were, however, not correct in this assertion: as for Orkney and Shetland, the Faroes were ruled by the Norse until *c.*1470 but were not part of the unified Kingdom of Great Britain from this date. In his map of 1804 (d), James Kirkwood, one of a family of engravers and map makers, provides what is possibly the best-balanced cartographic solution to the issue of the inclusion of the Northern Isles in terms of the symmetry of the map image as a whole. In doing so, however, he has placed Shetland well out of position.
Source: (a) William Hole, *Scotia Regnum*, [1607]. (b) Antonio Zatta, *Il Regno di Scozia* (1776). (c) Vincenzo Coronelli and Jean Baptiste Nolin, *Le Royaume d'Escosse divisé en deux parties . . .* (1689). (d) James Kirkwood, *This Map of Scotland, Constructed and Engraved from the Best Authorities* (1804).

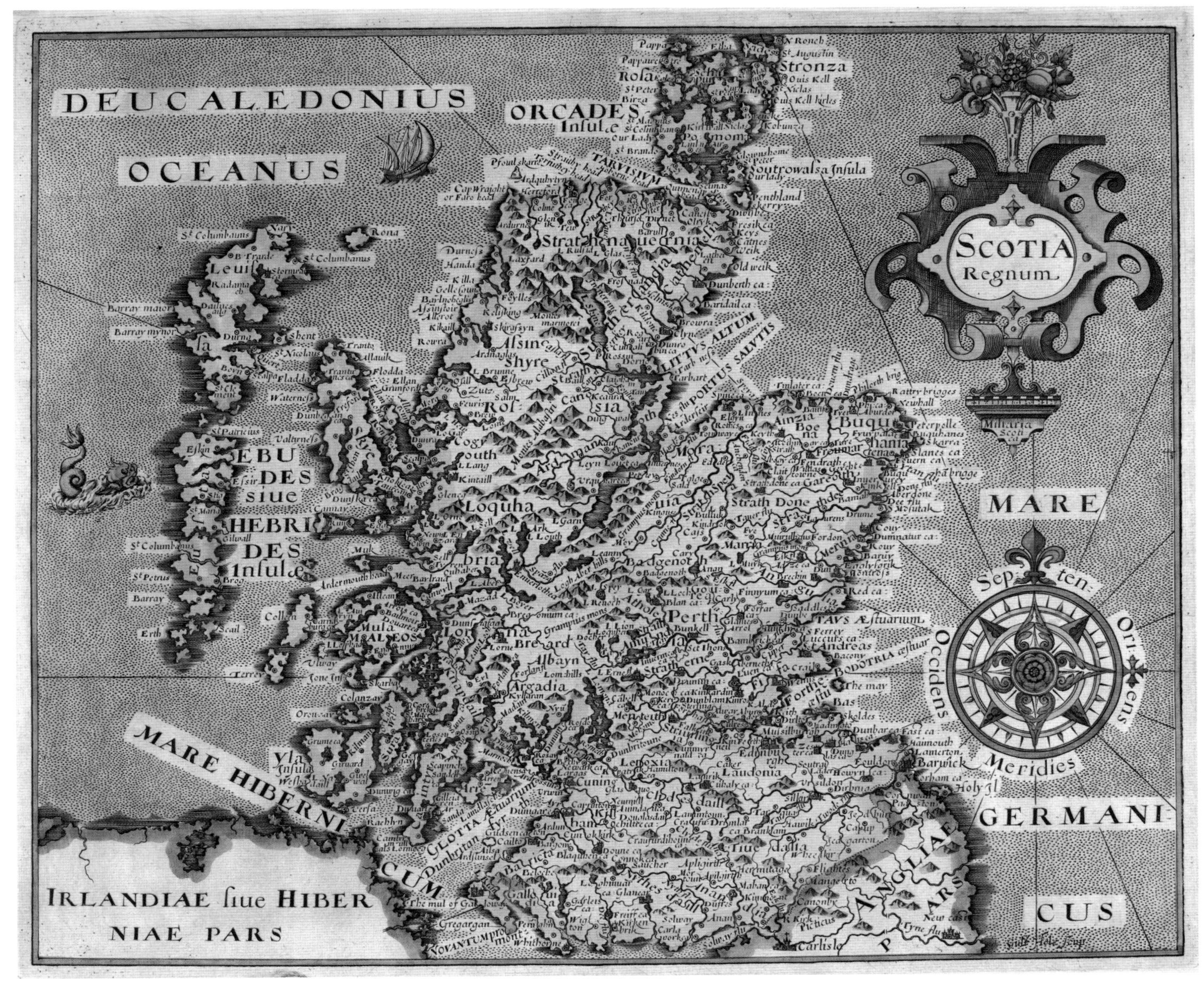
DEUCALEDONIUS
OCEANUS
ORCADES Insulæ
SCOTIA Regnum
MARE
GERMANI
CUS
EBUDES siue HEBRIDES Insulæ
MARE HIBERNICUM
IRLANDIAE siue HIBER
NIAE PARS
ANGLIAE PARS
Septentrio
Oriens
Occidens
Meridies
Stronza
Loquha
Perth
Laudonia
Argadia
Albayn
Strath Done
Badgenoth
Carlislo
a.

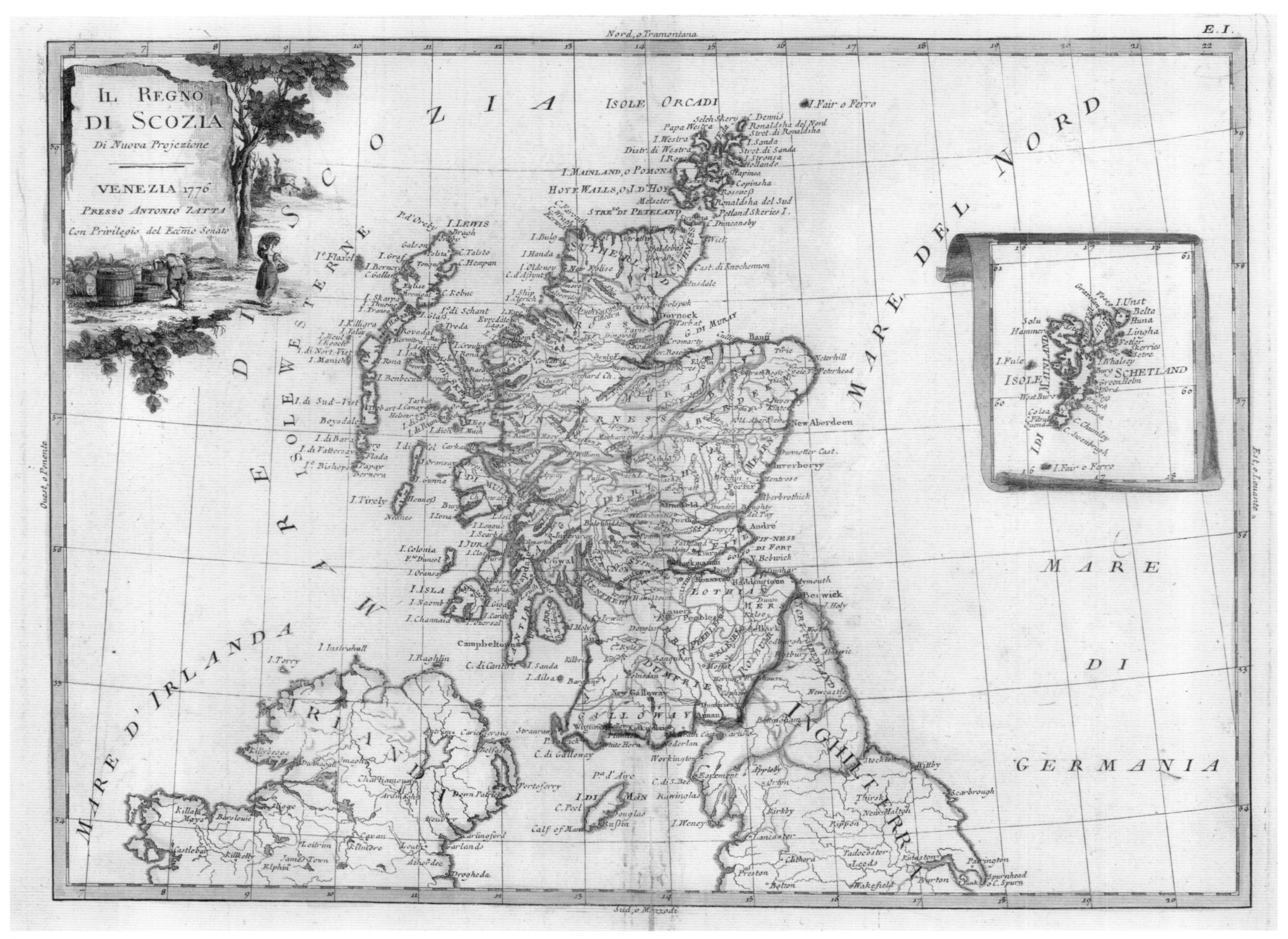
IL REGNO DI SCOZIA
Di Nuova Projezione
VENEZIA 1776
PRESSO ANTONIO ZATTA
Con Privilegio del Ecc.mo Senato
Nord, o Tramontana
Sud, o Mezzodi
Ouest, o Ponente
Est, o Levante
E. I.
SCOZIA
ISOLE ORCADI
MARE DEL NORD
ISOLE WESTERNE
MARE D'IRLANDA
IRLANDA
INGHILTERRA
MARE DI GERMANIA
ISOLE SCHETLAND
New Aberdeen
Campbeltown
Berwick
I. Fair o Ferro
b.

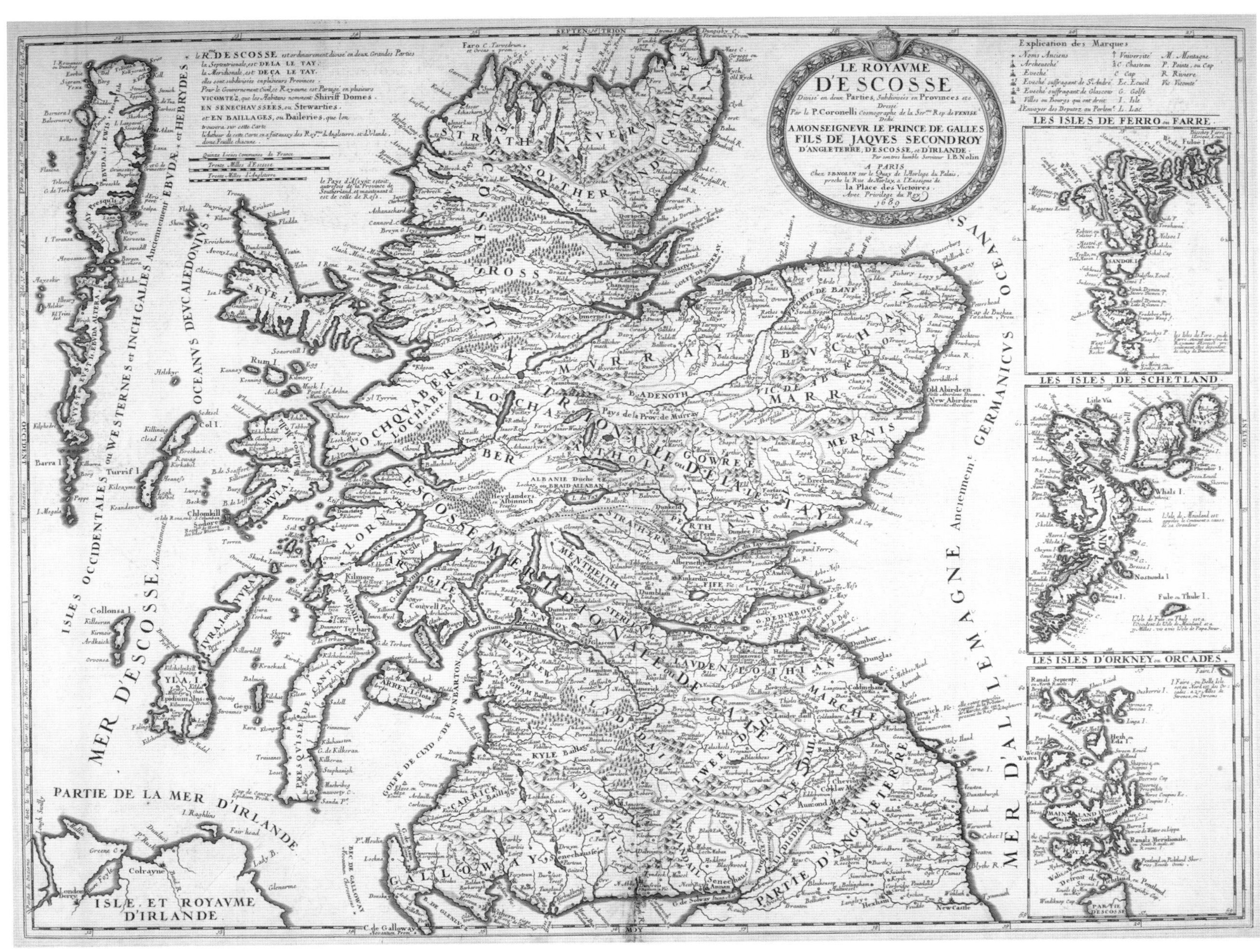

LE ROYAUME D'ESCOSSE
Divisé en deux Parties, Subdivisées en Provinces etc.
Dressé Par le P. Coronelli Cosmographe de la Ser.me Rep. de VENISE
Dédié
A MONSEIGNEUR LE PRINCE DE GALLES
FILS DE JAQUES SECOND ROY
D'ANGLETERRE, D'ESCOSSE, et D'IRLANDE.
Par son très humble Serviteur I.B. Nolin
A PARIS
la Place des Victoires
Avec Privilege du Roy
1689
LES ISLES DE FERRO ou FARRE
LES ISLES DE SCHETLAND
LES ISLES D'ORKNEY ou ORCADES
MER D'ESCOSSE
PARTIE DE LA MER D'IRLANDE
ISLE ET ROYAUME D'IRLANDE
MER D'ALLEMAGNE
OCEANUS DEUCALEDONIUS
ISLES OCCIDENTALES ou WESTERNES et INCHGALLES
SEPTENTRION
MIDY
OCCIDENT
ORIENT

c.

This Map of Scotland, Constructed and Engraved from the best Authorities by J. Kirkwood & Sons, is Respectfully Dedicated to Brig. Genl. Dirom of Mount Annan Deputy Quarter Master General for Scotland.
NORTH SEA
PENTLAND FRITH
ORKNEY ISLANDS
POMONA
PENTLAND FRITH
CAITHNESS
WESTERN ISLANDS
LEWIS
SKY
CAITHNESS
SUTHERLAND
DORNOCH FRITH
MURRAY FRITH
FRITH OF TAY
FRITH of FORTH
HADDINGTON
CLACKMANNAN
MILITARY DISTRICTS
CANALS
POPULATION of SCOTLAND
NAVIGABLE RIVERS
BRITISH OCEAN
JURA SOUND
FRITH OF CLYDE
NORTH CHANNEL
THE SHETLAND ISLANDS
IRELAND
SOLWAY FRITH
PART OF ENGLAND
EMS. A. 74.

d.

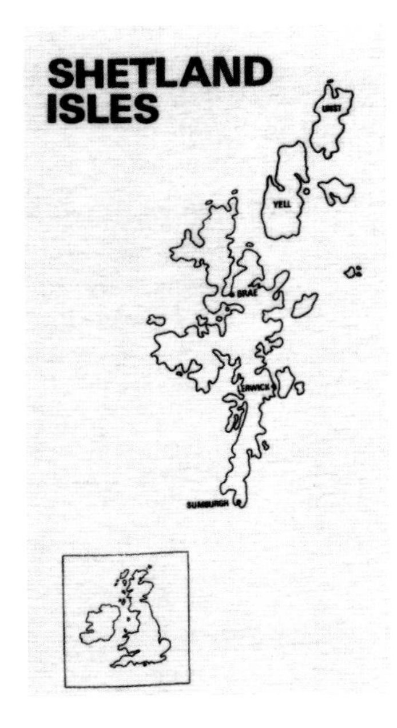

FIGURE 1.19
The idea for this T-shirt map – an act of 'cartographic embodiment' from the islanders' point of view – originated in the BBC Radio Shetland office in the mid 1980s. The conventional representation of Shetland in a box to one side has been subverted by treating the rest of Britain in this fashion (cf. fig. 1.18). The relative size of the Shetlands compared with the reduced size of the British Isles serves to remind us that, while our map conventions are commonplace, they are, nevertheless, conventions – established practices whose origins and forms we explore in this book.
Source: Shetland Isles T shirt (*c.*1980s).

The structure of the book

Scotland: Mapping the Islands is about the lure and 'pull' of maps and of islands in combination. It is about the ways in which islands attract – are made to attract – by being shown, in map form, to embody and to illustrate Scotland's geography and history in particular ways. We document something of the history of maps and of mapping, and examine something of the circumstances of individual islands as realised through maps. But this is absolutely not a work of 'nissological stasimetry' – to invent a category of island map study with reference to MacKenzie's 1774 classification – in which technical details are to the fore. The narrative focus is thematic. Rather than present an 'A–Z' gazetteer of islands and island maps, from Arran and Barra to Zetland (the official administrative name for Shetland until the 1970s) and show maps for each, and, in so doing, risk duplicating issues of chronology, geography and history, we have used eight themes through which to illustrate, in map form, the geography and history of Scotland's islands.

The thematic chapters within this narrative structure follow a broadly chronological order. Within each chapter, we address the ways in which an island's history and geography has been captured in maps at one time or another. We have also tried to reflect the sequences through which islands not only have appeared but also have come to exert their force and 'pull'. That is, the titling and ordering of chapters reflects those processes which, in the main, were followed in the past as islands were peopled, then named, then navigated to and from (or avoided as hazards), were defended, improved and exploited and so on. Each of the island map illustrations selected for these themes is given a caption which, as a minimum, identifies the source of the illustration in question. For the great majority of the map illustrations, we also provide a textual 'vignette', a particular brief story which extends that of the main text and embellishes the history 'behind' the map in question.

It is possible, then, to use and read the book in several ways. Read in conventional fashion from front to back, the thematic narrative will disclose how Scotland's islands have come into view through maps, and how island maps are themselves a key reason to explain the 'pull' of islands. It is also possible to use the map illustrations and the detailed captions selectively, either by island (using the index) or by map type, and so understand the changing representation of certain islands or island groups in addition to the evolution of map types. The book has been written by authors who, in various ways, know only too well the lure of maps and of islands. As others read it in ways of their choosing, it is our hope that it will be clear why maps and islands have the force they do.

Georgio Sideri (Calapoda/Callapoda), Portolan Chart of Europe (c.1560?).

lager
marre
praga
Alpe Alamanie
Ungaria
Ungaria
liburnia
Croua cia
Dalmacia
picardi a
burgonia
Genoa
venetia
francea
Francea
Italia
Colonia
olanda
Roma

Macphee's Hill

MINGULAY

Rua

School
(Boys & Girls)

Site of
St Columba's Chapel
Grave Yard

Rudh' an Droma

Landing Place

BARRA

MINGULAY BAY

Crois an t-Suidheachain

Landing Place

Aneir

Hecla Po

Hecla

Sloc Chremisgeo

CHAPTER TWO
PEOPLING

Maps document and describe more readily than they explain. The distribution of things may be clear enough, but, even when different maps are used to compare the same place at different times, the reasons behind any discernible differences are seldom clear from the maps themselves. Patterns may be apparent, causal processes much less so. Maps may show several forms of human settlement – sites currently occupied, the layout of streets, field boundaries, for example – but they do not do so of their own free will: map makers and those people upon whom map makers rely need to be able to locate, identify and interpret features in the landscape before they can translate that understanding into map form. For a variety of reasons, maps may deliberately omit or symbolically 'over-represent' what is known, but they cannot document what is there but not yet known. This 'cartographic discordance', if you will, is intrinsic to maps: they stylise and symbolise what is known according to particular purposes yet they may not show other things even when they exist, either because of their deliberate omission, or, more problematically, from ignorance of what else is there in the part of the world being depicted.

These issues are particularly relevant to understanding the peopling of Scotland's islands by way of maps, especially for the more distant past. Many islands were inhabited for millennia before the stamp of earlier human occupation was first recorded on maps, by individual map makers in the eighteenth century and, more commonly and systematically, by Ordnance Survey from the mid nineteenth century. There are a few exceptions to this: Joan Blaeu's 1662 map of Shetland is notable in showing the broch at Mousa (fig. 2.1). Scotland's rich map history records the presence of people in various ways: as town maps, as forts and castles (see chapter 5), in the location of religious sites (see fig. 3.2 for example), in the symbolic representation of villages, and so on. Yet, while this is true, we need to recognise a further and inconvenient truth: knowledge in map form of Scottish islands' peopling and un-peopling over thousands of years – their 'deeper' history and geography – is largely a product of only the last 150 years or so.

Opposite. Detail from Ordnance Survey, Six-Inch to the mile, *Inverness-shire (Hebrides), Sheet LXX* (surveyed 1878, published 1880).

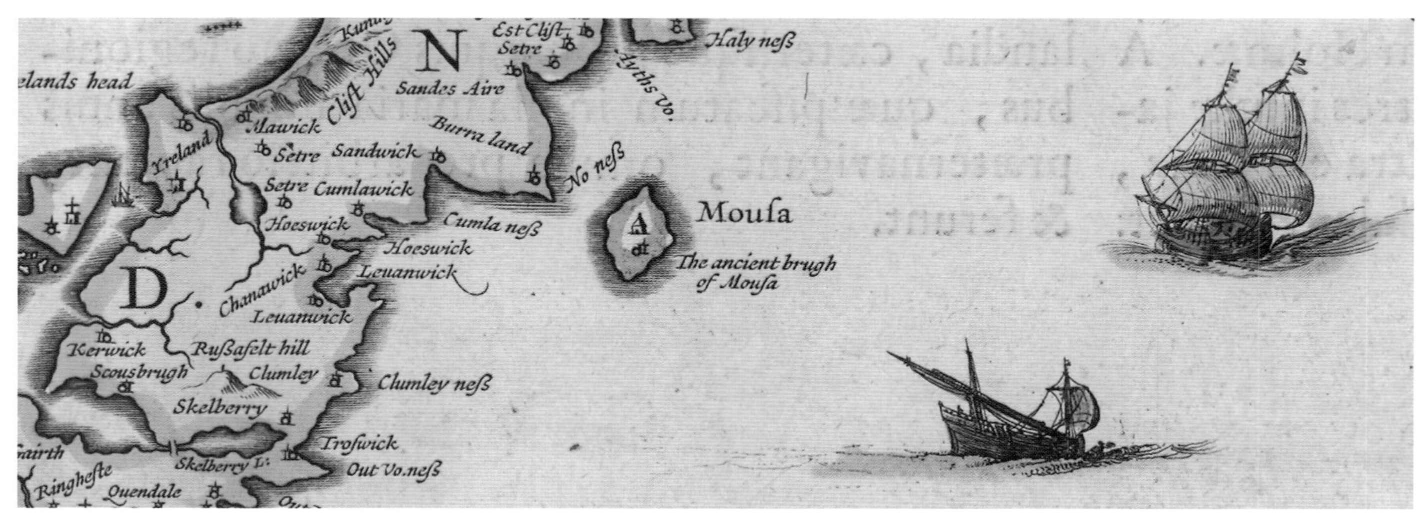

FIGURE 2.1
The broch at Mousa on the Shetland island of that name – 'the ancient brugh of Mousa' – as shown on Blaeu's map of 1662. It is unusual for early maps of Scotland to show settlement features from even earlier times. In the nineteenth century, as we shall see, Ordnance Survey began to develop a more systematic style to represent ancient settlement: see figs. 2.3 and 2.5 in particular.
Source: Timothy Pont/Joan Blaeu, *Orcadum et Schetlandiae Insularum accuratissima descriptio* (1662).

Early settlement geographies

Whatever the inhabitants of Scotland's islands may have called themselves in the past, we need to be careful about describing them as 'Scots' and their islands as 'Scottish'. The *Scotti*, originally Gaelic Irish, settled from the fifth century AD in what is now Argyll, but it was not until about AD 900, after centuries of political association and cultural assimilation, that Pict and Gael became Scot. In the Northern and Western Isles, place names testify to an enduring Norse presence between *c.*870 and *c.*1470 (see chapter 3). The result of this multiple peopling is that Scotland's genetic geography is a variable mix, perhaps especially in the islands: almost 60 per cent of the population in the Northern Isles is descended from the indigenous Picts, *c.*40 per cent are of Norse ancestry. For earlier periods, we have no way of knowing the names of peoples, nor should we associate islands' first human colonisers with later regional or national descriptors. Evidence of human presence survives in archaeological fragments and in built forms whose origins and dating are often subject to different, even contradictory, interpretation.

The first certain evidence for the peopling of Scotland dates to about 7500 BC, the Mesolithic. Because living off the mixed resources of land and sea was crucial, it is unsurprising that evidence of hunting–gathering–fishing communities is associated with Scotland's islands and coastal fringes. On Jura's east coast, for example, evidence of stone-set buildings, nut shells, and worked flints used to scrape animal hides suggests a community of foragers and hunters there between about 6250 and 6000 BC. On many islands, shell mounds – middens – testify not just to a diet based partly on limpet and other sea shells but, from the size and number of the mounds and the presence of harpoons and fish hooks, to a semi-resident population involved in foraging and fishing.

The features of Neolithic (literally, 'New Stone Age') settlement in Scotland between about 4000 and 2000 BC created, in part, a new settlement geography given the clearance of wooded areas and the development, albeit unevenly, of settled agriculture. In places where there was for obvious reasons a need for continuity of early settlement – access to the bare necessities of life – later evidence has sometimes obscured the earlier human presence. Understanding Neolithic Scotland reflects also the attention paid to its built features, such as chambered tombs. Archaeologists have identified five main types of chambered tombs in Scotland, based on their shape, structure and (rough) geographic spread. The so-called 'Clyde' tombs are found in Argyll and Arran with one or two examples in the Hebrides, in Perthshire, and in the south-west mainland. 'Orkney–Cromarty–Hebridean' type tombs are distinguished by this geographical label. The 'Maeshowe type' of passage-grave tomb is found only in Orkney, notably at Maeshowe itself, a chambered tomb dating from about 3000 BC and one of the most important Neolithic sites in Europe. 'Bargrennan' passage-graves are found only in south-west mainland Scotland. 'Clava-type' passage-graves and associated cairns are located only in Inverness-shire and adjacent counties. Spectacular as these sites for the dead are, Neolithic sites for the living in Scotland include, most famously, Skara Brae on Orkney Mainland and Knap of Howar on Papa Westray in Orkney.

Skara Brae well illustrates the slow, and sometimes uneven, 'emergence' of Scotland's archaeological heritage in terms of our knowledge of past settlement geographies. The site was first 'discovered' – uncovered, almost literally – as the result of progressive coastal erosion and a violent storm in 1850. Once thought to be a Pictish village and to date from some time between 500 BC and AD 500, the most recent of a series of excavations at the site now places its origins to about 3100 to 2450 BC and shows that the site underwent an almost continuous sequence of occupation, each leaving different traces of construction.

Where past island settlements such as at Skara Brae, at Rinyo on Rousay or at the Links of Noltland on Papa Westray survive largely below the present-day land surface, other distinctive features stand out in the landscape. The Stones of Stenness at Orkney, constructed about 3000 BC, and the nearby Ring of Brodgar, which has not proved possible to date with accuracy but probably dates to 2500–2000 BC, are Neolithic henge and stone circle monuments. Rather than see these monuments as individual sites, Maeshowe, the Stones of Stenness, the Ring of Brodgar and other smaller monuments in this part of Mainland Orkney are best understood as the surviving features of a complex Neolithic landscape. This is evident in the designation of these remains as a World Heritage Site from December 1999 and in the continuing excavation at the Ness of Brodgar: in combination, this is one of the most important ancient settlement sites in Europe.

In general none of this would be known from maps until, in the second half of the eighteenth century, antiquarians started to note and map such features (fig. 2.2). Even then, long-standing archaeological features were only sketched in, as is shown on William Aberdeen's map of 1769. Visitors did not know how to explain – and certainly not to date – these artefacts from earlier times. Reporting in 1792 upon his tour of the Orkneys, Principal Gordon of the Scots College in Paris recorded evasively how 'Different reasons have been assigned by different persons for the circular and semicircular form of these Scandinavian temples, for such they certainly have been.' 'Had I more time,' he concluded, 'I would beg leave to recommend it to any future traveller in those places, to pay a particular regard to the tumuli.' Later antiquarians did so, and more besides.

Early Scotland appears in more detail on maps only from the mid nineteenth century. Historians and antiquarians, including Principal Gordon, had speculated before then on Scotland's archaeological remains: Martin Martin and Sir Robert Sibbald in the 1680s; the Earl of Buchan's Society of Antiquaries begun in 1780; parish ministers writing the parish-based commentaries of the 'Old' *Statistical Account* under Sir John Sinclair between 1791 and 1799; and, later, Sir Walter Scott among literary figures. Some commentators, as we have seen for the Rev. Kenneth Macaulay writing about St Kilda in the 1760s, borrowed from earlier counterparts as they sought to explain earlier settlement patterns (see fig. 1.17). But the first formal surveys of early Scotland took place in the nineteenth century, many of them through the work of

a.

b.

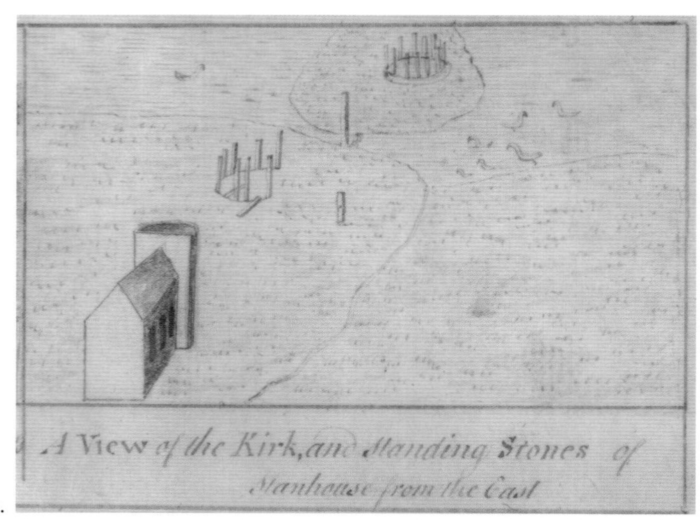

FIGURE 2.2

These two details, (a) and (b), taken from William Aberdeen's map of the Orkneys in 1769, show the stylised representation of an altogether older 'Scotland' on a map designed to show cadastral and administrative boundaries on the islands. Detail (a) shows two rather rudimentary stone circles: the one marked 'Stand^{g} Stones' indicates what is known today as the Ring of Brodgar; the other, south of the loch's edge and marked 'Standing Stones', indicates the site of the Stones of Stenness. Inset (b) shows the two henge and stone monuments in landscape style – what in the eighteenth century was called a 'scenographic view'. Scenography and topographic representation did not require accuracy. In his sketched depiction of the Ring of Brodgar (the uppermost monument in inset (b)), Aberdeen shows twelve standing stones. The site today has twenty-seven pillars, the remains of what, originally, was thought to have been sixty such stones.
Source: William Aberdeen, *A Chart of the Orkney Islands* (1769).

gentleman 'amateurs', natural historians and folklorists, or professional surveyors with antiquarian interests. The Ring of Brodgar was first surveyed by the naval hydrographer and chart maker Captain F. W. L. Thomas in 1849. The late Neolithic stones at Callanish on Lewis were known before then, but were only excavated from the surrounding peat in 1857 (fig. 2.3). As the officers of Ordnance Survey and the Hydrographic Office turned their skills to depicting Scotland's islands so they unearthed and mapped earlier Scotlands on the islands (fig. 2.4). Structures above ground, such as Scotland's many duns and brochs, obvious for centuries, only now began to appear on maps (fig. 2.5). One inhabited feature, the crannog, is an exception to this general cartographic appearance of Scotland's early settlement geographies only by the nineteenth century (fig. 2.6 and pp. 54–5).

Ordnance Survey did not have an archaeological officer until the appointment of O. G. S. Crawford in October 1920. With others, and in conjunction with bodies in Scotland such as the Royal Commission on the Ancient and Historical Monuments of Scotland, begun in 1908, and Historic Scotland (which two organisations came together in 2015 to form Historic Environment Scotland), Crawford was influential in promoting the use of aerial photography in archaeological work and in establishing standards for dating and representing on maps the presence of the human past in our landscapes. As new finds have been made, and new interpretations offered, so maps have changed in what they show and how (fig. 2.7). Excavations and aerial survey will continue to identify, unearth, and later map the Scotland beneath our feet.

Population and social change, *c.*1690–*c.*1841

Knowing how, when and why levels of population changed in the islands, either in aggregate or for individual islands and places within them, is hampered by the variability of sources. Before the first census of 1801, the parish registers, nominally begun in the mid sixteenth century, are the principal resource available by which to know Scotland's demographic trends

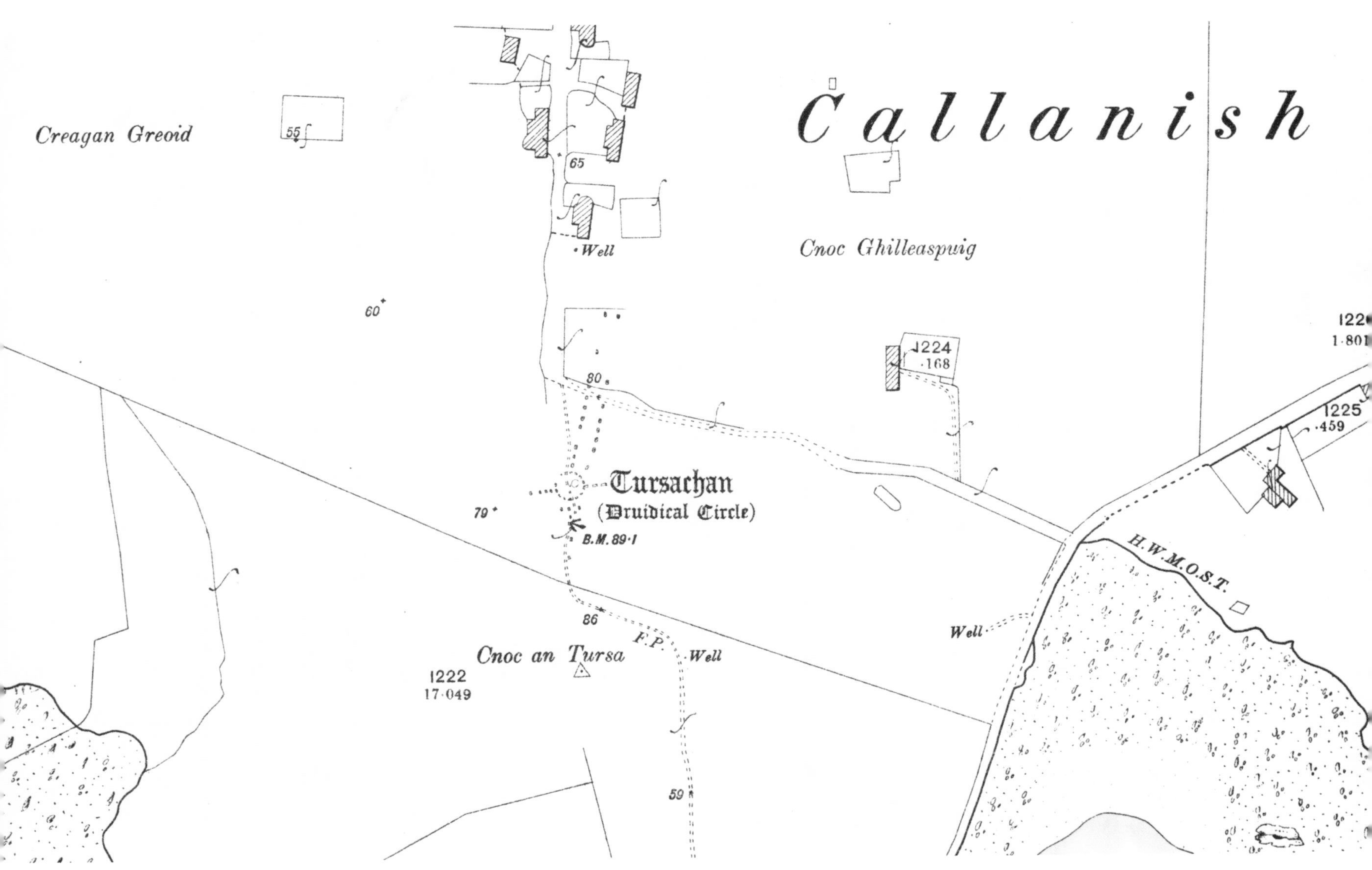

FIGURE 2.3

The standing stones at Callanish – Clachan Chalanais or Tursachan Chalanais in Scottish Gaelic – were visited in the late 1690s by the Gaelic-speaking historian and natural philosopher Martin Martin, as part of his enquiries for the Royal Society of London into the natural productions and beliefs of the Hebrides/Innse Gall and Western Isles/Na h-Eileanan an Iar, including St Kilda. These were published as *A Late Voyage to St Kilda* (London, 1698) and *A Description of the Western Isles of Scotland* (London, 1703). Martin, who accompanied John Adair during the latter's maritime surveying of the Outer Isles (see chapter 4), was informed by local inhabitants that Callanish/Calanais 'was a place appointed for people in the time of heathenism'. On this map of 1897, the epithet given is 'Druidical', a term first applied to the site by the eighteenth-century antiquarian William Stukeley. Suggestions made in the 1980s that the stones at Callanish/Calanais, which date from the late Neolithic (*c.*2900–2600 BC) and had fallen out of use by about 1000 BC, were aligned celestially so as to provide what, in effect, was a prehistoric lunar observatory, do not now have widespread credence.

Source: Ordnance Survey, 25 inch to the mile, *Ross & Cromarty (Island of Lewis), Sheet XXV.2* (revised 1895, published 1897).

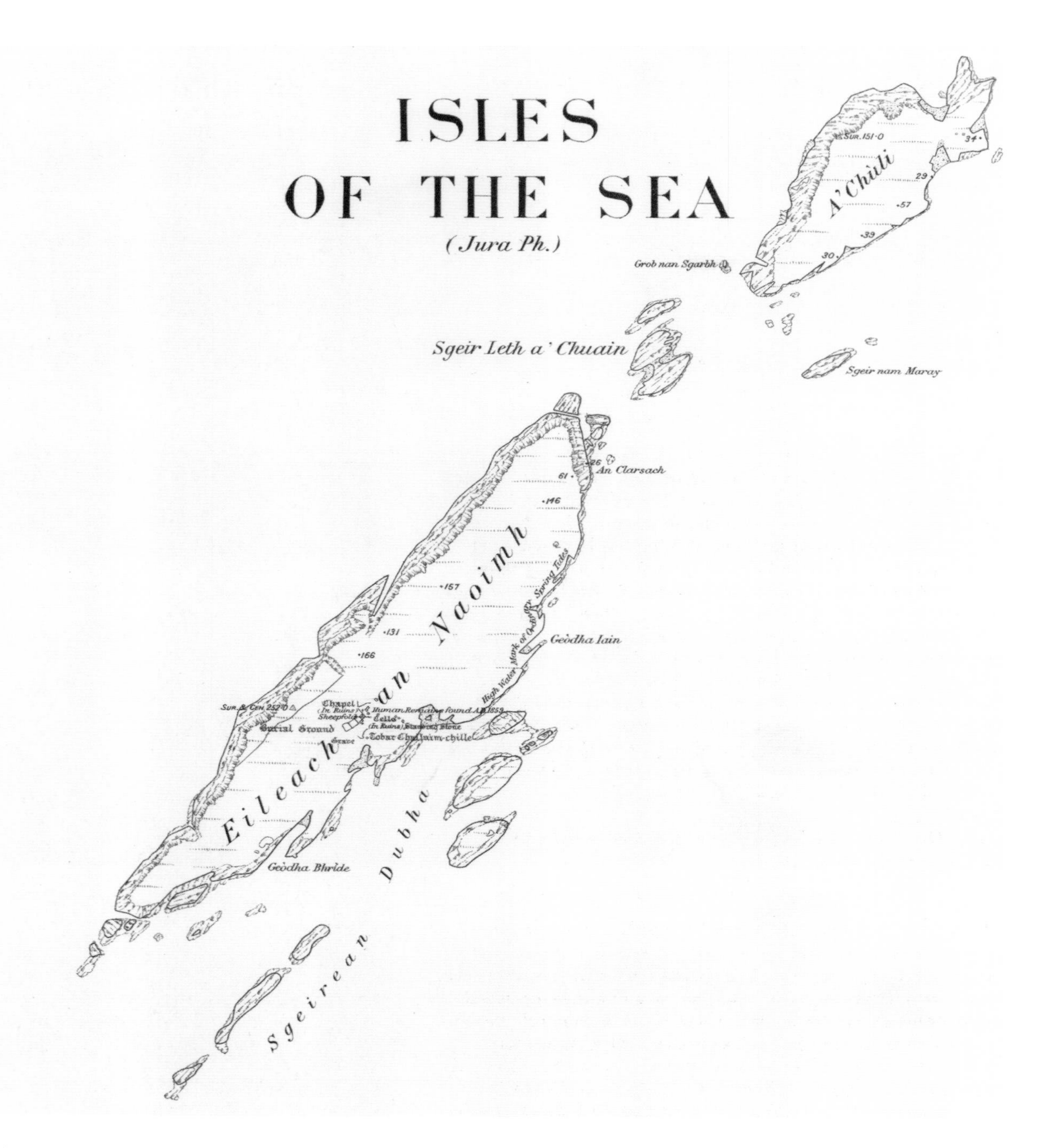

FIGURE 2.4

This detail of a map of Eileach an Naoimh ('Isle of the Saints'), the southernmost of the Garvellachs ('the Isles of the Sea'), shows how relatively recently the evidence of Scotland's early peopling was brought to light and placed on maps. Eileach an Naoimh has the remains of a Celtic monastery, thought to have been established by St Brendan in about AD 542, and numerous corbelled stone cells, which resemble large beehives, for the use of monks.

Source: Ordnance Survey, Six-Inch to the mile, *Argyllshire, Sheet CXXVIII* (surveyed 1875, published 1881).

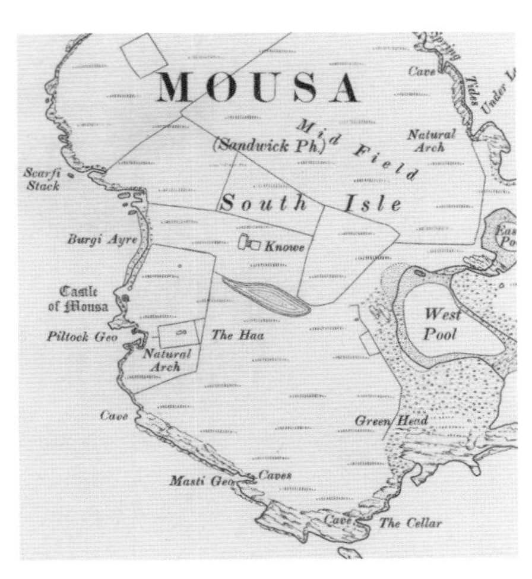

FIGURE 2.5

(*Above*) The map shows a plan view of the Broch of Mousa, or Mousa Broch, on the island of that name in Shetland. Dating from about 100 BC, it is one of the finest examples of its type in Scotland. Brochs, a type of Atlantic roundhouse, Iron Age in date and unique to Scotland, have been subject to different interpretations since they began to be the subject of serious study in the nineteenth century. Initial theories that they were simply communal defensive structures were later replaced by the view that they were buildings of high status, even having a symbolic function, a view itself replaced in the 1980s by a return, in modified form, to the defensive interpretation. Relatively few brochs have been the subject of excavation, and scholars are even divided over their exact definition: Historic Environment Scotland identifies 571 throughout Scotland; some archaeologists see the figure as little more than 100, arguing for the lower total on the basis of differences in form and structure. Mousa Broch certainly had a defensive role in the past, as mentioned in two Viking sagas. What Sir Walter Scott considered a 'Pictish fortress' was in 1877 described as the 'Castle of Mousa'.
Source: Ordnance Survey, Six-Inch to the mile, *Zetland, Sheet LXIII* (surveyed 1877, published 1881).

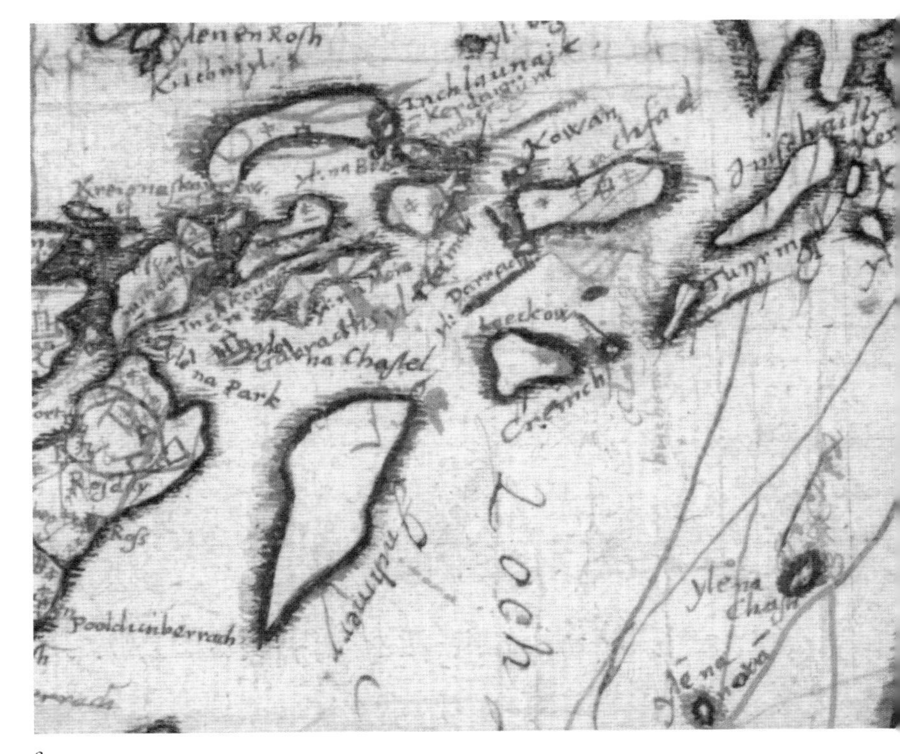

a.

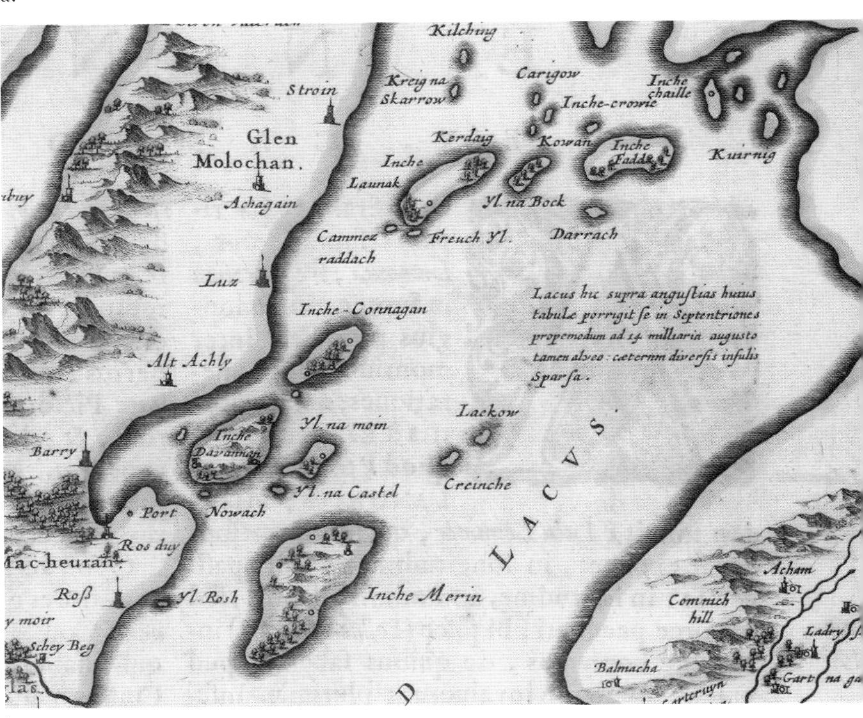

b.

FIGURE 2.6

(*Right*) Crannogs are partially or completely artificial islands. Many in Scotland, Ireland and continental Europe were occupied for considerable periods, a few for as long as 5,000 years, between the Neolithic period and the sixteenth century AD. As for brochs (see fig. 2.5), archaeologists and others are not agreed as to the definition of a crannog, principally on the basis of their varied form and methods of construction. As we show elsewhere (see chapter 5), some crannog sites were occupied until the early 1500s and were clearly marked on the manuscript maps of Timothy Pont in the later sixteenth century. While Pont may have marked these sites, his manuscript depiction did not always translate to the printed versions by Blaeu. Detail (a) shows Pont's maps of the islands of Loch Lomond. The printed map produced by Joan Blaeu in 1654 (b) looks rather different.
Source: (a) Timothy Pont, Loch Lomond, Pont 17 (*c*.1583–1614). (b) Timothy Pont/Joan Blaeu, *Levinia Vicecomitatus, [or], The Province of Lennox called the Shyre of Dun-Britton* (1654).

BUTT OF LEWIS
DUN CARLOWAY
CALLANISH
GARYNAHINE
DUN CROMORE
OUTER HEBRIDES
DUN AN STICER
VALLAY STRAND
DUN TORCUILL
BARPA NAM FEANNAG
BARPA LANGASS
DUN LIATH
DUN HALLIN
DUN SULEDALE
KENSALEYRE
DUN FIADHAIRT
DUN GERASHADER
DUN BEAG
REINEVAL
LOCH A' BHARP
DUN RINGILL
RUDH' AN DUNAIN
DUN BORERAIG
DUN GRUGAIG
DUN BHARPA
AN SGURR
SRON AN DUIN

a.

Dun Carloway
Callanish
Garynahine
Dun an Sticer
Clettraval
Dun Torcuill
Barpa nam Feannag
Tigh Cloiche
Barpa Langass
Dun Hali
Annait
Dun Boreraig
Dun Fiadhai
Reineval
Kilpheder
Loch a' Bharp
Dun Cuier
Dun Bharpa
Sgor nam Ba
Sron an Duin

b.

c.

FIGURE 2.7
These details (a–c) of the Western Isles/Na h-Eileanan an Iar, taken from Ordnance Survey's small-scale series mapping of Ancient Britain, exemplify not only the discovery and geography of earlier settlement sites but also the different ways in which such discovery has evolved over time, differentiated by symbols showing type of feature, relative age, and so on. Detail (a) is from the map of Ancient Britain produced in 1951; (b) from the equivalent map in 1982; and (c) from 2005.
Source: Ordnance Survey, Ten Miles to One Inch series, Maps of Ancient Britain, 1951, 1982 and 2005. Reproduced with the permission of Ordnance Survey.

and their geographical variation. In the islands especially, these were unevenly kept until civil registration – enumeration of the vital facts by officers of the state – replaced the Church-based collection of population statistics from 1854. Before 1755, there is no single source from which it is possible to derive even an estimate of Scotland's population as a whole, far less so for the islands where, often, the population was scattered, and systems of parochial administration and population enumeration did not work. Even when we do have figures such as those derived by the Edinburgh minister Alexander Webster, in his 1755 'An Account of the Number of People in Scotland' – known today as 'Webster's Census' – they must be treated with caution: on the island of Gigha, for example, Webster lists the island's population as 514, of which total '102⅘' persons were 'fighting men'. On Tiree at the same time, there were '301 and four-fifths' fighting men in a total population of 1509.

Allowing for variations in cause and timing, the population of Scotland's islands increased steadily from the end of the seventeenth century and did so for several reasons. Famine – occasioned by harvest crises (the result of wet summers, harsh winters, and delayed growing seasons) – became less common after the so-called 'Ill Years' of the 1690s and, where and when it did occur, could be relieved by the import of grain or the export of black cattle. Epidemic disease grew less common and less fatal. Mortality was in consequence gradually lower throughout the eighteenth century although infant mortality remained high. There were exceptions, of course: the people of the Outer Hebrides/Innse Gall suffered during the harvest failures of 1782–83 in particular (the result of a prolonged winter and a short, wet, cold summer). The high death rates of the very young on St Kilda relative to the rest of Scotland, even by the late nineteenth century, were a source of recurrent concern to inhabitants and to outside commentators. What is known of fertility rates, marriage rates and age at first marriage is too fragmentary for us to be able to say anything definitive before the nineteenth century. Yet we do know, on numbers alone, that Scotland's population experienced a growth rate of 28 per cent between 1755 and 1801. Since this was also a period of considerable emigration, the actual rate of increase may have been as much as 32 per cent or 33 per cent. Rates of increase were similarly great in the opening decades of the nineteenth century until between 1831 and 1841, when Scotland's population growth rate slowed, largely as the result of increased mortality in its cities and larger towns. Scotland's population increased again from the later nineteenth century, the result of mortality decline which was itself the consequence of the better administration of public health and the control of epidemic disease in the nation's burghs.

For Scotland's islands, three further matters must be noted in relation to this all-too-brief population history: marked regional and inter-island differences, extensive migration, and, in the Western Isles/Na h-Eileanan an Iar until the later eighteenth century, the dominance of the clan system of tenurial holding and land management. Migration involved temporary Highland–Lowland, islands–mainland, population displacement for harvest labour and to other forms of seasonal employment in the south. In times of stress, this acted as a safety valve demographically and economically. Emigration drew off the people permanently. More so in the past than today, the Northern Isles differed from the Western Isles/Na h-Eileanan an Iar in their geography and in the systems of landholding. Shetland's upland interior is largely unproductive land, even today, except for the Tingwall valley in the mainland and Dunrossness to the south. On the Orkneys, by contrast, most islands – with the exception of Hoy – are low-lying and have relatively large proportions of land suitable for arable cultivation. The Western Isles/Na h-Eileanan an Iar have considerable geographical variation in the availability of land for arable agriculture. Lismore and Tiree, for example, are generally fertile and low-lying; parts of inland Lewis/Leòdhas are moorland yet the island has a western fringe of machair, fertile sandy grassland, which provides some of the richest land in all the islands. Eastern Harris/Na Hearadh is different yet again in its acid soils and lack of good ground.

To these facts of terrain we must add differences in the ways land was held and managed. Norse rule in the Northern Isles may have formally ended by about 1470, but the Norse system of odal land ownership, which was based on absolute possession of the land and the sharing of the land amongst sons and daughters on the death of the odal holder, lasted until at least the early seventeenth century. The result was a social and agrarian landscape of small-to-medium independent proprietors. In the Western Isles/Na h-Eileanan an Iar, the clan system placed a particular emphasis on the kin group, upon the senior ruling family or sept within it – the clan chief especially – and upon the mutual fulfilment of obligatory duties: military service on behalf of the clan or chief should it be required, or the chief's support for one's kindred and for the customary ways in which land was inherited and worked.

The working-out of these demographic, geographical and cultural circumstances was different on different islands. In the Western Isles/Na h-Eileanan an Iar, the gradual easing of mortality and of epidemic diseases, together with the dominant clan system, allowed the build-up of population after about 1690, especially after the introduction of the potato from 1740. Crofting as a form of land management is not an archaic feature of the islands, Highland mainland and Northern Isles; it was largely created, from the second half of the eighteenth century, by landlords keen to retain a paying tenantry but on smaller holdings. This also created a new landless class of people, the cottars. By using seaware (seaweed) as fertiliser and making use of the *cas chrom*, or wooden plough, even small patches of arable land could be put to work to sustain a largely subsistence economy where rent was commonly paid in kind and where cash income came from the sale of black cattle or local fishing.

Where island soils were particularly fertile or the interests of clan chiefs prevailed, population numbers could increase to levels unimaginable today, especially where temporary migration or by-employments such as weaving or fishing acted to supplement the economy and where the cottar class remained to work the land, but not own it or rent it directly. One 'census' in 1764–65 of the Small Isles parish – the islands of Canna, Muck, Rum and Eigg – prompted by the Rev. Dr John Walker's enquiries into the economy and population of the Hebrides, enumerated the total population as 1,159 persons, the great majority of whom were aged under 39. Near-contemporary maps (fig. 2.8) give no clue as to the balance of crofter, cottar, or other population category within this 'over-crowding'. Strictly, by the standards of the time, we should not describe it thus. This was a populous island parish, one of many, inhabited by loyal tenants working to established customs, and dependent upon the potato. By 1821, the population of the Small Isles was 1,620 persons. In 2011, the combined population total of these islands was 161 residents. On Tiree, renowned for its fertility and well populated in consequence, grain payments recorded in seventeenth-century rentals accounted, on average, for between 40 per cent and 50 per cent of the total rent paid. In contrast, on Mull by the 1740s, most rent was made up in livestock produce (fig. 2.9).

Since islands could thus be different in their resource base, we should similarly resist easy generalisations in explaining the collapse of this system from the mid eighteenth century. Market forces certainly spread from the south, but the people of the islands were always connected commercially with the world beyond themselves. Political loyalties were tested and, for some, found wanting, either because of that spirit of independence endemic to clanship or, after 1707, and the Jacobite rebellions of 1709, 1715 and 1745, because others saw that system as backward, inimical to standards of Britishness and civilisation. The large-scale forcible clearance of people to make way for commercial sheep farming was a feature of the nineteenth century, but people were leaving the islands for Lowland towns, London or the Americas from the 1760s and earlier. Reports from Tiree estate managers in the 1730s spoke of what we would now term 'environmental degradation': livestock numbers were too many for the island's pastures. Agricultural improvers in the eighteenth century railed at 'the prejudice of the people' and their adherence to established ways. Rather than see change from north to south or, even, island by island, it is better understood in terms of estate by estate as, one by one, chiefs became landlords, and kinsmen became rent-paying crofters or landless cottars.

Today, offshore banking often involves islands' status as tax havens for capitalism's elite. From the 1730s, Scotland's

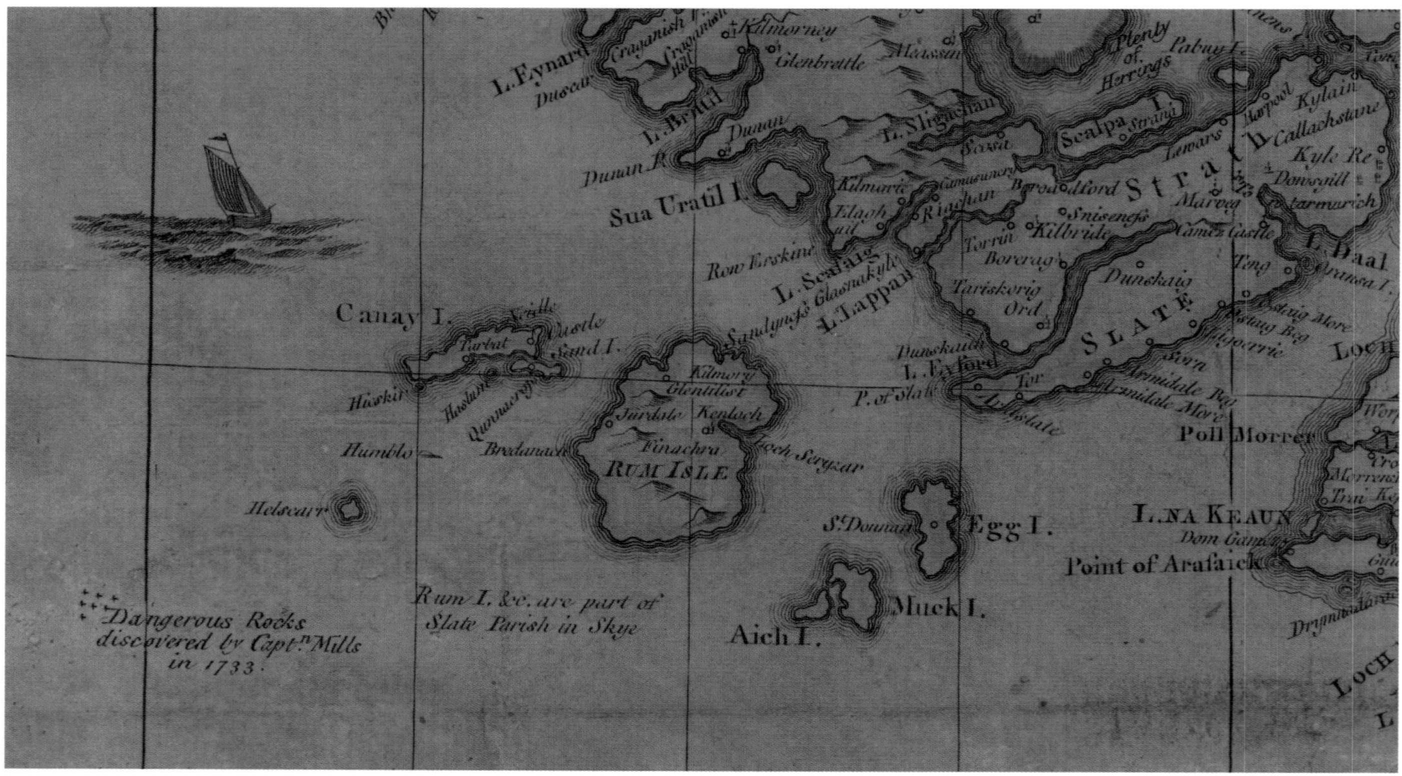

FIGURE 2.8

(*Above*). From this map of the Small Isles (Canna, Eigg, Muck, Rum), taken from the north-west section of Thomas Kitchin's map of 1773, we can see that Kitchin provides information of use to mariners regarding the navigational hazards to the west of this island group as well as the location of several townships on the islands (with the exception of those on Eigg). Thomas Kitchin was a central figure in the map-engraving and publishing world of eighteenth-century Britain, chiefly in London, where in 1732 he was apprenticed at the age of thirteen to the leading map maker Emanuel Bowen. Kitchin was involved with several maps of Scotland, including John Elphinstone's map of 1746, and was behind the then-novel inclusion of maps in eighteenth-century periodicals and magazines, notably in the *London Magazine*. Although he was renowned as an engraver and for the quality of his map work, there is no evidence that Kitchin visited Scotland to undertake detailed original survey; he died in 1784.
Source: Thomas Kitchin, *A New and Complete Map of Scotland and Islands Thereto Belonging* (1773).

FIGURE 2.9

(*Opposite*). The island of Mull in the mid eighteenth century, taken from the south-west section of James Dorret's map of 1750. The impression given here is of a populous island: smaller townships are shown, and brief comments made about the location of mineral resources. Tobermory – here 'Tober Morey' from the Scottish Gaelic 'Tobar Mhoire' (the Well of Mary) – is marked, although the modern settlement dates from 1788, when it was established as a fishing port by the British Fisheries Society following preferential financial terms offered by the Duke of Argyll, one of the Society's directors. Despite some early misgivings about the establishment of Tobermory as a fishing village given its distance from what were then the main fishing grounds, the town prospered and it is today the main centre of population on Mull and a popular base for tourism (but see also fig. 2.11). The similarity in title between Dorret's map of 1750 and Kitchin's map of 1773 (fig. 2.8) is interesting. James Dorret was employed by the Duke of Argyll: his map of 1750, which made use of Roy's Military Survey of 1747–55 and other contemporary maps, was used as the 'base' map for much of Scotland's mapping over the next forty years. Although there is no direct evidence to support the connection, it is possible that Kitchin's choice of title for his later map was a deliberate homage to the map maker and to his patron, the influential Duke of Argyll.
Source: James Dorret, *A General Map of Scotland and Islands Thereto Belonging* (1750).

P.t of Ardnamurchan
ARDNAMURCHAN
Kilchaan
Mingary Castle
Acharacle
SUNA
Loch
Row na Killich
Penmoir
Tober Morey Bay
Tober Morey
I. Calve
Sound of Mull
Dutchman's Cap
Carnburgh I.
Pladda
Mingary
Pemoloch
QUINES
Killvore
L. Ryfa
Drumsin
Aros C.
MULL
Glackuggary
Lefford
Kilmolowig
Rowintleve
Gomedra
Kilbranan
Lagganulva
Loch na Gaul
Keaul
Ulva
Staffa I.
Arnskerry
Gribon Part of
ISLAND
L. Ba
Glenkanner
Bradilaltach
Scallasdell
TOIR
Torghormaig Kilpatrick
Ardchoirk
C. Dowart
Balnahard
Kilummer
Bailmeanach
Balnahays
Rosal
Ringael
Ringross
Kilinaig
Torran
Ormsaig
BROLOS
Burg
Ron Burg
Loch Sereedan
Kilpatrick
Ardtoun
Bunessan
L. Lye
ROSY
Carfwick Bay
Glenbair
Karowick
Inigart
Karval
TONTIRE
L. Buie
C. Moy
L. Spelve
Crogan
Drumnatayan
Tychadba
Iona or Ikolmkill
Haynish
Soy
L. Don
Kilean
Ina Gaun
Mac Allasters Bay
Point of Ardintride
Lismore
Carrick a Rock
Kerrera
Back
Loch
Ardincaple
Obain
Carnbaan
Clachan
NETHER
Eysdale I.
A fine Slate quarry
Seil
Balechuan
Dunmore
Melford
L. Melfort
Lung Isle
Orlarack
Shuna
Searba I.
Ardtornish
L. Aylin
Kinlochalen
Morvern
Ternat where Copper Mines are wrought
Glendow where Lead Mines are wrought
Kilcolmkil
Knock
Drumnin
Auliston
Laggan
Sallachan
Kenloch Teagus

FIGURE 2.10
The island of Gigha, from the south-east section of George Langlands' *Map of Argyllshire* (1801). At the time this map was drawn, the population was 556 persons in an island which, at its maximum, is only about 6 miles in length and 1.5 miles in breadth with limited areas suitable for arable. This number fell to 398 by the 1891 Census and to 110 by 2001. Today, the island has a resident population of 163, a total supplemented considerably in the summer by tourists. Since 2002, the island has been owned and managed by its inhabitants through the Gigha Heritage Trust, thus arresting a series of failed absentee landlord and estate owners' schemes which had served only to progressively depopulate the island. The Gigha buy-out followed another island purchase by its inhabitants, that of Eigg, in 1997.
Source: George Langlands and Sons, *This Map of Argyllshire . . .* (1801).

islands and their inhabitants were themselves the commodities of mercantile capitalism, increasingly bought and sold as assets according to landlords' whims and market values. For some islands, these facts have been reversed only in the late twentieth century (fig. 2.10). Yet, even in the face of such changes, island populations could still be retained if chiefs-cum-landlords continued to privilege customary obligations over capitalist advantage, if there were sufficient farming or other resources to work with – and if the potato harvest could be depended upon.

Leaving and re-peopling after *c.*1841: the traces on maps

The people of the islands were no stranger to food shortages and scarcity. The nineteenth century opened with several instances of harvest failure – in 1806–07, 1811 and 1816–17 – but, commonly, income from the sales of black cattle was used to purchase grain and offset severe difficulty. Agrarian societies are always at risk – from the vagaries of climate, from fluctuations in price and, in the Western Isles/Na h-Eileanan an Iar especially from the 1820s and 1830s, from the demands of landlords for greater levels of rent. In 1836 and again in

1837, many island communities faced near starvation as partial but successive failures in the potato crop and in the oat harvest reduced many people to destitution. Worse was to come.

On Skye in February 1846, it was reported that the potato crop was failing, though its cause, blight, was not yet understood. On Harris/Na Hearadh in the following month, the potato crop was 'all but a complete failure'. By December of that year, the Free Church of Scotland, which established a Destitution Committee to aid the people of the islands and Highlands, reported that 79 per cent of the potato crop in the crofting districts had failed utterly. Approximately 200,000 people were reckoned 'destitute of food'. In 1847, the potato crop failed again, and grain harvests were much reduced 'by severe and boisterous weather'. On Tiree, officers of the Destitution Committee reported desperate scenes in November 1848:

> The island, although the most fertile, is, notwithstanding, most impoverished, in consequence of the excessive numbers of population. Of 4266 people, 272 families (or 1360 souls) are cottars, nearly all of whom are entirely dependent upon the noble proprietor and the public for support ever since the first failure of the potato crop. Besides the cottars, there are a very large number of crofters, who are, for several months in the year, very destitute of food . . . in the month of July last [1848], there were 2115 people (and these exclusive of paupers on the poors' roll) receiving their relief meal, and, who, without that relief, would in all probability have perished for want of food. Nothing could be more unfortunate or deplorable than the condition of these people.

We cannot be sure of the exact details, and the effects were locally variable, but at least half and in some cases much more of the potato crop failed in Scotland's western island parishes each year between 1847 and 1850, with serious failure also in 1852 and 1854. Portree on Skye, which served the northern Hebrides/Innse Gall, and Tobermory, which provided for the more southerly affected areas, became relief depots and temporary homes for the dispossessed (fig. 2.11). For a decade from 1846, the peoples of the islands faced, almost daily, a struggle for existence.

Island parishes where crofting was dominant were the hardest hit: on Lewis/Leòdhas and Harris/Na Hearadh, Skye, Tiree and Coll, Mull, the Small Isles, Benbecula/Beinn na Faoghla and Barra/Barraigh. The effect was exacerbated by the relative failure of temporary migration to provide sufficient cash income, by the policies of clearance-minded landlords, and by the decline in earnings from cattle sales given a downturn in prices. Emigration was already underway but the combined result of these circumstances was to sharply accelerate the existing rates of out-movement. On Lewis/Leòdhas, half the population of Uig parish left between 1841 and 1861; almost a third from Jura and from Barra between 1841 and 1851; perhaps a little over a quarter of the people from each of Skye's parishes in the ten years from 1851. Yet the effects were also very varied.

Mingulay/Miùghlaigh, one of the Bishop's Isles, was near collapse in 1849: 'a more miserable place, or a more wretched population, can scarcely be conceived'. But because it was part of the MacNeil estates, several families evicted from nearby Barra/Barraigh chose to live there rather than emigrate: population on this small island – about 2.5 square miles in area – increased from 113 in 1841 to 142 by 1891 and 135 by 1901. From 1907, however, the island began to lose its people permanently: fishing was hazardous in the absence of a sheltered landing, returns from the land were poor, attempts to seize land on adjacent islands by forcible occupation failed. By 1912, the island was deserted (fig. 2.12).

Perhaps because it was so relatively recent and has been seen as an especially poignant manifestation of wider and longer-term circumstances, the departure of people from the archipelago of St Kilda and its main island, Hirta, has a particular place in the Scottish cultural and geographical imagination. The island's permanent residents left in 1930 – the last

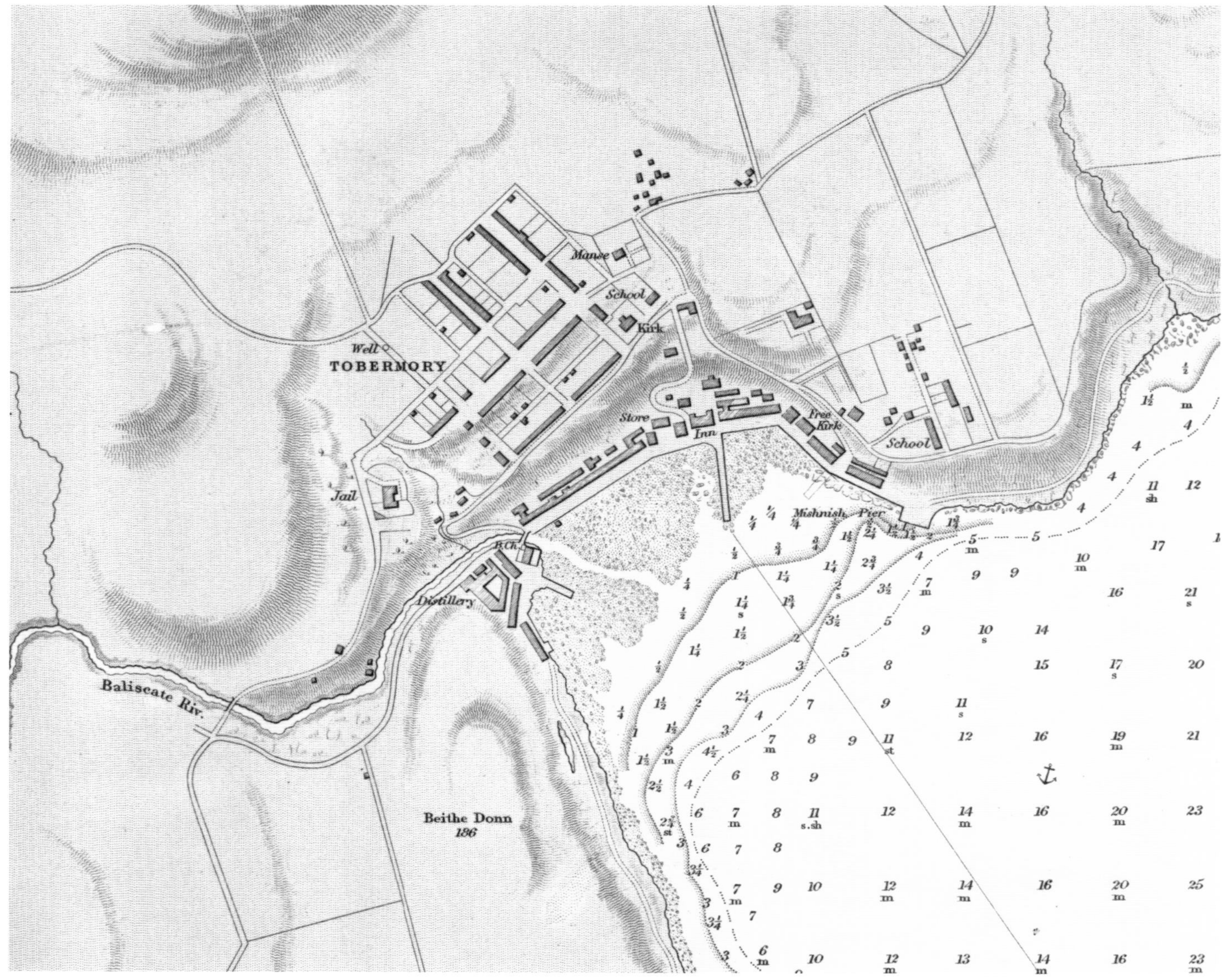

FIGURE 2.11

This image, an inset in plan view of the village of Tobermory and its harbour, is from the 1847 Hydrographic Office chart. It is something of an irony that the surveyors of the Hydrographic Office should have mapped the town and its harbour and given this impression of neat order at just the time Tobermory was becoming a famine relief depot for the people of the southern Hebrides and a temporary transit camp for destitute islanders. Between 1847 and 1852, Sir James Matheson, proprietor of Lewis/Leòdhas, obtained summonses of removal for 1,367 tenants on his estates and, in 1850, 660 persons (132 families) were removed from Barra/Barraigh alone. Many were shipped to Tobermory, thence south to the urban Lowlands. Tobermory was also the collection point for islanders cleared from the island of Ulva between 1847 and 1850. Not all moved on: both as a consequence of the opportunities to be had from managing others' misfortune and from outsiders staying, the population of Tobermory increased by 11 per cent between 1841 and 1861.
Source: Hydrographic Office, *Tobermory Harbour.* Admiralty Chart 1836 (surveyed 1847, published 1848).

a.

b.

FIGURE 2.12
The presence (and absence) of people which is inferred from the mapping of human settlement – and the ways in which maps capture the traces of such occupation but may not always accurately date or explain them – is clear from these details of maps of Mingulay/Miùghlaigh. Detail (a) shows the outline and distribution of dwellings by the late 1870s (the island's population in 1881 was 150 persons) together with the school house. By 1910, the year in which the school closed, there were only six families resident on the island. They left in 1912. The island was briefly but only seasonally occupied in the summer of 1930 and 1931, the latter the year in which Ordnance Survey published its 'Popular edition' map of Mingulay/Miùghlaigh (b). No trace of that earlier human settlement was marked. The island was acquired by the National Trust for Scotland in 2000.
Source: (a) Ordnance Survey, Six-Inch to the mile, *Inverness-shire (Hebrides), Sheet LXX* (surveyed 1878, published 1880). (b) Ordnance Survey, One-Inch to the mile, *'Popular' edition, Scotland, Sheet 33* (1931).

surviving of them, Rachel Johnson, then an eight-year-old girl, died only in April 2016 as this book was being completed. As we have illustrated in this chapter, maps can help capture the new lives of islands as, also, and for a variety of reasons, they cannot always document the dynamic ebb and flow of human life there. Nearly thirty years after it was abandoned, St Kilda was once more permanently inhabited following the advent of the Cold War, the fear of nuclear attack from the Soviet Bloc and the decision by the British Government to install a rocket range on Benbecula/Beinn na Faoghla on the Hebrides/Innse Gall (see chapter 5). Initially, fears were expressed that the military would look to occupy, certainly modify and perhaps demolish some of the buildings in the village on Hirta. In May 1957, D. R. MacGregor of the Department of Geography at the University of Edinburgh spent a few days on the island, his project intended as a sort of 'rescue cartography' (fig. 2.13). In the event, fears over any major changes wrought by the military re-occupation of St Kilda were unfounded. And the population of many of the Northern Isles rose swiftly in the 1960s and 1970s as, first, military activities and, second, oil-related employment brought people to islands that had been in danger of losing their population and the infrastructure that goes with it. Tourism considerably supplements resident numbers. The effects and scale of these population movements is not always apparent on maps – even if it is the case that a much older and populated Scotland began to appear in the work of Ordnance Survey in the nineteenth century and in antiquarian sketches by men like William Aberdeen in the later 1760s.

FIGURE 2.13
Map of Hirta, the main island of the St Kilda archipelago, produced by D. R. MacGregor of the University of Edinburgh, May 1957. MacGregor taught map production and interpretation as part of his teaching duties in the university's Department of Geography: his skills with the plain table and theodolite are clearly evident here. The village which features centrally here was built in 1834, replacing the medieval settlement. MacGregor captures the archaeological and landscape features as well as the form of the settlement.
Source: D. R. MacGregor, *St Kilda Village* (surveyed May 1957).

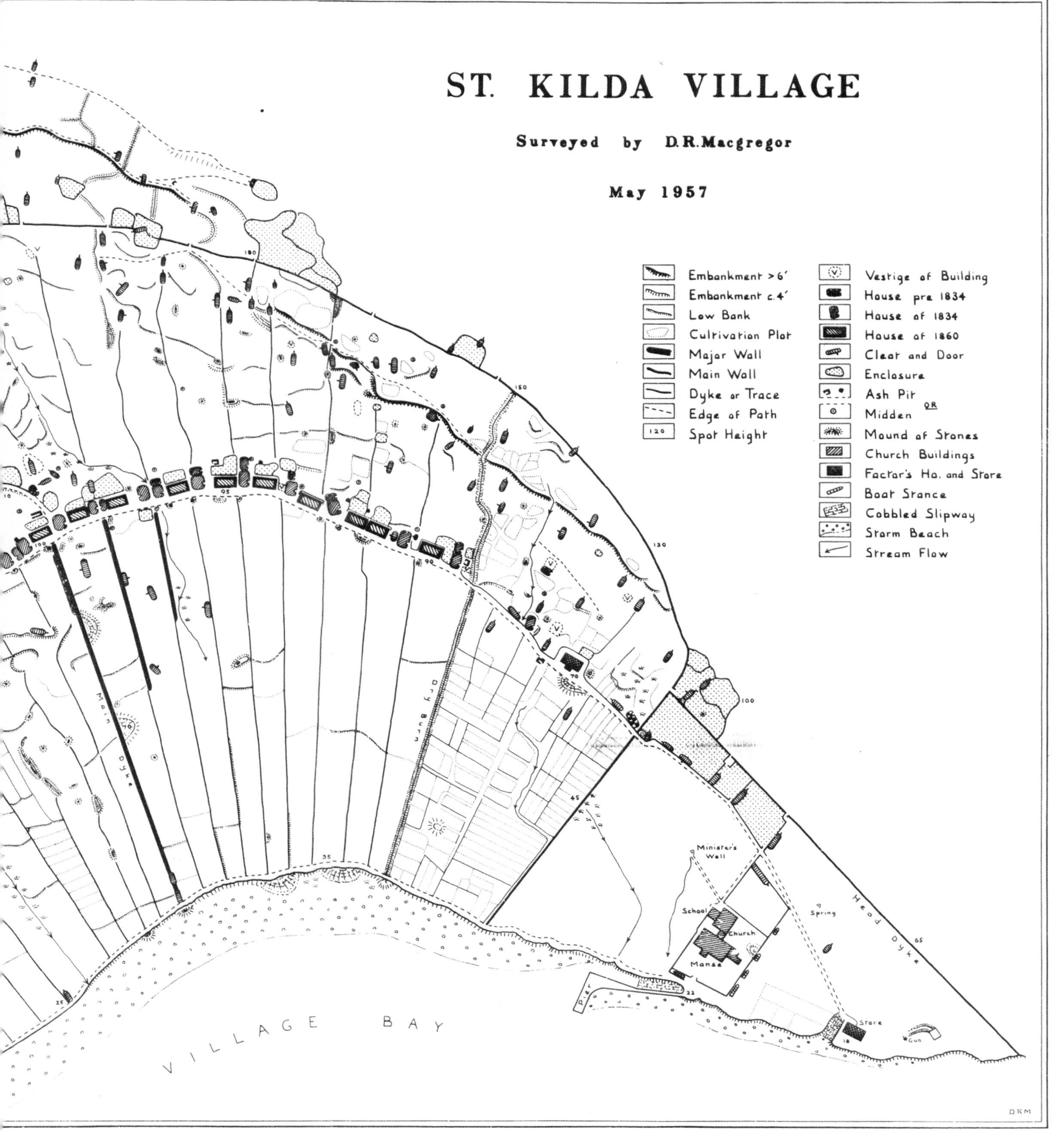

...uted in eight days by one person using theodolite and plane-table

C18 : 21 (2)

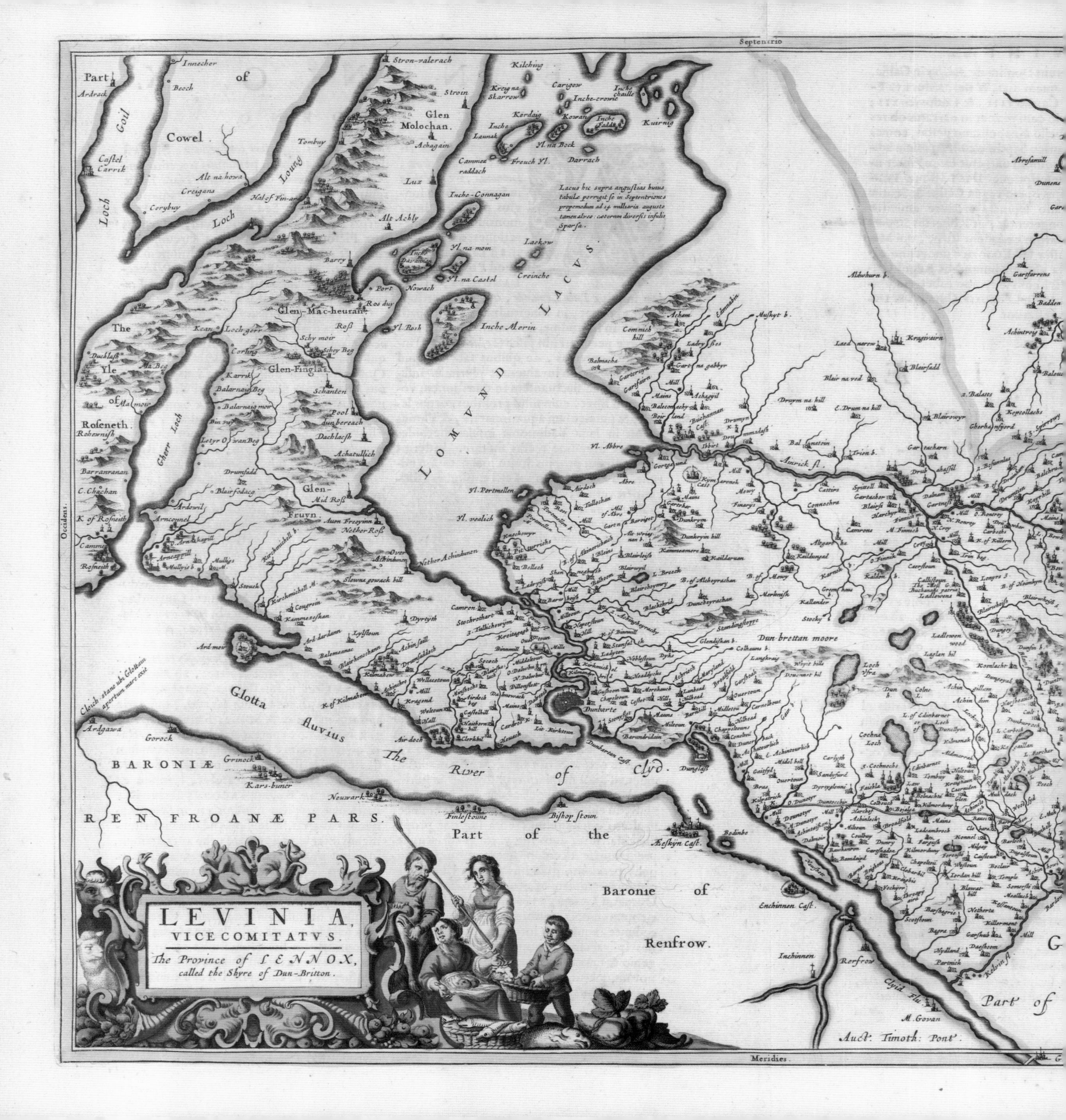

Septentrio
Meridies
Occidens.
LEVINIA, VICECOMITATVS.
The Province of LENNOX, called the Shyre of Dun-Britton.
Auct: Timoth: Pont.
Part of Cowel
Loch Goil
Loch Loung
The Yle of Roseneth.
Gherr Loch
LOMOND LACVS
Lacus hic supra angustias huius tabulæ porrigit se in Septentriones propemodum ad 14 milliaria augusto tamen alveo: cæterum diversis insulis Sparsa.
Glotta fluvius
The River of Clyd.
BARONIÆ RENFROANÆ PARS.
Part of the Baronie of Renfrow.
Clyid Flu.
Dun-brettan moore
Dunbarton
Glen-Mac-heuran
Glen-Fingla
Glen Molochan.
Inche Merin
Inche-Connagan
Loch Vsra
Cochna Loch
Ard-mor
Ardgawa
Gorock
Grinock
Kars-buner
Newwark
Finlestoune
Bishop stoun
Æeskyn Cast.
Enchinnen Cast.
Inchinnen
Renfrow
M. Govan
Partnich
Kelvin fl.
Anrick fl.
Buchannan Cast.
Roseneth
Cloich-stane ubi Glotta in apertum mare exit

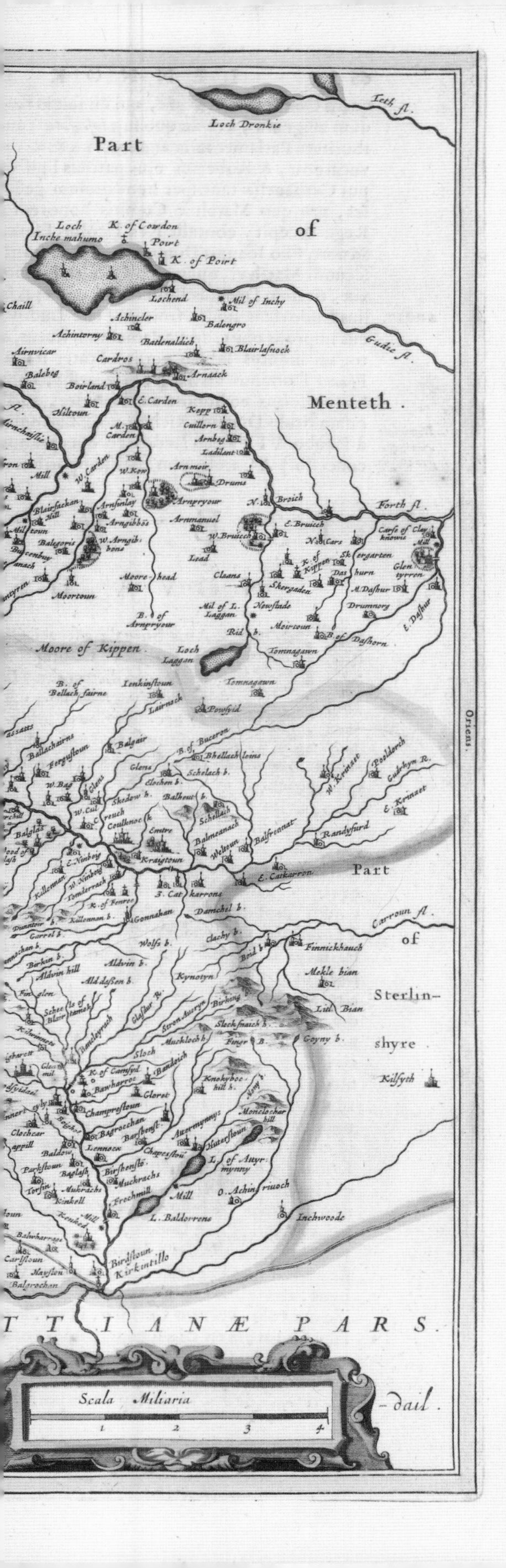

Joan Blaeu, *Levinia Vicecomitatus, [or] The Province of Lennox*… (1654).

Chapaelton
Kilburn
Dockra
Manse
Largs
Alpest
Gogo
Burnside
Side
Lady Stone
White Bay
Douncraig
Battle in 1263
Killing Craig
Here is a safe Road steed for Ships of any Burden & good Anchorage
Balaquin
Portray
GREAT
CUMBRA
O F
Scate Bay
Fortune
Kelburn
Kirktoun
Keplefoot
Belchetly
Sherrifs Port
Red Castle
Fairley
Millport
Watch House
Castle ruins
Southanan
South Point
Clashcraig
ruins
Southannan
C
Ness
Gurroch Head
Brig of Urd
Glenside
LITTLE
Light House
CUMBRA
Hunterston
Biglaw
Kilgriskin
Campleton
Corsby
Y D E
Porting Cross
Campbleton
Carlung
ruins
Arneal
Springside
Orchard
ruins
Kilbride
Law
NB. The two Cumbraes belong to the Shire of Bute; and Cowal to Argyleshire, they were not Surveyd but taken from the best Charts, only the great Cumbra is laid down from an Actual Survey of Lord Glasgow's.

CHAPTER THREE

NAMING

The naming of things matters – for islands no more or less than for anything else. In August 2015, this fact was brought home, a little rudely perhaps, to local council officials on the island of Bute following their attempts, using Scottish Gaelic, to welcome visitors to Rothesay, the island's largest town. It was intended that the sign which greeted ferry passengers as they disembarked should read, in Gaelic, 'Fàilte gu Baile Bhòid' – 'Welcome to the beauty of [the town of] Bute'. But failure to include the required grave accent over the 'o' on 'Bhòid', thus making it 'Bhoid' (pronounced 'Bod'), Gaelic for the male member, rendered the welcome message altogether different: 'Welcome to Penis Island'. Diacritical marks are there for a reason. The connections between maps, mapping and place names, toponyms, are of considerable historical and geographical importance, whether they speak to issues which give offence – as in the case of several place names and natural features in the United States of America, for example, such as 'Nigger Hill' or 'Squaw Butte', many of which have been renamed by legislation – or are simply the result of a sign of writer's carelessness.

This is true equally for the names of islands, the names of places on islands, and even the names for different types of islands. To the Dutch map maker Joan Blaeu, for instance, as he worked in the mid seventeenth century trying to make sense of the many islands making up the Orkneys (fig. 3.1), the islands as a whole presented a hierarchy of names to be used in relation to size:

> Now the whole land of the Orkneys is divided into three parts by reason of their names, viz. largest, smaller, and smallest. The largest parts are called islands, which are inhabited and cultivated. The smaller parts are called holms, a local word, that is, grassy plains set in water, most of such small extent that they would scarcely maintain one or two settlers, if they were cultivated; and because they produce only grass, they are deserted and left for pasturing animals. The smallest parts are called Rocks (with the vernacular name of *Skerries* among the inhabitants of the Orkneys), in which nothing

Opposite. Detail from Andrew Armstrong, *A New Map of Ayr Shire, Comprehending Kyle, Cunningham, & Carrick* (1775).

Eglis Øy
Galts neß
Holms
Sound
Elwick
Sands
Dert Sound
Theeues holm
Heiller holm
Cares neß
Woerk
Wiland
Tankerneß
Kirsia or Hia
Scok neß
Vaquoy
The head
Stamsgord
Skyrwyck
Roous Øy
Cogairth
Bellabreck
Trumland neß
Sourwick
Hucla
Banks
Cors
Wasbuster
Lubadel Hils
Quendale
Brugh
West neß
Wyer
Hen of Gersa
Gres Øy
Tingskery
Contreneß
Wytfelt hill
Dams Øy
Greenbuster holm
Crukneß
Myrsetter
Rendale
Woodwick
Akerneß
Akerneß
Alhallow
Costa
Costa head
Burgar
Sweno
Setre
Engelser
The Brugh of Byrsa
The Heronstak Byrsa
Breck
Banks
Marwick
Marwick head
Mousland
Arquoy
Flanes
Quoy
Cott
Gorn
Kingsea
Ellabuster
Hacla
Wale
Harra
Starth
Howland
Cowbuster
Clarth
Rydlane
Burswel
Stonesthous
Orfer
Dale
Burnes
Rambuster
Greenbuster
Fyrth
L: Somerdale
Grunawater
Breck
Linxnes
Widlocknes
Wytfort
Papdale
Eßenquoy
Kirkewale
Yensta L:
Yensta
Langaskel
Kirkabusterneß
Holland
Brecks
Kirk
Brabuster
Skell
POMONIA, or MAINLAND
Foubuster
Sand
Ouer Sand
Stameskel
At Papedale
Hoom
Papley
Clet
Rose neß
Over Scapa
Kirkbuster
The Houp of Kirbuster
Setre L.
Twa
Lukfat
Griney
Aithster
Chumley
Sadwick
Stonhous L.
Ytebust
Kirbuster L.
Lie
Skel
Skel L.
Gairth
Whom
Stromneß
Lie
Kirbuster
Brig of Wairth
Outregil
Claistra
Cowbuster
Orfer head
Hanga Bak
Orefer holm
Glump
Hunda
Labholm
Burra
S. Margaret H
Quoyanes
Oxanes
Quoynnes
Galtnes
Calf of Flotta
Hariths neß
Stangerhead
Oxa
Lippa
S. Peters
S. Andrewes
Braga
Ireland
Carestone holm
Carestone
Grams Øy
Carlingskerry
Ryce Øy
Fara
Yeskinnabey
The Bowon
Brabuster
The Skerries of the Bow
The Comb of Hoy
West Hill
Mones
Caua
Lienes
Cruknes
Thur Vo.
S. Nicolas
Bow
Warthbuster
Troul
glen
Rootberry
Lyrwa B.
Patgil B.
Rysa
Ore
Aire
The most magnificent natural Fort, called Brabrugh
Warth Hil
Sads V.
Hogglins V.
HOY
Manclet
Heldale
wittin
Brims
Rora head Stour
Rackwick
Somerhoy
Sneuktourn
Rackwick litle
Hyberry
Waes
Tornes
The Stour, wher buildet that excellent Foul, caled the Lyer.
Wards
Kirkbuster
Ayth
Snelster
S. Colms K.
Fara
clet
Houscow
Bow
Wydwa Soud
Flotta
Warth head
Cara
Wydwa
Sandwick
Lady K.
Burwick
Barts head, or Balts head
The Leuther
Breecks
Halero
Southa
Cantop tyd
Souna
The Souna
The Heppers
OCEANUS

FIGURE 3.1
This detail from a map of Orkney and Shetland by Willem and Joan Blaeu shows the name commonly used by map makers for the Orkney Mainland – 'Pomonia', by others 'Pomona' or 'Pomania' – which stems from a mistranslation by the late sixteenth-century Scottish historian and humanist George Buchanan. The mainland island of the Orkneys was never commonly understood by this name, by local inhabitants at least, although the name survived on maps into the eighteenth century. The Pentland Skerries, which Joan Blaeu considered might also be thought of as a holm by virtue of its size, is here also termed 'Pichtland Skerries'. 'Skerry', plural 'skerries', is a corruption of the Gaelic *sgeir* (rock).
Source: Timothy Pont/Joan Blaeu, *Orcadum et Schetlandiae Insularum accuratissima descriptio* (1654).

> grows, even grass, with the exception of one Rock, viz. Pentland Skerry . . . which is worthy of the name of Holm rather than Skerry.

This example speaks to more general matters. As Blaeu's remark about the terms 'Rocks' and 'Skerries' suggests, the names given by map makers to inhabited places or to geographical features may not always accord with how that place or feature was known and named by locals. Nor should we assume that there is only one local name, or even only one language, at work in those places for which names are recorded on the map. Further, because place names on maps may change over time (consider, on fig. 3.1, the variant forms 'South Ranals Øy, or South Ronans Øy' for what, today, is South Ronaldsay, one of the Orkney islands), it is important (though it is not always possible) to know the earliest form of place names and to be able to trace variant forms between the first known name and the most recent form of a place name. Knowing the form of a name is not always to know why it has that form. Who has the authority to give a name form to places on a map: local inhabitants or the map's maker? Do the names by which a place or feature is known always easily 'translate' to the map, as it were, either from one language to another or from words spoken to words written? Put another way, who is a name's 'author' (and who the map's?); who has the 'authority' to name islands and the places on them? This chapter explores these and other issues of islands' naming.

Naming and the completeness of maps

The easily posed question 'What's a map for?' does not have a straightforward or single answer. Because maps simplify and symbolise the reality they purport to depict, we often have to interrogate the map for clues as to its maker's intent. Consider, for example, John Speed's depiction of places on Skye and on Lewis/Leòdhas and Harris/Na Hearadh and the southern Hebrides/Innse Gall as part of his 1610 map *The Kingdome of Scotland* (fig. 3.2). There, as elsewhere, Speed uses three

FIGURE 3.2
The Outer Isles and northern islands of the Inner Hebrides from John Speed, *The Kingdome of Scotland* (London, 1610). Note the place names 'S. Patricius', 'S. Maria', 'S. Columbanus' and 'S. Petrus', the locations of early Christian sites associated with these particular saints. The inclusion by Speed of King James VI and I in the frame of the map is a deliberate piece of political and Royalist propaganda. Here, Speed is saying, is the ruler and monarch of that newly created geographical entity 'Great Britain', the result of the Union of the Crowns of 1603. *Source*: John Speed, *The Kingdome of Scotland* (1610).

sorts of symbols for settlements, two of which were associated with named places. One is a stylised settlement in profile, used for the main castles, fortified dwellings and the larger inhabited places: this is employed for 'Dunbegan' (Dunvegan) and 'Tranternesca' (Trotternish) on Skye, and 'Stornwaye cast' (Stornoway Castle) on Lewis/Leòdhas, for example. The second is a red-coloured circle, sometimes with a black dot at its centre. Some of these symbols are linked to names, chiefly, it is thought, to names of chapels and other sites of religious significance; other examples of this symbol have no name attached (see South Uist/Uibhist a Deas, for example, on fig. 3.2). Speed's naming of places, at once political and religious, but always partial, tells us nothing of the dwellings of most of the local inhabitants. He would not have wanted to feature inhabitants' dwellings, nor would his patrons have expected him to. Never having visited the islands, Speed could not have known from first-hand encounter where the islands were or what they contained.

The names of places on Scotland's islands, and names of the islands themselves, those in the north and west particularly, incorporate elements from Norse, Scottish Gaelic and English, most commonly either as personal names, as descriptive terms for a natural or prominent feature, or a combination of both. In the Northern Isles, the languages spoken before the arrival of the Norse towards the end of the ninth century were Old Gaelic and, we may presume of a few persons, Latin. Norse, or its dialectal form, Norn, was spoken together with Scottish Gaelic and Latin throughout the Northern Isles and the western seaboard even after the cessation of Norse political rule. Along the western seaboard and in the Outer Hebrides/Innse Gall especially, this followed the Battle of Largs in 1263 and the related Treaty of Perth of 1266 which formally marked the end of hostilities between Norway, under King Magnus VI, and Scotland, under Alexander III. In the Orkneys and Shetland, the earldom of Orkney passed into the hands of the Scottish crown, to King James III, in February 1472. Battles and treaties do not directly affect the languages people speak. It is helpful, then, to think of different language 'stratae', as it were, with the principal languages of Norse or Scottish Gaelic differently 'thick' or 'thin' in different islands, according either to the longer-run presence there of one language community or to the actions of one dominant group in naming the places and features of that island.

Yet what was spoken may not be the language used to record on maps the names of places.

Let us return to South Ronaldsay (fig. 3.1). In its origins and earlier variants, this combines the Norse for island, '*a*' or '*ay*', with the personal name 'Ronald' or 'Rognvald', brother to one of the first earls of Orkney. Blaeu has two variants on his map of 1654, and the name is given as 'Roghnalsœy' in early fourteenth-century charter evidence. Native pronunciation would commonly have it as 'Rinnalsay'.

While individual maps published at one date may thus show one or more name forms – even when, as here, those forms were not the only ones used of that place – maps from different dates may be used to show something of the emergence of the 'modern' form of names. Consider in this case the islands which make up the Small Isles – Canna, Rum, Eigg and Muck – and the variant name forms given to those islands and to Coll and Tiree to the south of them (fig. 3.3a–c). The example of the shift from a near-Gaelic name form, 'Ylen na Aich', to 'Horse Island' for the small island off Muck is illustrative of that widespread shift from Gaelic names of places to English forms that has characterised the mapping of Scotland as a whole, not just of its islands. Further south, in the Firth of Clyde, the islands today known as Great and Little Cumbrae appear in different although recognisable forms in early written evidence: 'Cumberays' in 1264, 'Kumbrey' in about 1270, 'Litill Comeray' in 1515. On later maps, they appear as 'Kumbra I.', 'Lit[tle] Kumbra' on Herman Moll's map of 1745, and, thirty years later, more definitively, as 'Great Cumbra' and 'Little Cumbra'. By the early twentieth century, the modern name form had long been established (fig. 3.4). The names – which mean, in Norse, the isles ('*a*', '*ay*' or '*ey*') of the Welsh (*Cymri*) – suggest the past presence of another language group in an earlier Scotland, the Brittonic inhabitants of the kingdom of Rheged in southern Scotland and northern England.

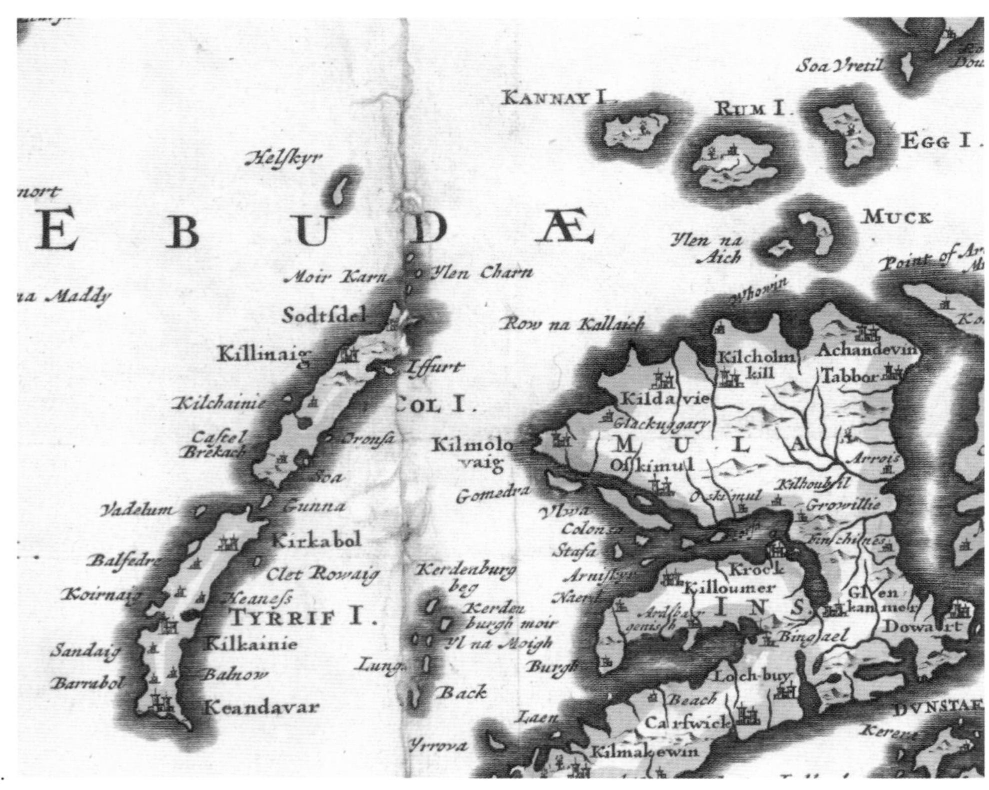

a.

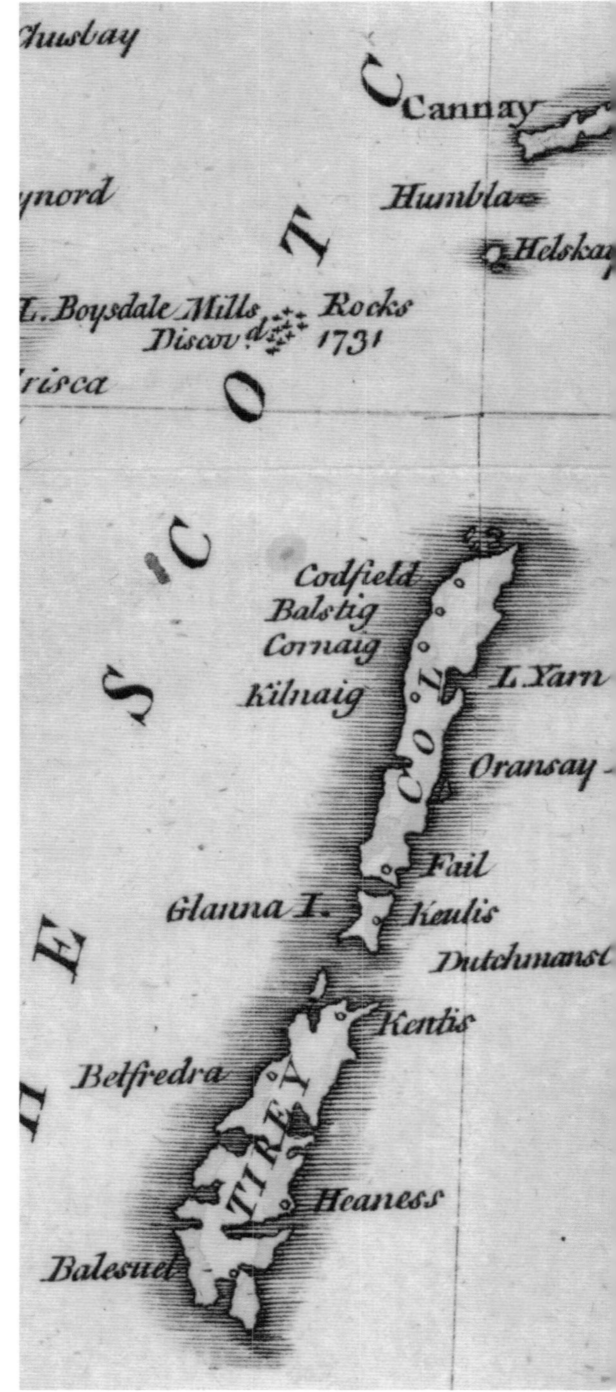

b.

FIGURE 3.3

The details from these three maps show, from left to right, the different forms and spellings of the names for the four islands making up the Small Isles, and for Coll and Tiree, between the end of the seventeenth century and the end of the nineteenth century. In de Wit's map from around 1680 (a), the constituent islands making up the Small Isles are spelled 'Kannay', 'Rum', 'Egg' and 'Muck'. Adjacent to Muck, de Wit has the name 'Ylen na Aich', a near transliteration of the Scottish Gaelic 'Eilean nan Each' (the 'Island of the Horses' or 'Horse Island'). The form of words used suggests that de Wit heard it from an informant or was told it rather than saw the name in written form. In Bowen's 1769 map (b), this small island is simply named 'Aich' (a corruption of the Gaelic *each*, horse) despite the map maker's claim to have based the map on up-to-date surveys. His use of a Gaelic term, albeit incorrectly spelled, was consistent with the strength of the language on the island at this time: the 'Old' *Statistical Account of Scotland* parochial report upon the Small Isles in 1796 notes simply that 'The language principally spoken, and universally understood, is Gaelic.' By the later 1870s, however, although Gaelic was still the language of the inhabitants of Muck and neighbouring islands, Johnston's map names the adjacent island in English – 'Horse Island' (c). Note, too, the changing name forms for Tiree in particular: 'Tyrrif' in *c.*1680 (perhaps a printer's error for 'Tyrrie'); 'Tirey' on Bowen's map, and 'Tiree' by 1879. Tiree may derive from the Gaelic *tìr etha* (land of corn), reflecting its fertility, or, perhaps, may be a personal name since 'Ith' refers to a man: the Gaelic *tìr* and Latin *terra* (land) are cognate.

Source: (a) Frederick de Wit, *Scotia Regnum divisum in Partem Septentrionalem et Meridionalem Subdivisas in Comitatus, Vice Comitatus Provincias Praefecturas Dominia et Insulas* (*c.*1680). (b) Emanuel Bowen, *A Map of North Britain or Scotland, from the Newest Surveys & Observations* (1769). (c) Alexander Keith Johnston, *Library or Travelling Map of Scotland* (1879).

Ora
Sora
L. Lyford
Arslate
Egg
Pt. of Arasa
Muck
Aich
Pt. of Ardnamurchan
Glendrain
Ardnam
Ormain
Penmoir
Ardmore
Tobermore B.
Belseat
Lagan
Killvore
Letter more
Hawn
Torsa
Ferg
Arros
Inve
Achare
Kilbranan
Ulva I.
Dysaig
Rosal
Kilnachan
Ringross
C. Moy
L. na Gaul
Ormsaig
Karowick
Kilpatrick
Ardalanish

Boisdale
L. Bhreatal
Dunan Pt.
Soa
Garbshen
Kilmaree
CUCHULLIN SOUND
L. Strath Aird Pt.
Canna
Compass Hill
Tarbert
Garresdale
Harb.
Sandy I.
Kilmory B.
BARRA
Small Isles K.
L. Scresort
Sleat Pt.
Huma Rk.
Bridianoch Pt.
RUM I.
Has Keval 2667
Longitude West of Greenwich
Longitude 9° 30' W. of Paris
Hysker I.
Ben More 2310
Harris
Pappadil
Glendebble
Parish of Small Isles
S E A
Mills Rks.
Ru Stoir
Eigg I.
Scuir of Eigg 1272
Kildonan K.
Sandavore
Castle I.
O F T H E
Horse I.
Muck I.
B R I D E S
Ardnamurchan Pt.
Sana B.
Great Pt.
I. More
Torrastan
Grisapoll
COLL I.
West End
Taylun B.
Gunna I.
Soa Ardnish
PASSAGE OF COLL
Caliach Pt.
Tobermory
Calgary B.
Treshnish Pt.
Kilmore
Balliphetrish
Kelis
Ru na Kelis
Dusker
Ken na varah Pt.
Vassipoll
Scorinish
PASSAGE
TIREE I.
OF
TIREE
Travay Bay
Haynish
Ken na varah Pt.
Sandaig
Pladda
L. Tuadh
Treshnish
Lunga I.
Gometra
Ulva
Blackmore
Sound of Ulva
Staffa
Fingals Cave
L. na Keal
L. Sunart

c.

FIGURE 3.4

(*This page and opposite*). The details from these three maps show, from (a) to (c), the different anglicised forms taken in the names of Great and Little Cumbrae between 1745 and 1933. Moll's 1745 map (a) also shows, in the Gaelic place name 'Ballach Martin' – from the Gaelic *Bealach* for a 'pass' – a variant of today's 'Ballochmartin', but without locating it accurately. Andrew Armstrong (b), whose use of symbols for the larger settlements recalls that of John Speed over 150 years earlier (see fig. 3.2), appears to position some features in the Firth of Clyde: they are, perhaps, the names of rocks (his 'Explanation' [of the symbols on his map] makes no mention of them). Note, too, his reference to the 'Battle in 1263', south of Largs, siting the event which diminished Norse political authority on Scotland's western seaboard. *Source*: (a) Herman Moll, *The Shire of Renfrew with Cuningham: The North Part of ye Shire of Air* (1745). (b) Andrew Armstrong, *A New Map of Ayr Shire, Comprehending Kyle, Cunningham, & Carrick* (1775). (c) John Bartholomew & Son Ltd, *Half Inch to the Mile Map of Scotland, Sheet 7: Firth of Clyde* (1933).

a.

b.

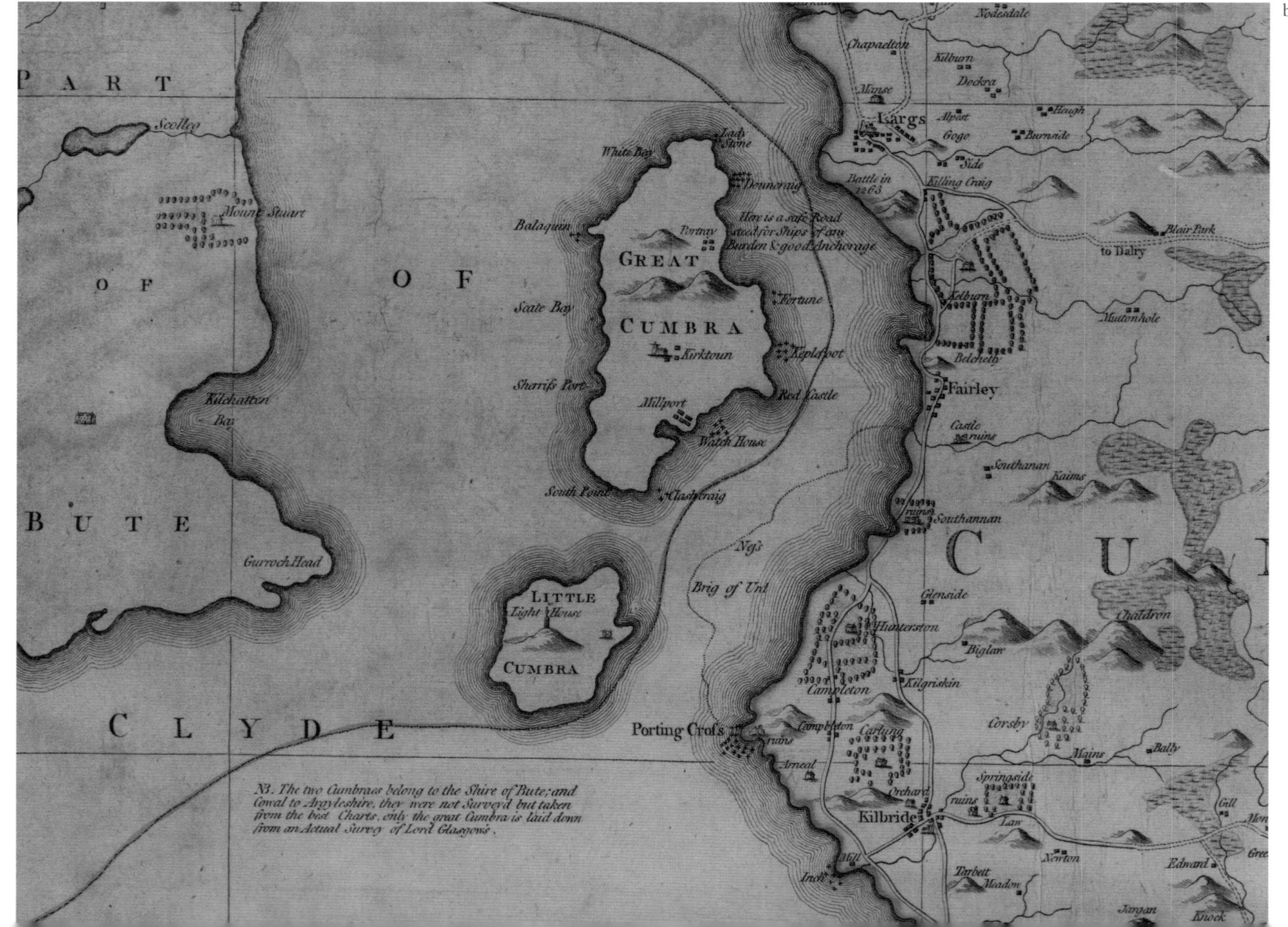

Kerrycroy
Scoulag Ch.
Scoulag Pt
Mountstuart
Kerrylamont
Kerrylamont B.
Bruchag Pt
Kerrytonlia
Kerrytonlia Pt
Kilchattan B.
Kilchattan
Pier
White Port
Glen Callum
Torr Mor
485
Garroch Hd
Largs Bay
LARGS
Tomont End
White B.
Skate Pt
Bell B.
Fintray B.
Great Cumbrae
CUMBRAE
Island
Sheriff's Port
Millport
Millport Bay
Portachur Pt
The Tan
Farland Pt
Fairlie Roads
Keppel Pier
Fairlie
Fence B.
Shauniwilly Pt
Long B.
Little Cumbrae
Island
Beacon
Castle I.
Gull Pt
Inner Brigurd Pt
Hunterston
Portencross
Farland Hd
Ardneil B.
West Kilbride
WEST KILBRIDE
Glentane Hill
Caldron Hill
Kaim Hill
Blaeloch Hill
Green Hill
Gogo Water
Girtley Hill
Bushy Bog
A 760
Broomfields
Downcraig
Bowencraig Walk
Clashfarland Pt
Underbank
Black Hill
Poteath
Goldenberry Hill
Thirdpart
Carlung Ho.
Ardneil
B 7048
B 782
Sea Mill
Crosbie
Castle Knowe
Biglees
T
H
O
F

c.

What these examples do not so readily disclose are the ways in which names were arrived at for the features and places on islands and then 'fixed' on maps. For insight into this question, we have to turn to the work of Ordnance Survey.

Naming by authority: Ordnance Survey at work in the nineteenth century

Established in 1791, Ordnance Survey's first published sheets were of Kent, in south-east England, in 1801. The place names which its officers and field surveyors initially and principally had to deal with were primarily Anglo-Saxon in origin. But, as the Survey moved gradually northwards and westwards in mapping the British Isles, so the languages and place names of the Celtic nations and the presence of Norse name elements increasingly presented difficulties over the correct orthography of the names to be included on the Survey's maps. To know what these difficulties were, and how they were overcome, is to be afforded insight into how and why Scotland's islands are named as they are.

The emergence of what we might think of as Ordnance Survey's 'ground rules' for naming places began in Lincolnshire and, for Celtic names, may be traced to early nineteenth-century Wales. Between about 1811 and 1820, Ordnance Survey mappers in Wales accepted as their authorising source the standard names given in printed documents such as the Census or other equivalent civil documents. The proof sheet of each map prepared was then sent to local clergymen and to landowners thought capable of correcting the orthography. By the 1820s or so, however, it was clear that such reliance upon the local gentry alone was generally ineffective, and a different procedure was adopted. In essence, names collected by the surveyors on the ground were verified by a mixture of sources. Primarily, responsibility for verification of names shifted 'back' from the proof stage – that is, after the map had been drafted and drawn up – to the survey stage, the stage of the map's in-the-field compilation. In this sense, authority for the correct name shifted from the local or distant expert to the surveyors on the ground.

In 1825, following his work in Ireland, Thomas Colby, the then Superintendent of Ordnance Survey (Colby was earlier employed, in 1817–18, not altogether happily, in mapping Shetland – a story we turn to in chapter 4), issued more detailed instructions to those doing the Survey's work:

> The persons employed on the survey are to endeavour to obtain the correct orthography of the names of places by diligently consulting the best authorities within their reach.
>
> The name of each place is to be inserted as it is commonly spelt, in the first column of the name book: and the various modes of spelling it used in books, writings &c. are to be inserted in the second column, with the authority placed in the third column opposite to each.

Colby's instructions gave further guidance over who, exactly, was an 'authority' in the eyes of the Survey:

> For the name of a house, farm, park or wood, or other part of an estate the owner is the best authority. For names generally the following are the best individual authorities and should be taken in the order given: Owners of property; estate agents; clergymen, post-masters and schoolmasters, if they have been some time resident in the district; rate collectors; borough and county surveyors; gentlemen residing in the district; Local Government Board Orders; local histories; good directories. Assistance may also be obtained from local antiquarian and other societies, in connection with places of antiquarian and national interest.
>
> Respectable inhabitants of some position should be consulted. Small farmers and cottagers are not to be depended on, even for the names of the places they occupy, especially as to the spelling. But a well-educated and independent occupier is, of course, a good authority.

22

Parish of Stornoway

List of Names to be corrected if necessary	Orthography, as recommended to be used in the new Plans	Other modes of Spelling the same Name	Authority for these other modes of Spelling when known	Situation	Descriptive Remarks, or other General Observations which may be considered of Interest
33 Rutha na Monach	Rudha na Mònach	Rutha na Monach Rutha na Monach " "	John McKenzie John Mackay John McKa Allan Ross	In Broad Bay about 20 Chains East of Aird Thunga village	A small point or headland on the sea shore. Rudha na Mònach signifies Peat Point.
34 Traidh na Faing	Tràigh na Fainge	Traidh na Faing Tràigh na Fainge " "	John McKenzie John Mackay John McKa Allan Ross	About 20 Chains South West of Rudha na Monach 11 Chains East of Aird Thunga Village	A small bay or Creek on the Shore of Broad Bay – Tràigh na Fainge signifies Pen-fold Shore

FIGURE 3.5
On this manuscript page from the Original Object Name Books, the printed headings to the columns were a standard feature, ratified by Major-General Sir Henry James, Director-General of Ordnance Survey, in his *Account of the Field Surveying, and the Preparation of the Manuscript Plans of the Ordnance Survey* (Southampton: Ordnance Survey, 1873). Entries, which were usually completed in the field by surveying staff although sometimes completed back at the Survey's head office, would show variant forms of names (if such existed) in the column 'Various modes of Spelling the same Names'. The column marked 'Authority for those modes of Spelling' is often the most informative with regard to the source cited (when it was not direct personal testimony) in establishing the name to be used. Earlier maps were commonly referred to as supporting evidence, without any recognition as to how the places on those earlier maps were in turn named and by which authority. The column 'Descriptive Remarks, or other General Observations which may be considered of interest', allowed further comments to be added, sometimes documenting local lore or historical events associated with the place in question. The Original Object Name Books are thus a rich resource for studying Scotland's ethnography, geography and history.
Source: Original Object Name Books for the Parish of Lochs on the Island of Lewis, Ross and Cromarty. Volume 104, page 22 (1848–52). Crown copyright, National Records of Scotland, OS1/27/53 page 22.

This system, which became established practice by about 1850, was re-affirmed later in the nineteenth century by subsequent directors of Ordnance Survey, and endorsed in its *Instructions to Field Revisers* (1932).

Several points may be made in relation to these rules of engagement over names and naming authorities. The 'name book' referred to into which names were inserted was known as the 'Original Object Name Book' (fig. 3.5). Mapping – accurate and 'complete' mapping in any case – was dependent upon accurate naming, accurate reporting and accurate

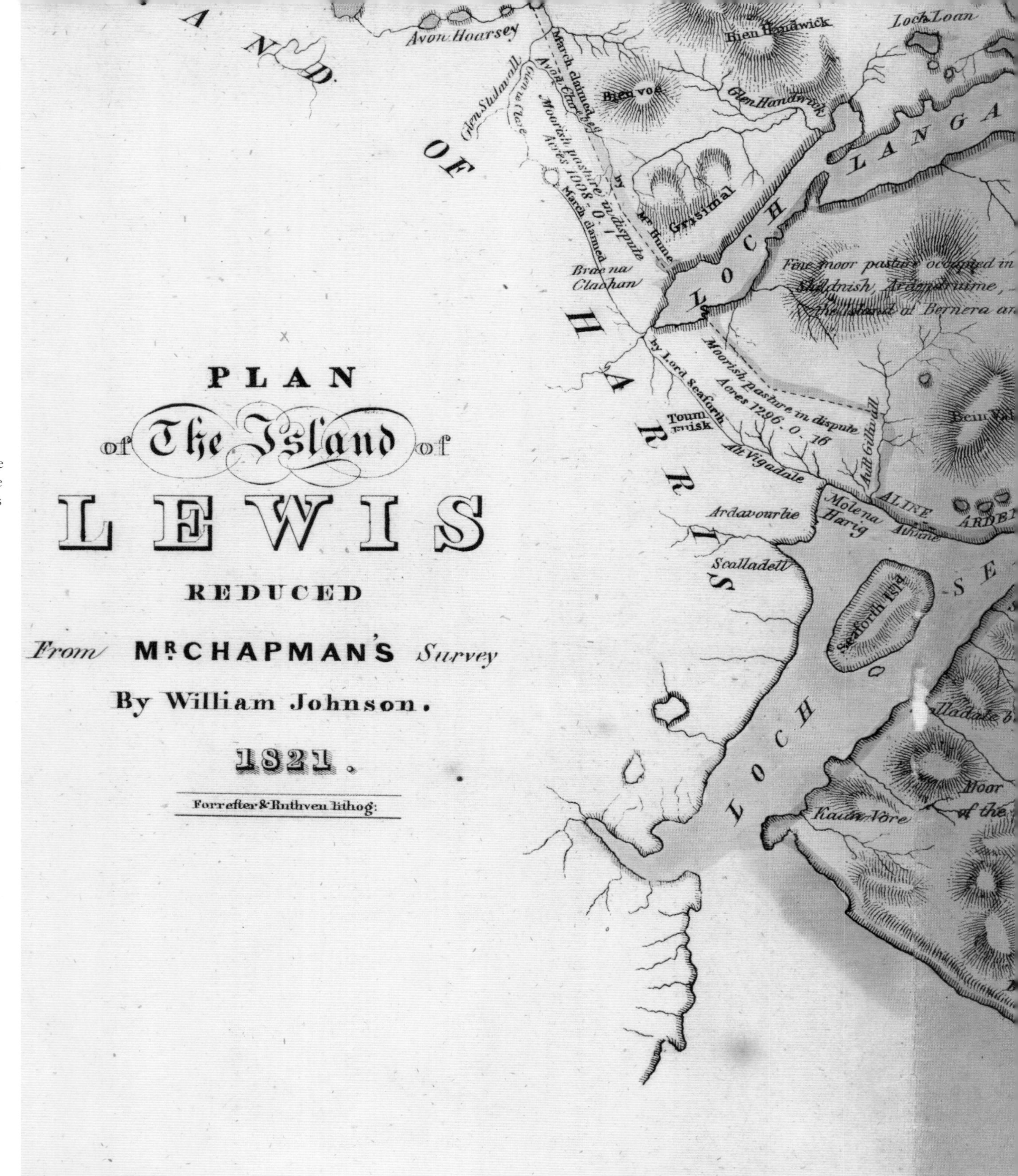

FIGURE 3.6
William Johnson's map of Lewis/Leòdhas was produced in 1821. This map, with its descriptions of the quality of land as well as of place names, was probably available to later Ordnance Survey 'sappers' as they worked on the island in the second half of the nineteenth century. Note, too, that in addition to disputes over names, there were also disputes over estate boundaries, shown here between Lewis/Leòdhas estate and the Harris/Na Hearadh estate to the south. Action over these boundaries was raised in the Court of Session in 1804 by Hume MacLeod against the Earl of Seaforth and not finally settled until 1853.
Source: James Chapman/William Johnson, *Plan of the Island of Lewis, Reduced from Mr Chapman's Survey* (surveyed 1807–09, published 1821).

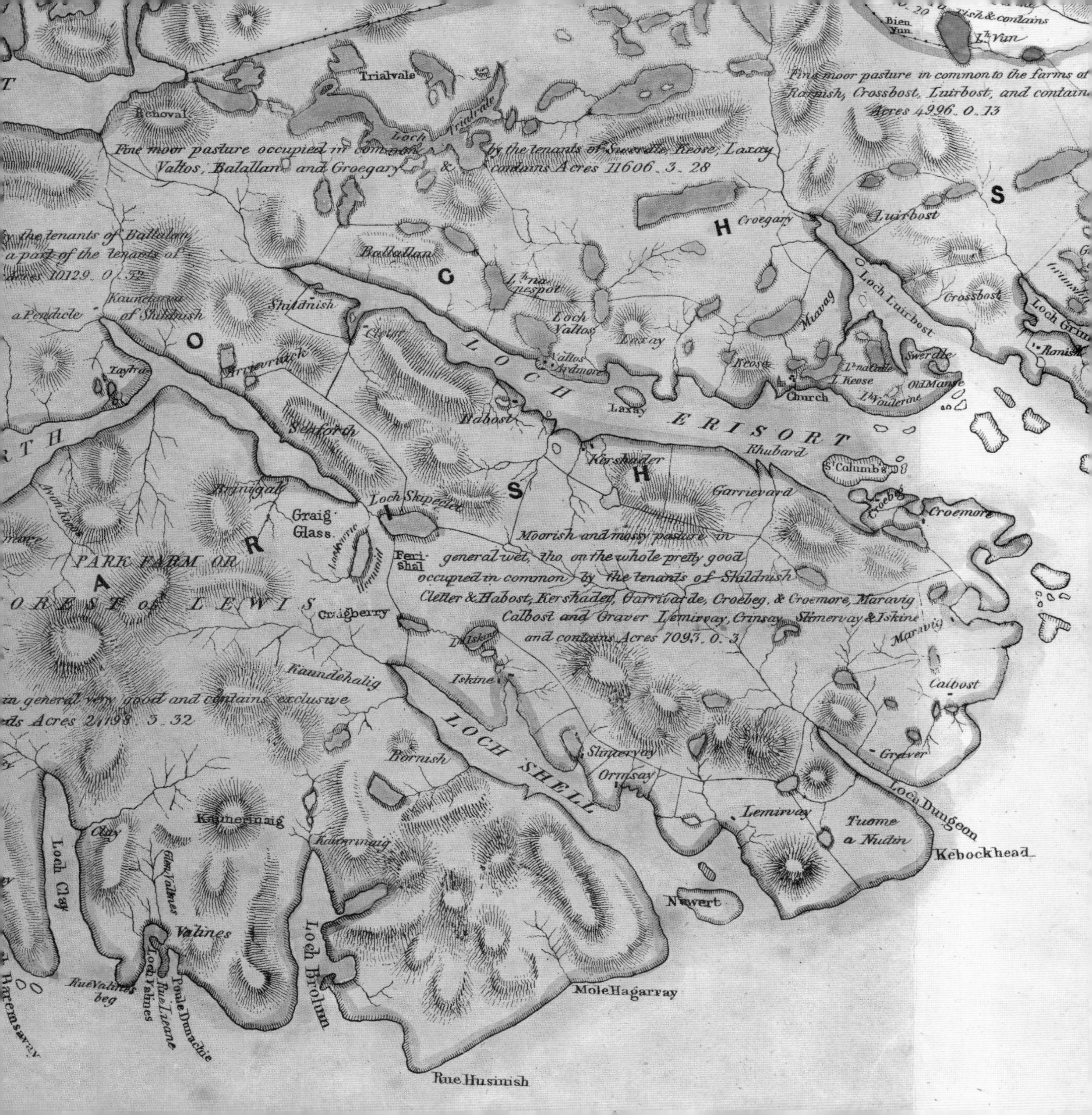

Trialvale
Loch Trialvale
Rehoval
Fine moor pasture occupied in common by the tenants of Swerdle, Keose, Laxay
Valtos, Balallan and Groegary & contains Acres 11606. 3. 28
Fine moor pasture in common to the farms of
Ranish, Crossbost, Luirbost, and contains
Acres 4996. 0. 13
L^h Vun
Bien Yun
Luirbost
Croegary
Balallan
Shildnish
Kaunetarra of Shildnish
a Pendicle
Loch Valtos
Valtos
Ardmore
Laxay
Keose
Church
Swerdle
Crossbost
Loch Luirbost
Old Manse
L^h Keose
Miavag
Ranish
LOCH ERISORT
Habost
Seaforth
Rhubard
St Columbs
Kershader
Garrievard
Croebeg
Croemore
Loch Skipeetee
Brinigal
Graig Glass.
Ferishal
Moorish and mossy pasture in
general wet, tho on the whole pretty good
occupied in common by the tenants of Shildnish
Cletter & Habost, Kershader, Garrivarde, Croebeg, & Croemore, Maravig
Calbost and Graver Lemirvay, Crinsay, Slimervay & Iskine
and contains Acres 7093. 0. 3
PARK FARM OR
FOREST of LEWIS
Craigberry
Iskine
Maravig
Calbost
Kaundehalig
in general very good and contains exclusive
Acres 21198 3. 32
LOCH SHELL
Bornish
Slimervay
Orinsay
Graver
Loch Dungeon
Lemirvay
Tuome a Nutin
Kebockhead
Kaimerinaig
Clay
Loch Clay
Valines
Loch Valines
Loch Brolum
Newert
Mole Hagarray
Rue Valines beg
Poule Dunachie
Rue Husinish

recording. Maps might accurately locate and symbolise natural features and inhabited places in a variety of ways, but, without the correct names of such things, the map's value was limited. There is a presumption that there was a correct orthography to a name and that it could be commonly spelt. And there is, importantly, an association between the correct form of the name and the social status of the informant. Authoring a map, that is, authenticating the names that were to appear on it, was thus a matter of its authorisation from people of the right standing – at one end of the social scale property owners, who were commonly to be sought; at the other small farmers and cottagers, who, for the purposes of the Survey's naming of places, were 'not to be depended on'.

The facts of geography often mean, however, that rules cannot be straightforwardly followed in all circumstances. The social structure of nineteenth-century Hebridean Scotland in particular – an island society then in a state of transformation (see chapters 2, 6 and 7 in particular) – meant that Ordnance Survey had to compromise its own naming principles. There were few landowners. Of those few, several were absentee. In 'rationalising' their estates by displacing the native inhabitants in favour of sheep or deer forests, neither landlords nor their estate managers evinced much interest in the names of places or in the associations between land, lore, language and life as embodied in Gaelic's largely oral culture. On Harris/Na Hearadh and Barra/Barraigh, as in many other parishes, 'shepherds' and 'boatmen' were often called upon as reliable authorities for the island's names. On Lewis/Leòdhas, although two teachers from village schools on the island were used in this respect, the majority of informants were crofters or cottars. For each parish, one man – the informants were nearly always men – was appointed as interpreter, acting either in a literal sense or as the interlocutor and 'go-between' for native inhabitants and the men of the Corps of Royal Sappers and Miners (known by the locals simply as 'the sappers'). Earlier maps of the island were used to help the surveyors, despite the fact that such maps might themselves not have provided detailed coverage of the places named (fig. 3.6).

Ordnance Survey was aware of the problems caused by the presence of different originating languages in naming Scotland's islands. In March 1891, the then Director General, Colonel Sir Charles Wilson, laid out something of the issues faced:

> In the case of the Norse names which prevail along the western coasts and islands, and in the counties of Sutherland and Caithness, great difficulty arose as to the proper mode of spelling. All the local authorities, or nearly all, were in favour of spelling the names in the Gaelic fashion. . . . Therefore the question was should Fiskavaig (Fish Bay) be written as pronounced, or in its Gaelic form Fiscabhaig? Should Stornoway be written as pronounced, or in its Gaelic form of Steòrna bhaigh, where *bh* represents *v*, and *gh* represents *y*? Should Roneval be written as pronounced, or in its Gaelic forms of Ronebhal, as written to all mountains (*bh* for *v*) in the Lewis Survey, or Ronemhal (*mh* for *v*) as generally written in other districts?

While this suggests the Survey was attentive to these issues, on the ground and thus on the map, its English-speaking surveyors often worked differently. The Gaelic folklorist Alexander Carmichael, who worked with the Survey on Harris /Na Hearadh, North Uist/Uibhist a Tuath, South Uist/Uibhist a Deas, Benbecula/Beinn na Faoghla and Barra/Barraigh in the late 1870s, considered that 'the system pursued by the Ordnance Survey in Scotland in regard to taking up place names is altogether erroneous. Non-Gaelic-speaking men go about non-English speaking people to take down Norse-Gaelic names with their English meanings!' Carmichael claimed to be thorough in his own work: 'I have gone to the locality and in every instance corrected the place-name from the living voice on the spot. From these corrections I have written out each name in correct Gaelic and have revised and re-revised my own work. I have adhered strictly to the local sound and pronounciation [sic] of every word.' He was mortified to be told that one of the sappers involved, a Captain Macpherson, intended to adopt neither Norse nor Gaelic theory in spelling 'but to give the name in phonetic spelling', and to learn from Wilson that, for reasons of expense, the names of numerous smaller landscape features had been omitted from the draft maps being prepared for publication.

Ordnance Survey maps provide one of the most important sources for the geography of island Scotland from the mid nineteenth century (figs 3.7 and 3.8). As policies towards Gaelic in Scottish life have swung from antipathy towards the language in the 1800s to recognition of its place in national culture in the twentieth and twenty-first centuries, so Gaelic naming practices on maps have followed suit. Ordnance Survey has prepared and revised its Gaelic Names Policy five times between November 2000 and April 2016. Working with the Ainmean-Àite na h-Alba (AÀA) partnership (Gaelic Place-Names of Scotland), the Survey's stated policy is to support the National Gaelic Language Plan 2012–2017 and the Gaelic Language (Scotland) Act (2005) through the depiction of Gaelic names on its maps, and by working where it can to ensure consistency in the orthography of Gaelic place names within its different map series. This applies to proper names only, within the natural environment, the man-made environment and of administrative divisions, but does not include descriptive names in the landscape ('cattle grid', for instance). The effect in terms of map styles and content and in terms of the presence of Gaelic as a living language in the landscape can be quite dramatic (fig. 3.9). In its policy, Ordnance Survey has helped 're-Gaelicise' modern Scotland in ways which would have struck Sir Charles Wilson and others as odd. It has not been alone in this policy. The Bartholomew firm of map makers in Edinburgh was similarly attentive to Gaelic names in its topographic maps in the early twentieth century (fig. 3.10). Yet the processes of name authoring and of authorisation that lie behind and 'beneath' the many place names recorded and reported by map makers are not themselves evident from the paper documents. Maps, commonly, are silent witnesses to their own complex making.

Naming above and below ground

As Ordnance Survey worked with others to name islands, holms and skerries in the nineteenth century, other agencies were at work surveying and naming the very rocks themselves. The Geological Survey of Great Britain and Ireland was formally begun in 1845, but work in and on Scotland was not properly begun until 1854. The Geological Survey of Scotland was formally begun only in 1867. Work on the islands of the north and west, and on the Highlands, did not start in detail until the later 1870s and 1880s. What emerged from the slow study of Scotland beneath the surface was considerable geological complexity at a local scale and, across the islands and the north and west Highlands generally, evidence of some of the oldest rocks and tectonic processes in the world (fig. 3.11).

Rocks endure, their names – 'Gabbro, Granite, Gneiss, Quartz-Porphyry' – a poetic taxonomy of sorts to enduringly ancient geographies. On the land's surface, names may endure even when the human things they signify do not. At much the same time that Geikie's surveyors were bringing Raasay into colourful life (fig. 3.11), others were transforming the island's geographies forever. In 1845, the island was sold to George Rainy who, amongst other acts of pious over-lordship, forbade his tenantry to marry. Between 1852 and 1854, Rainy oversaw the clearance of the population of Raasay, ninety-four families in all, in twelve townships, events later recalled in evidence given to the Napier Commission's investigation into the condition of the crofting and cottar population of islands and Highland Scotland in 1883. One of the townships cleared, Hallaig, lay on Raasay's eastern shore, beneath the hill of Dùn Cana (a basaltic outcrop, given as 'Dun Caan' by Geikie in 1876: see fig. 3.11).

Of Hallaig on the ground only stones remain. The township had been deserted for nearly two decades by the time the sappers of Ordnance Survey brought Raasay to shape and Hallaig to empty life (fig. 3.12). From Hallaig on the map, there is an enduring resonance. 'Hallaig' is the title of a poem, in Gaelic, by Sorley MacLean, born on Raasay in 1911, in which he uses Hallaig, and Raasay, to meditate upon the clear-

ORDNANCE SURVEY OF SCOTLAND

TIREE — SHEET

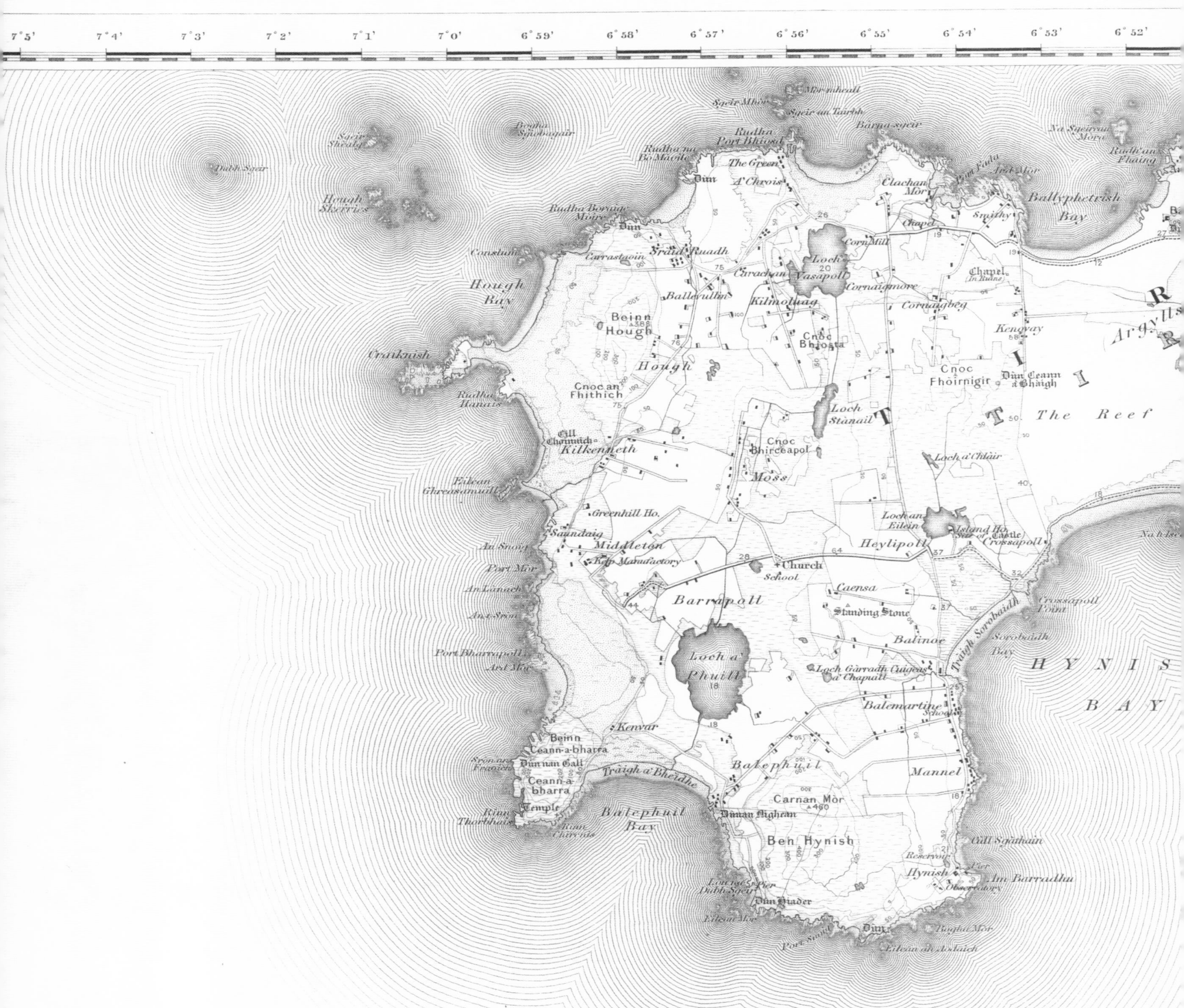

FIGURE 3.7
The island of Tiree from Ordnance Survey's one-inch first edition series. The Ordnance Survey's first edition maps of Scotland at this scale were published between 1856 and 1895, and based on surveys conducted between 1843 and 1878. Attention is paid to the present settlement patterns and, by the use of different lettering, to former inhabited places such as duns (forts) and chapels. *Source*: Ordnance Survey, One-Inch to the mile, *Scotland, Sheet* 42 (surveyed 1875–76, published 1885).

FIGURE 3.8

(*Right*) The inclusion of three different name forms on this Ordnance Survey One-Inch map from the mid nineteenth century for what are today commonly known as the Shiant Islands suggests no easy compromise was reached by the Survey's sappers and other naming authorities over the 'correct' name for the islands.
Source: Ordnance Survey, One-Inch to the mile, *Scotland, Sheet 99* (surveyed 1848, published 1858).

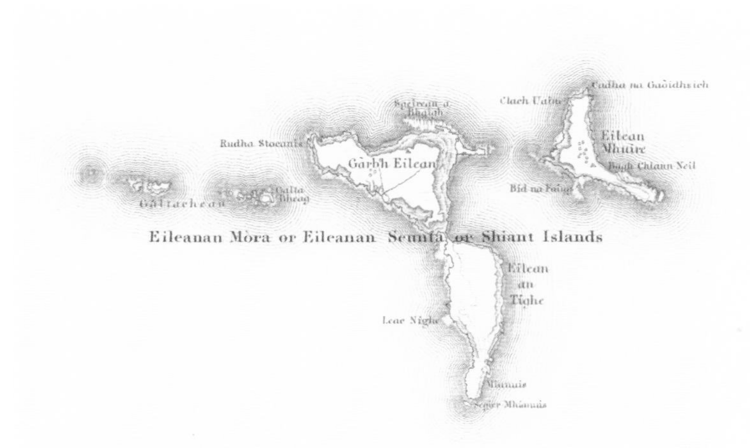

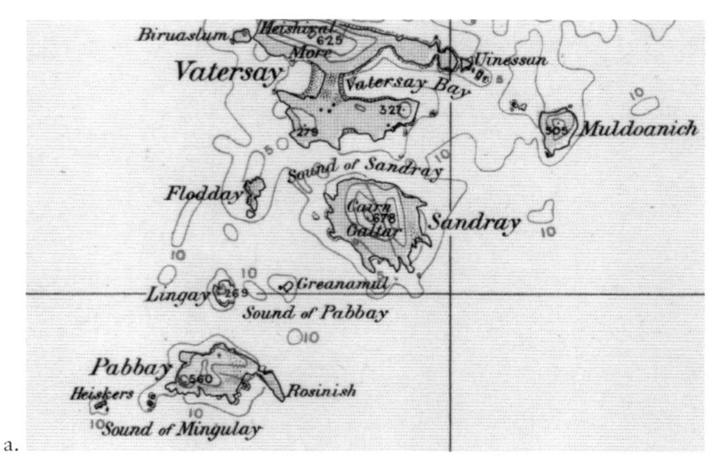

a.

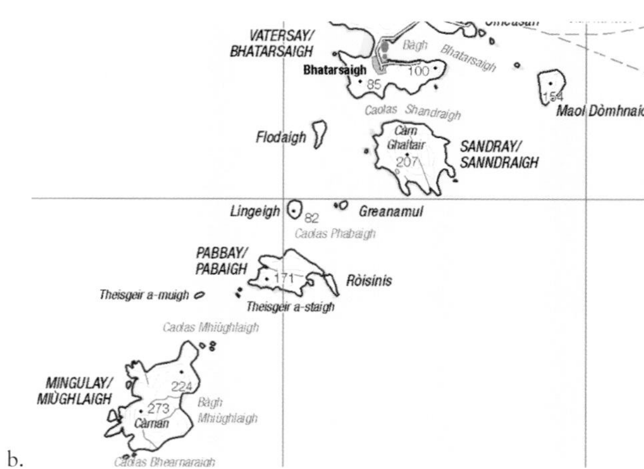

b.

FIGURE 3.9

(*Left*) Even allowing for the different map styles, the emphasis given to Gaelic place names and proper names in recent Ordnance Survey maps of the Western Isles/Na h-Eileanan an Iar (b) is in marked contrast to its maps of the islands from the first half of the twentieth century (a). Ordnance Survey is clear today in its use of Gaelic for the different hierarchy of names. For names in the natural environment, normally only one name will be shown, the locally and/or historically accepted form (Gaelic or English), with exceptions made for some principal features. For names in the man-made environment, dual names are shown unless this is difficult for cartographic reasons (of space or layout, for example), in which case only the English name is given. For administrative names, the Survey follows the practice determined by the authority in question.
Source: (a) Ordnance Survey, Quarter-Inch to one mile, *Scotland, 3rd edition, Sheet 6* (1922). (b) Ordnance Survey, *1:250,000 Raster* (2016). Contains OS data © Crown copyright and database right (2016).

FIGURE 3.10
These two map details, from Ordnance Survey (a) and from the Bartholomew firm (b), both show a commitment to Gaelic place names, here of parts of west Lewis/Leòdhas, but in maps of different style and purpose. The range of shading to symbolise height is characteristic of the work of the Bartholomew map-making firm. Ordnance Survey has employed graduated and numbered contours to signify elevation. It is likely that John George Bartholomew of the Bartholomew firm discussed the question of Gaelic orthography and names on maps with officers of Ordnance Survey and of the Royal Scottish Geographical Society (which latter body Bartholomew was influential in establishing in 1884); the Society's Committee on Gaelic Place Names, established in 1891, advised Ordnance Survey and others until 1899.
Source: (a) Ordnance Survey, One-Inch to the mile, *Scotland, Sheet 104* (revised 1895, published 1897).
(b) John Bartholomew & Co., *Half-Inch to the Mile Map of Scotland, Sheet 23* (1902).

a.

b.
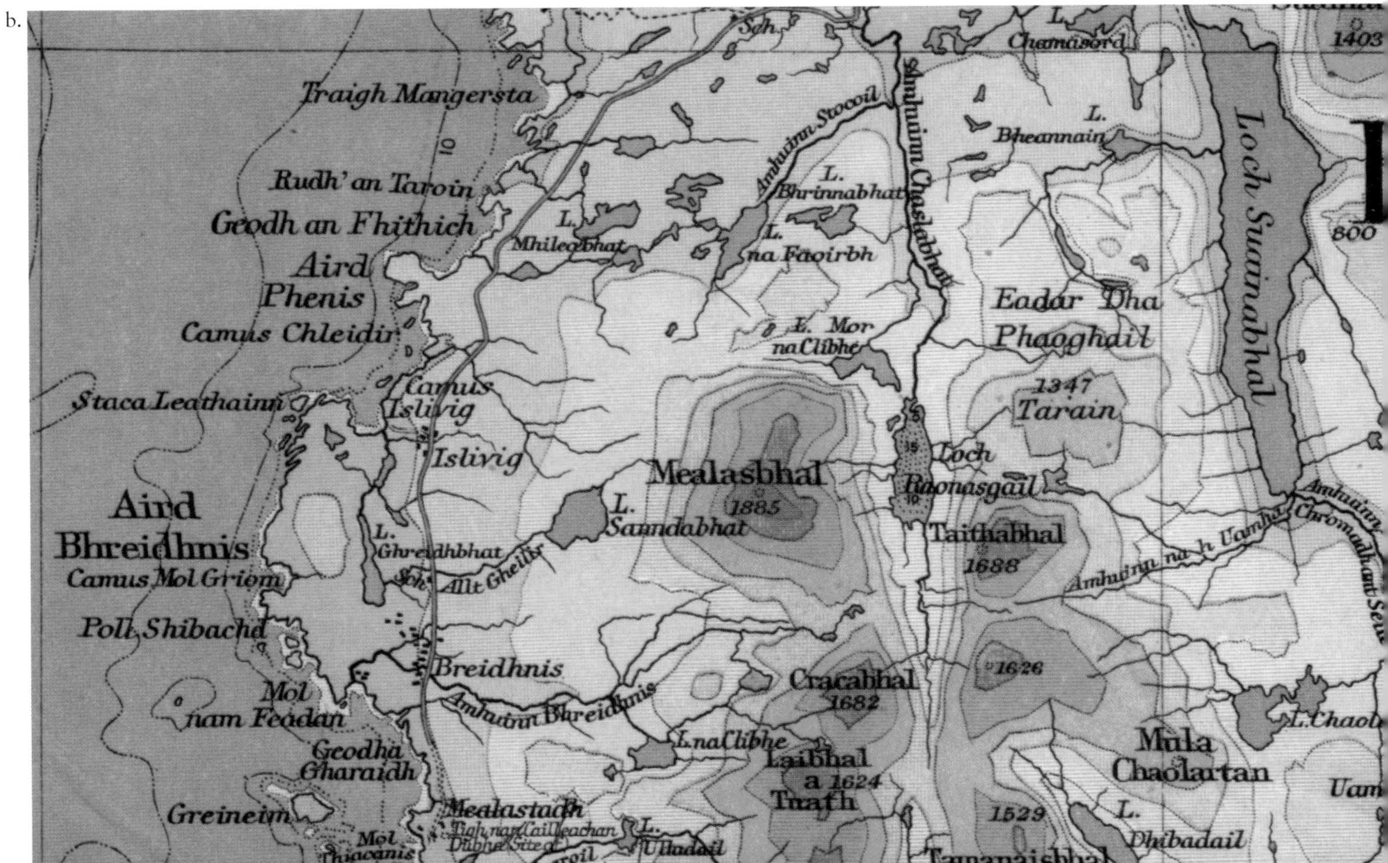

FIGURE 3.11
Archibald Geikie was made Director of the Geological Survey of Scotland at its formal inception in 1867. The pale grey shading to the north of Raasay, marked by the letter 'a', signifies sandstone. The pale shading marked with the capital 'A' is gneiss and those outcrops marked with a 'Q' in orange shading are the quartz-porphyry rocks. Little account is here taken of the 'drift geology' or surface deposits. Because of this, and because geological survey work did not cover all areas with equal thoroughness (it was impossible to do so given the limitations of time and money), this is in several respects a map of inference, with shadings by area drawn on the basis of relatively few point-based observations, and the lines suggesting sharp demarcations between rock types when, in reality, this was often not the case. Even so, Geikie's work and that of his fellow geologists was enormously influential in bringing an ancient 'Scotland' to light. *Source*: Archibald Geikie, *Geological Map of Scotland* [Topography by T. B. Johnston]. (1876).

FIGURE 3.12
Raasay, showing to the south of the island the location of the cleared township of Hallaig. 'North Fearns' and 'Beinn na Leic' feature in Sorley MacLean's poem 'Hallaig', as does 'Dùn Cana' – 'Dùn Caan' on the Ordnance Survey map – beneath whose shaded slopes the poet imagines the comings and goings of the islanders in the past. *Source*: Ordnance Survey, One-Inch to the mile, *Scotland, Sheet 81* (surveyed 1873–77, published 1882).

ances of people from their land. Writing of Hallaig's 'vivid speechless air', MacLean imagines there the absent presence of his forebears.

Island landscapes are commonly imagined, named and mapped in the mind. As documents which depend upon written authority, maps are simply one way of recording and reporting in order to name. But different sorts of maps are possible, of islands and island life as for other features of the world. Names on maps point to physical features as well as to inhabited places. Yet, as numerous Gaelic speakers and poets have shown, and as the landscape writer Robert Macfarlane has recently noted, the names of many landmarks on islands such as Lewis/Leòdhas and Harris/Na Hearadh are remembered, not recorded: they are held fast in the memory, and so passed down from one generation to the next, as alternative, and perhaps more detailed and vibrant, versions of what others have inscribed in ink on paper. Where it is undertaken, as it is in Shader and elsewhere on Lewis/Leòdhas, community cartography with a view to 'memory mapping' exists not to 'fix' names through the agency of social authority but to serve much more as projects in the heritage of memory, keeping alive names which embody island landscapes as living things, shaped and named by the people who inhabit them. These things matter greatly when the very rocks themselves – valued by many in the language of home and belonging – may be seen by others only as sources of road aggregate and offshore profit, as was the case for the super-quarry planned (but not authorised) for Rodel/Roghadal and Roineabhal on Harris/Na Hearadh in the 1990s (see chapter 7 and fig. 7.8). Would the name endure if the thing it stood for was only a memory? Maps are custodians of memory and acts of authority in scripted form and, perhaps also, a means to erase names in a landscape lest care be taken to record what is mapped in the mind.

THE
Uïg Parish
Parish of Loch
THE FOREST
PART OF HARRIS
Mangastay
Brenish
Malisty
Duskere Rocks
L. HAMNEVAY
SCARP I.
Skinet
L. Rhesort
Tirescull
Kenrhesort
Dushenish
Clorigin
Cleasamul
Soa
L. Miavag
TARANSAY
Langarod
Bouspick
Ilanisa I.
L. TARBOT
L. BUNOANETTER
Luskender
Linshader
Calvay
Kirabust
Briasclet
Balallin
Valtis
Garvard
Brenigal
Aline
Skeladale
L. Seafort
Marwick
Brolum
Ru. Ushenish
L. Newrt
L. SEAFORT
L. CLAY
L. VALUMIS
L. BROLUM
Diroclet
Urga
Knocknidcod
Trinniset
SCALPAY or ILAN GLASH
Skergriatich R.
Ru. Grebanish
Ru. Chlner
L. CREOSAVAGH
RA-SCATAVAGH
L. NAHUAGH
L. STOKENIS
Skerinoe
Garivelan
SHIANT
Ilanakily
VIII
VI

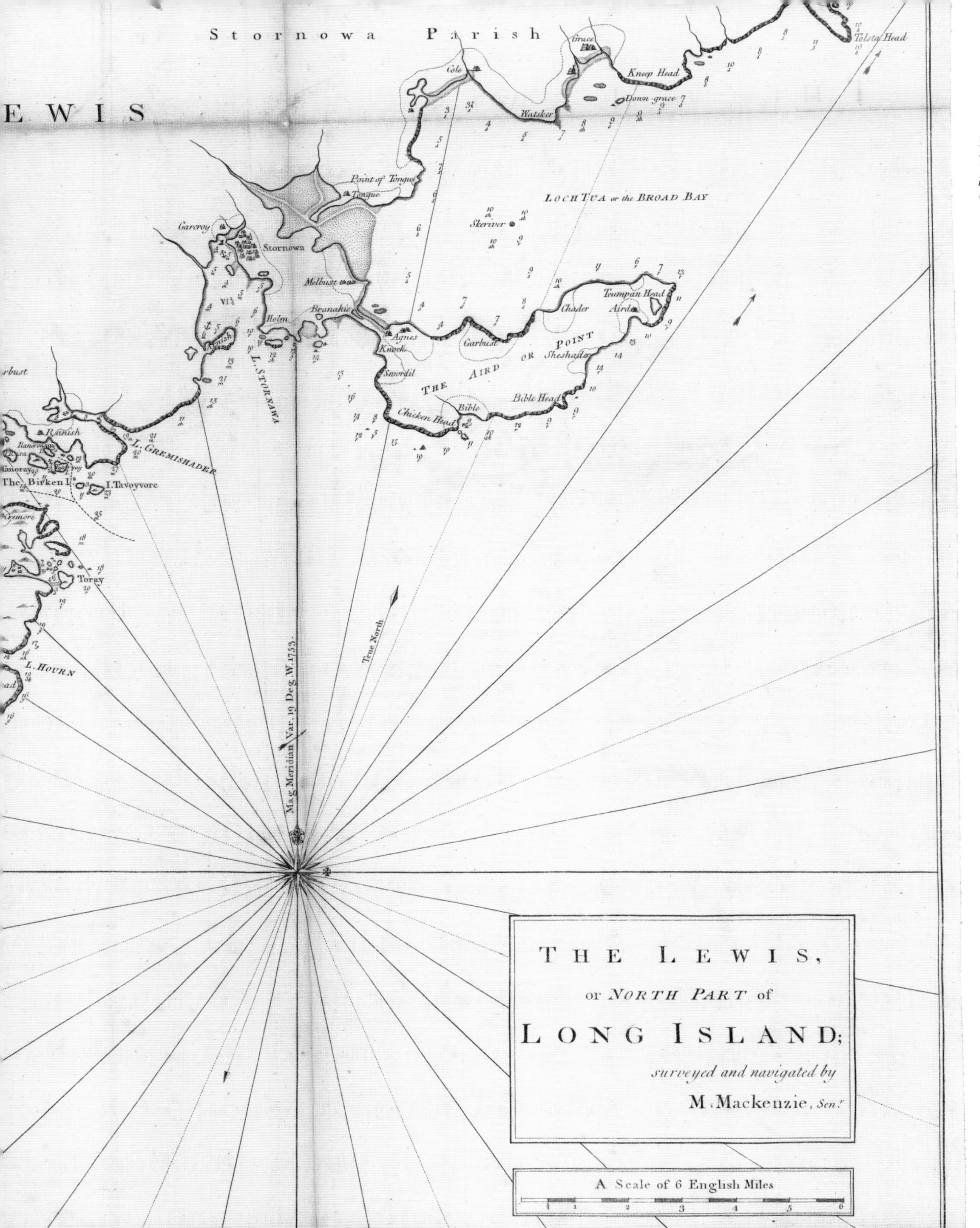

Murdoch Mackenzie, *The Lewis or north part of Long Island* (1776).

Croulin I.
Loch Indaal
Kenlochdaal
Caniscrois
Ruinteuoch
Sandick
Malawicky
Pabbay I.
Longa I.
Skerinderick
Skerintarson
Broadford
Knock
Scalpa I.
Strolumis
Dumarua
Loch Eynort
Duncann
Kilmore
Armidale
Heast
Tormore
Ord
Borereg
Tarskevag
Talevile
RAZAI I.
Brea
Sconser
Glamocascard Hill
Balmenoch
Tormore Hd.
Pendinavig
Penmore
Culester
Porbree
Airderachig
Loch Stigachan
Torrin
Loch Slepin
Loch Eyfort
Kilmaree
Aird of Strath
Ruinaiscan
Ilanahardah
The Point of Slate
Egg
Loch Nagan
Rises 11S. 6N. Tides
VI
D
N
A
L
S
I
SKY ISLAND
Culen Hills
Loch Scavig
Soay I.
Loch Skresort
Colinanne
Ruindowin
Loch Brittle
Kenloch
Camisviaseg
Slaninesha
Herries
RUM I.
Kilwiry
Guaridil
Rocky
Shells
Snisort
Grakenish
Crule
Eynort
Housdale
Arriharan
Duskere
Loch Eynort
Drynach
L. Harport
Gesto
Bracadale
Fernalea
Ruinaglach
Taliscar
Ulimish
Coraghan P.
Rocky
Sandy I.
Cana Island
Loch Harlosh
Colbust
Balmenoch
Griserniſh
Loch Bracadale
Orensa I.
Wia I.
Haverser I.
Harlosh I.
Balimore
Loch Pulroag
Loch Nagulin
VI
Humela Rock
Mud
Dunvegan Castle
Kilmore
Claggen
Loch Varksag
Macleods Maidens
Halival Hill
Itrigil
Greenok
Ilanskyandin
Very little Stream
Rocky
Hyſker
Debidil
Skeratt
Lowrgell
Gravel
Mud
Boreraig
Ramsake
Hamar
Waterstene
Copnahow Head
Rocks
Pulteel
Niest
Betamore
Sand
Rocky

CHAPTER FOUR

NAVIGATING

The maps most familiar to us in modern context are those which we use to find our way from place to place – maps for wayfinding or navigating. In the history of maps and map use as a whole, this is a relatively recent feature: what is today the commonest use of the commonest types of maps is not what maps were used for in pre-modern times. For local travel, we often do not require maps at all: the maps we employ are mental, stored in our heads and understood through everyday usage. For places unfamiliar to us through daily encounter, however, the key questions of 'Where do I want to get to?' and 'How do I get there?' are usually answered through recourse to a map. For islands, the act and art of navigating depends greatly upon the sorts of maps we produce of them and how maps show what lies in between the islands in question. This chapter examines the representation of Scotland's islands in marine charts and other maps, and the importance of islands in navigating Scotland's seas.

Scotland's islands figure in marine charts both as the subject to be mapped – the aim being to determine an island's shape and depict its 'internal features' – and, more importantly, as objects of navigational importance. Siting islands by their correct position in terms of latitude and longitude and delineating their dimensions accurately is of vital importance to those at sea who need to sight them, to know where the island lies, where safe anchorage may be had, and so on. But Scotland's islands were not easily positioned or shaped, either in navigational terms or as the objects of the map maker's attention. One of the earliest types of map, the portolan chart, first appeared in the thirteenth century. In their depiction of coastlines and in the commonplace use of rhumb lines – lines corresponding to a compass bearing for use in navigating – portolan charts are in general a remarkably accurate form of marine wayfaring map. As we saw in chapter 1, however, the positioning and representation of Scotland's islands on Georgio Sideri's (Callapoda) portolan chart of 1560 (fig. 1.6) was little more than symbolic. As we have also seen, when Scotland first comes more fully into recognisable view in map form from the 1560s, numerous islands are shown to the west and north of the mainland (figs 1.2, 1.3). Several have identifiable names, but, to modern eyes, a much less recognisable shape. In their position, outline and content, these islands bear little relation to what is known of them today.

Opposite. Detail from William Heather, *A New and Improved Chart of the Hebrides, or Lewis Islands and Adjacent Coast of Scotland* (1804).

After Scotland had been circumnavigated in the 1540s and a map produced from this work in the 1580s (fig. 1.3), Scotland's islands begin to show a more 'modern' geography in terms of their dimensions and position relative to one another. But they were not themselves the map's subject. That Gerard Mercator in 1595 and Joan Blaeu in 1654, both map makers of European renown, could disagree about the shape and orientation of major islands such as Skye, for example, or the position and size of the small island of Rona (see fig. 4.1), is interesting. But it should not come as a surprise. In navigational terms and as mapped objects, Scotland's islands and their surrounding seas were not the direct concern of its map makers until the end of the seventeenth century.

Bringing Scotland's islands into view: the work of John Adair, 1686–*c.*1713

Scotland's islands as a whole were first depicted in detail on printed maps in Joan Blaeu's 1654 *Atlas Novus*. This work, a combination of maps and written descriptions of the country, was largely based on manuscript maps and texts compiled by Timothy Pont at the end of the sixteenth century, together with revisions to Pont's work and additional material by Robert Gordon, minister of Straloch in Banffshire. Other sources incorporated into Blaeu's 1654 *Atlas Novus* – effectively, Scotland's first atlas – included that of the Scottish churchman and historian George Buchanan from his *Rerum Scoticarum Historia* (1582), and the English antiquary William Camden. Camden's major work, *Britannia*, first published in 1577, went into numerous and ever-larger editions in later years (that of 1607 included a full set of county maps of England and Wales, as well as a map of Scotland, fig. 1.18a). *Britannia* provided the first county-based chorography, or regional description, of those countries. It was, however, relatively uninformed about Scotland, even to the point of being dismissive.

Maps and geographical descriptions can reach their 'use-by' date quite rapidly. This was certainly the view of Sir Robert Sibbald, Scotland's first Geographer Royal. In 1683, Sibbald

FIGURE 4.1
(*Opposite and overleaf*). These two maps show how different leading map makers in the early modern period depicted the islands on the west coast of Scotland. Consider, in particular, the relative orientation given to the Isle of Skye and, in Mercator's map, the positioning and dimensions of Rona. Map (a), produced at a scale of about 25 miles to the inch, is in part based on the Nicolay map of 1583 (see fig. 1.3), but also includes information from Mercator's map of 1564 and from other contemporary sources. The main islands are correctly positioned in relation to one another, more or less, but that is not so in terms of their size: consider Rona, for example – what is today North Rona – which lies north of the Outer Hebrides/Innse Gall rather than between Lewis/Leòdhas and the north-west mainland as is shown here (but compare this representation of the island with that of Forlani in *c.*1561: fig. 1.2). Elsewhere, 'Collen' (Coll) and 'Terrey' (Tiree) and 'Mula' (Mull) feature if not (yet) with their current toponomy.
Source: (a) Gerard Mercator, *Scotia Regnum* (1595). (b) Robert Gordon of Straloch, *Scotia Regnum cum insulis adjacentibus* (1654).

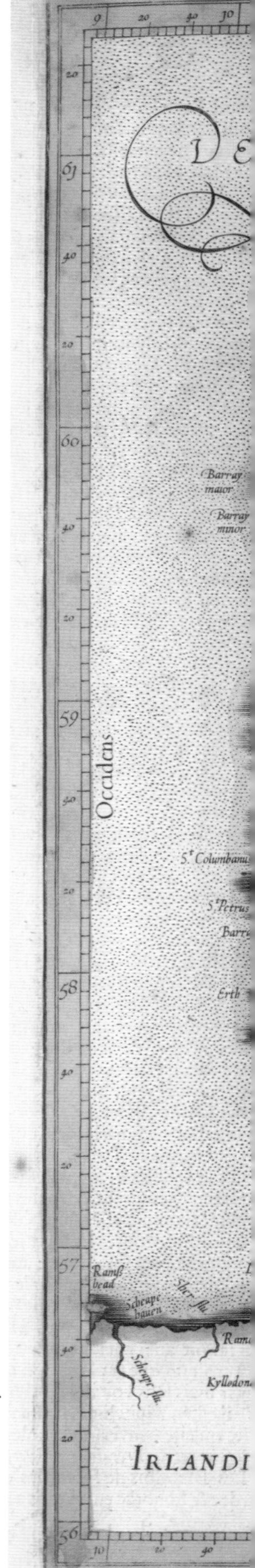

a.

SCOTIA, Regnum.
Miliaria Scotica
Per Gerardum Mercatorem Cum Priuilegio
Septentrio
Meridies
Oriens
ORCADES insulæ
HEBRIDES insulæ
MARE GERMANICUM
HIBERNIAE
Yla insula
Mula
Pomonia
Strathnauernia
Sutherlandia
Cathenesia
Rossia
Loquhabria
Lorna
Argadia
Morauia
Buquhania
Perth
Atholia
Fifa
Laudonia
Marchia
Cuningham
Kyll
Carricta
Gallouidia
Clidesdaill
Teuidalia
Anandia
Lennoxia
Menteith
Albayn
Cantyr
Arren
ANGLIAE PARS

SCOTIA
REGNVM
cum insulis adjacentibus.
Robertus Gordonius a Straloch descripsit.
Septentrio
Meridies
Oriens.
Occidens.
Orcadum Insularum pars
PENTLAND FYRTH
Dungisby head
Parro head
Terbaet-Ness
Reed-head
THE FYRTH OF FORTH
THE FYRTH
OF CLYD
SULWAY FYRTH
CUMBERLAND
PART OF ENGLAND
NORTH HUMBER LAND
PART OF IRELAND
Lewis
Skye
Rum
Egg.
Muick
Mul
Iona
Iura
Yla
Arren
Ross
Kintail
Glen Elg
Morvern
Laern
Braid Al
Oth oll
Mernis
Angus
Strath Ern
Menteth
Fyf
Edinburgh
Lothian
Twed dail
Selkirk
Clyds dail
Annan dail
Eskdail
Liddisdail
Caithnes
Sutherland
Strathnavern
Lochaber
Badenoch
Murray
Universæ insulæ hæ, in hoc mari late sparsæ (Boëthio Hebrides, Buchanano Æbudæ dictæ) hodie communi nomine Occidentales insulæ dicuntur.
Miliaria Scotica quorum 50 respondent uni gradui coelesti.
Miliaria Anglica quorum Sexaginta respondent uni gradui.

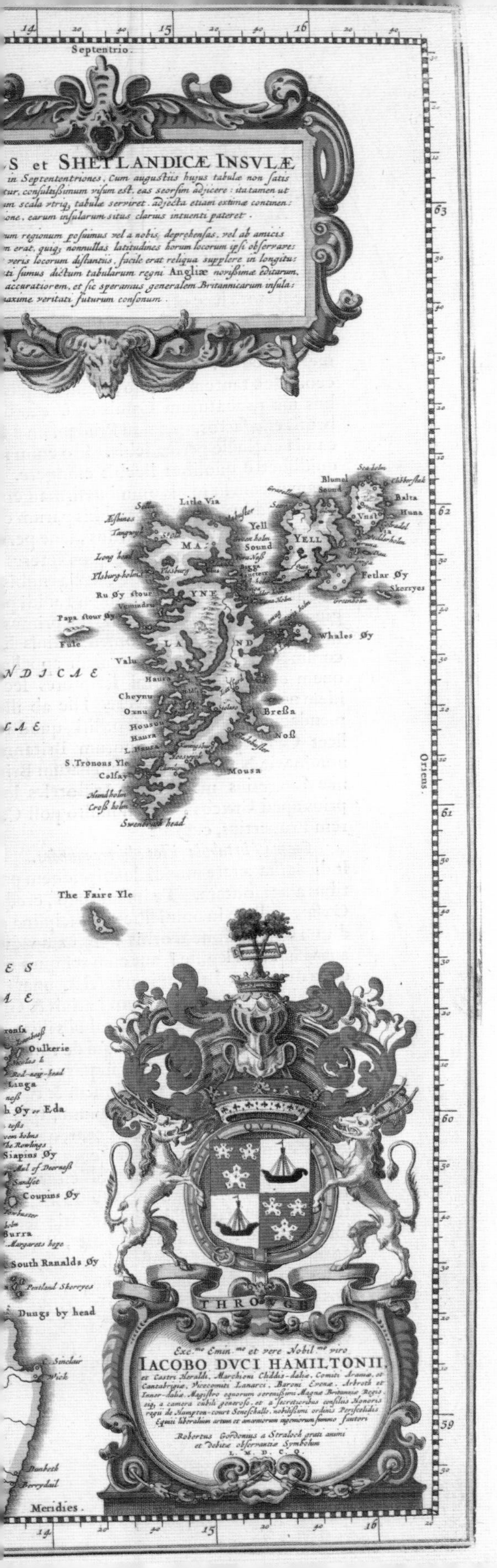

b.

laid out plans for a new 'Scotish [sic] Atlas', what he called a 'description of Scotland ancient and modern'. In justifying this proposed plan of action, Sibbald took aim at Blaeu's 1654 work and its sources. 'For the theater of Scotland published by Bleau [sic]', wrote Sibbald, 'except it be the Description of some few shires by the learned Gordon of Straloch, . . . containeth little more than what Buchanan wrote, and some few Scrapes [scraps] out of Cambden'. The latter, Sibbald further reckoned, 'is no friend to us in what he writeth'. Sibbald also had more immediately utilitarian ends in mind: 'And in respect that there are many Islands around this ancient Kingdom of Scotland, and many impetuous and contrary Currents and Tides, and in several places the Coast is full of Rocks, or Banks of Sand, which ought to be exactly described for the security of Trade.' The man chosen to undertake this mapping work, for Sibbald and for Scotland, was John Adair.

Born in 1660, John Adair – 'Mathematician and Skilful Mechanick' as Sibbald described him before the two fell out over maps and money – received support from Scotland's Privy Council for his national mapping work as early as May 1681. Further governmental support for him was forthcoming from June 1686 when the Scottish Parliament passed an Act 'In Favours of John Adair, Geographer, for Surveying the Kingdom of Scotland, and Navigating the Coasts and Isles thereof'. Funding for this was to be provided by a tonnage levy applied differentially upon Scottish and foreign ships (one shilling a ton on Scottish ships, four shillings a ton on foreign vessels).

If we may see Adair as Scotland's first government-funded map-making civil servant, he had counterparts elsewhere. In England, John Seller's *The English Pilot*, a marine atlas partly based on Dutch charts, had been published in 1670, and the hydrographer Greenvile Collins' *Great Britain's Coasting Pilot* first appeared in 1693. Collins surveyed Orkney in 1688 as part of this work (fig. 4.2). In Holland, the Dutch led the way in marine charts and sea atlases. Maps for wayfaring at sea were thus a shared imperative for Europe's maritime nations. The 1686 Act in support of Adair makes this clear for Scotland: 'the Hydrographical Description of the Sea-Coast, Isles, Crieks [sic], Firths, and Lochs, about the Kingdom, are not only

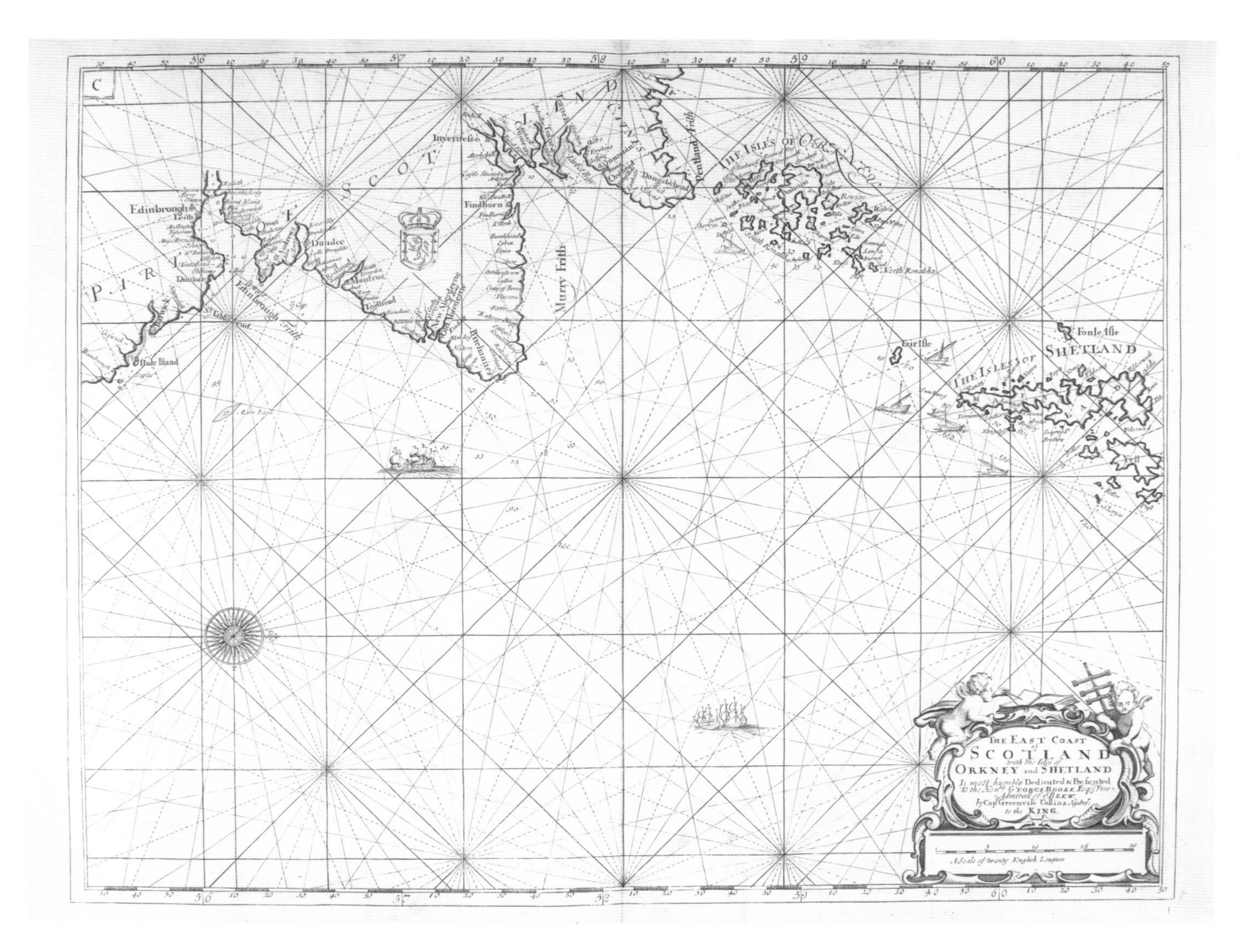

FIGURE 4.2
Greenvile Collins' map of 1693 combines something of the earlier mapping features of the *isolario* map form (cf. Callapoda's map of *c.*1560, fig. 1.6) with later forms of the chart. He has identified selected names on the mainland and placed them at right angles to the coast so as not to clutter other sections of the map (very much in the *isolario* tradition). Collins had extensive experience as a hydrographer. His 1693 *Great Britain's Coasting Pilot* was the first such work by a Briton. *Source*: Greenvile Collins, 'The East Coast of Scotland with the Isles of Orkney and Shetland' from *Great Britain's Coasting Pilot* (London: Richard Mount, 1693).

Honourable and Useful, but most necessary for Navigation . . . The want of such exact Maps, having occasioned great losses in time past: and likewise, thereby Forraigners may be invited to Trade with more security on our Coasts'. Accurate maps of Scotland's islands and seas would allow Scotland to know itself better and to be open for business.

Adair began his charting work from north to south, first on the east coast, then on the west. By 1692, he had surveyed parts of the east coast between Perth and Aberdeen. In 1696,

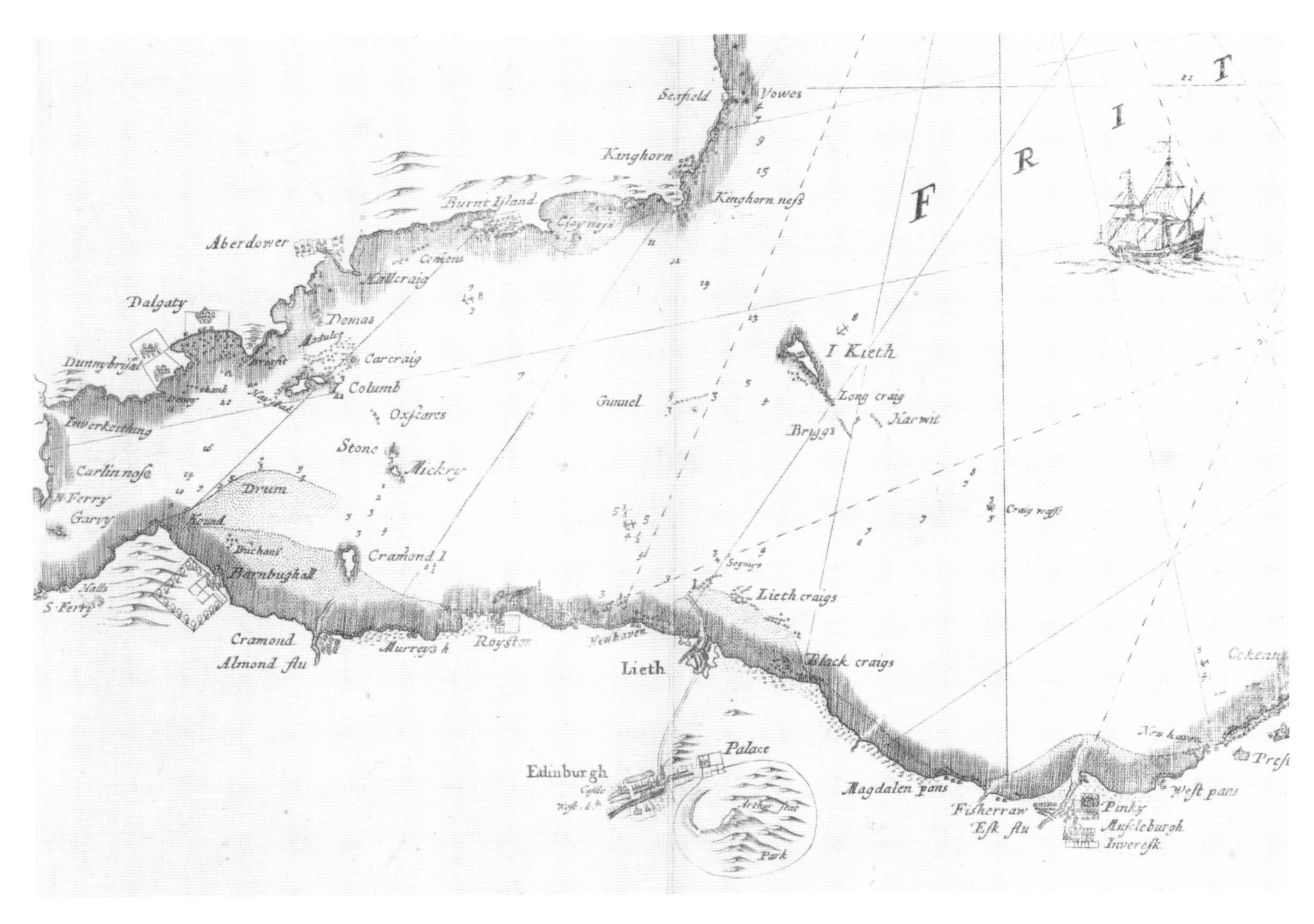

accounts show him hiring a vessel for use around the 'small Islands [on the western seaboard] and along the south coast'. In 1697, he was 'presently to goe and make mapps and Descriptions of the coasts of Argyle, Kintyre, Yla [Islay], Arran, Bute and the inner pairts of Clyde'. Between May and November 1698, he was surveying the Orkneys, then the Western Isles/Na h-Eileanan an Iar. Published charts exist of the east coast, for the Aberdeen area, and for the Firth of Forth (fig. 4.3). By 1713, he had nineteen charts in manuscript, most

FIGURE 4.3
This map shows the western stretches of the Firth of Forth and its islands, marked both for their relative position to one another and as hazards to the coastal shipping whose tonnage (symbolised in the ship shown here) was a source of revenue for Adair himself: it was thus very much in his own interests to draw up good maps of 'Inch Kieth', 'Ox Scares' and other islands.
Source: John Adair, 'The Frith of Forth from the Entry to the Queen's Ferry With all the islands . . .', from *The Description of the Sea-Coast and Islands of Scotland* (Edinburgh, 1703).

of them drawn up by 1704. These surviving manuscript charts form four distinct geographical groups: the Hebrides/Innse Gall and the north-west coast; the south-west coast; the Northern Isles of Orkney and Shetland, and the Aberdeenshire coast. In his *The Description of the Sea-Coasts and Islands of Scotland, with large and exact maps, for the use of seamen* (1703), Adair combined charts with textual descriptions – tides, harbours, routes for safe navigation. Adair's work demonstrates for the first time in Scotland a direct association between government initiative, island mapping and the demands of navigation.

Adair never completed the systematic mapping of Scotland's islands and coasts. There are several reasons for this. Proficient surveyor no doubt, Adair was slow. He was slow because, until released from his contract in 1691, Adair was obligated to produce maps for Sibbald's own never-completed Scottish atlas project. He was also hindered by lack of funds: the tonnage levy was never sufficient for his needs. Survey instruments were expensive. And he faced competition for this income from John Slezer, whose 1693 *Theatrum Scotiae* presented topographical views of Scotland (see fig. 5.4) as Adair hoped to provide hydrographical and navigational ones. His work was more than once hampered by bad weather. Not least, as one man faced with mapping all of Scotland's islands, Adair faced a near impossible task. And, as was the case for Seller and Collins, Adair did not use a standard and accurate hydrographic surveying technique. Such a thing did not then exist. This problem would not be addressed until the 1740s.

'Island fixing': Murdoch Mackenzie (senior) and the development of accurate marine survey

> The Lives and Fortunes of Sea-faring Persons, in great measure, depend on the Accuracy of their Charts. Who-ever, therefore, publishes any Draughts of the Sea-coast, is bound in Conscience, to give a faithful Account of the Manner and Grounds of the Performance, impartially pointing out what Parts are perfect, what are defective, and in what Respects each are so; that the Public may be enabled to judge of its Sufficiency, and how far the Draughts are to be relied on.

With these words, the Orkney-born hydrographer Murdoch Mackenzie, one of Britain's leading hydrographers and island mappers, introduced his *Orcades: or a Geographic and Hydrographic Survey of the Orkney and Lewis Islands in Eight Maps* (1750). Mackenzie owed his employment as an island map maker to the patronage of James Douglas, 14th earl of Morton, who had succeeded to his estates, which included the Orkney Islands, in 1738, and to Colin MacLaurin, professor of mathematics at the University of Edinburgh, who taught the principles of mapping as part of his classes. In this and later work, Mackenzie established new standards for the mapping of Scotland's islands and seas – and brought the Orkneys in particular into sharper focus (fig. 4.4).

Mackenzie began his Orkney Islands survey in August 1744. He completed it in 1748, at which point he turned, rather more hurriedly, to Lewis/Leòdhas in the Outer Hebrides/Innse Gall (fig. 4.5). As he was at pains to point out, the work was a new departure in island mapping:

> The Method made choice of in taking this Survey of the Orkney and Lewis Islands, differs from the usual Way, of Sailing along the Land, taking the Bearings of the Head-lands by a Sea-Compass, and guessing at the Distances by the Eye, or Log-line; and also from a very common, tho' less certain Method, of constructing Charts, without either surveying, navigating, or viewing the Places themselves; but only from verbal Information, copied Journals, or superficial Sketches of Sailors casually passing along the Coast.

Mackenzie's remarks may be read in different ways. Contemporaries would have understood them as an affirmation of maps' utility and as a caution about maps' accuracy. The credibility of maps as documents to be relied on in navigating depended on more than others' word of mouth or guesswork. Other maps and map makers at the time were no less anxious to be seen to be up-to-date but words on the map itself could not necessarily be guaranteed to be of benefit to navigation if the map were not used at sea (fig. 4.6). Today, map historians would see Mackenzie's declaration as typical of Enlightenment chart and map making: in an accompanying narrative or memoir, the map's maker would make clear the methods used to produce the maps. To map readers and users then and now, Mackenzie's words are also statements about his own standards and usefulness: trust was embodied mutually, in the maker and in the map.

Mackenzie's importance rests in his innovative combination of methods, beginning with his use of triangulation from a measured baseline. 'Beacons, or Land-marks' were built on the summits of all the 'remarkable Hills and Eminences'. Taking advantage of a hard frost, he erected his theodolite (lent to him, with other instruments, by the Navy) and measured his baseline on the frozen 'Loch Stenhouse' (today, the Loch of Stenness) on Orkney Mainland. From there, he built up a network of triangles. He also went round the coast, marking on rough drafts the 'true Shapes of the Heads, Points and Bays'. Boats were positioned 'to lye at the Extremities of Rocks and Shoals, till their Positions and Dimensions were determined by the requisite Observations on Land'. When a 'complete Map of an Island was made out in this Manner', Mackenzie went round the island in question by boat, taking soundings for depth, measuring the rate of tides and taking 'Notes of the Land-marks, for avoiding Rocks, sailing thro' Channels, &c' (fig. 4.7). This whole process, he tells us, was repeated island by island, 'till the Survey of all the Orkney Islands was completed'.

Mackenzie's *Orcades* set a new benchmark for hydrographic and island mapping. The many names of subscribers listed to the front of the book – merchants in Rotterdam, Kirkwall, Inverness, London and Leith, members of Scotland's gentry, the East India Company and many others – testify to the high regard in which it, and he, was held. From 1751, he was employed by the Admiralty to undertake survey work of Scotland's west coast and Ireland. This work appeared as *A Treatise of Maritim* [sic] *Surveying* (1774) and as *A Maritim Survey of Ireland and the West of Great Britain* (1776, two volumes). The first work was strongly instructive, as Mackenzie laid out his methods for others to follow: 'How to Measure a Straight Line on the Surface of the Sea', 'How to Survey an Island', 'Circumstances to be noticed in Describing Rocks and Shoals', and so on. The second work was essentially illustrative, with Scotland's islands and related navigational remarks appearing in detail (fig. 4.8).

Mackenzie's work was not a reflection in the Scottish context of what others were doing elsewhere. It is more appropriate to see him as a pioneer, leading the way in new forms of mapping and using Scotland's islands to do so. Mackenzie's island mapping established Scotland's islands, the Orkneys especially, as 'test sites' for new, mathematically informed, cartographic procedures. New levels of accuracy were realised, principally for islands' margins, shape and relative position. He sought to provide what, in modern terms, we might call a language of consistent cartographic symbolisation (fig. 4.9), a 'symbolic shorthand' for the features of islands and their margins. He even kept island mapping in the family – his nephew, Murdoch Mackenzie junior, went on to become Maritime Surveyor at the Admiralty.

Safety and survey: government initiatives, 1795–1855

There are important parallels between the mapping of land and of the sea in late eighteenth-century Britain. Ordnance Survey, founded in 1791 to begin the systematic mapping of the nation, has its origins in part in the work of William Roy and David Watson in the Military Survey of Scotland undertaken between 1747 and 1755. The maritime equivalent, the Hydrographic Office, was founded in 1795. It has its origins in the work of the two Mackenzies and others from mid century, from a concern that island maps and sea charts were being produced but not coordinated in any systematic way,

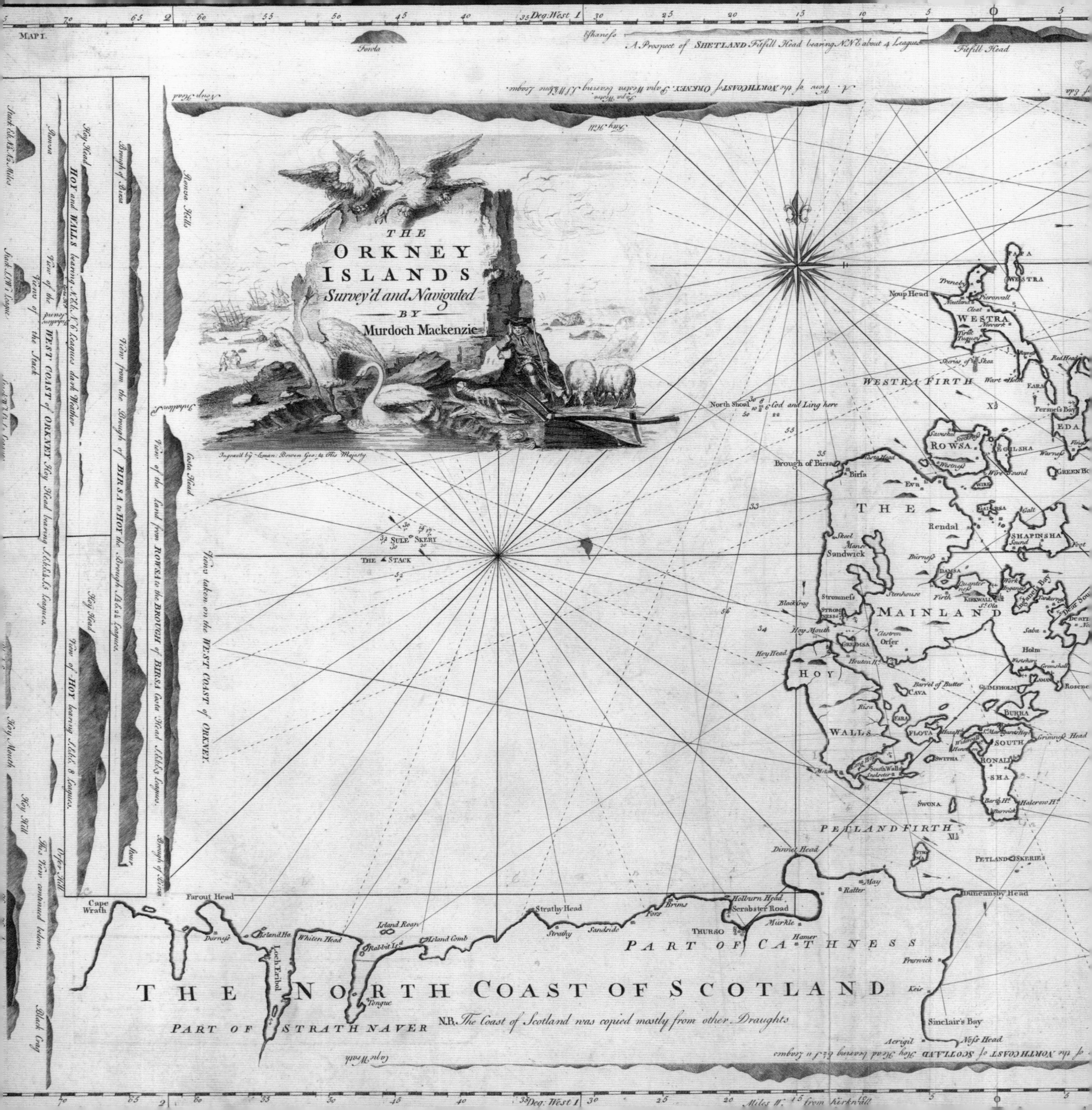

MAP I.
THE ORKNEY ISLANDS Survey'd and Navigated BY Murdoch Mackenzie
Ingrav'd by Eman: Bowen Geo: to His Majesty
A Prospect of SHETLAND Fitfill Head bearing N.N.E about 4 Leagues
Fitfill Head
Fowla
THE NORTH COAST OF SCOTLAND
PART OF CATHNESS
PART OF STRATHNAVER
N.B. The Coast of Scotland was copied mostly from other Draughts
THE MAINLAND
HOY
WALLS
WESTRA
ROWSA
EDA
SHAPINSHA
SOUTH RONALD SHA
BURRA
FLOTA
FARA
CAVA
SWONA
PETLAND FIRTH
WESTRA FIRTH
PETLAND SKERIES
SULE SKERY
THE STACK
North Shoal
Cod and Ling here
Brough of Birsa
Noup Head
Stromness
Sandwick
Kirkwall
Cape Wrath
Farout Head
Strathy Head
Hollburn Head
Scrabster Road
THURSO
Dunnet Head
Duncansby Head
Sinclair's Bay
Noss Head
Tongue
Loch Eribol
Whiten Head
Deg: West
Miles W. from Kirkwall

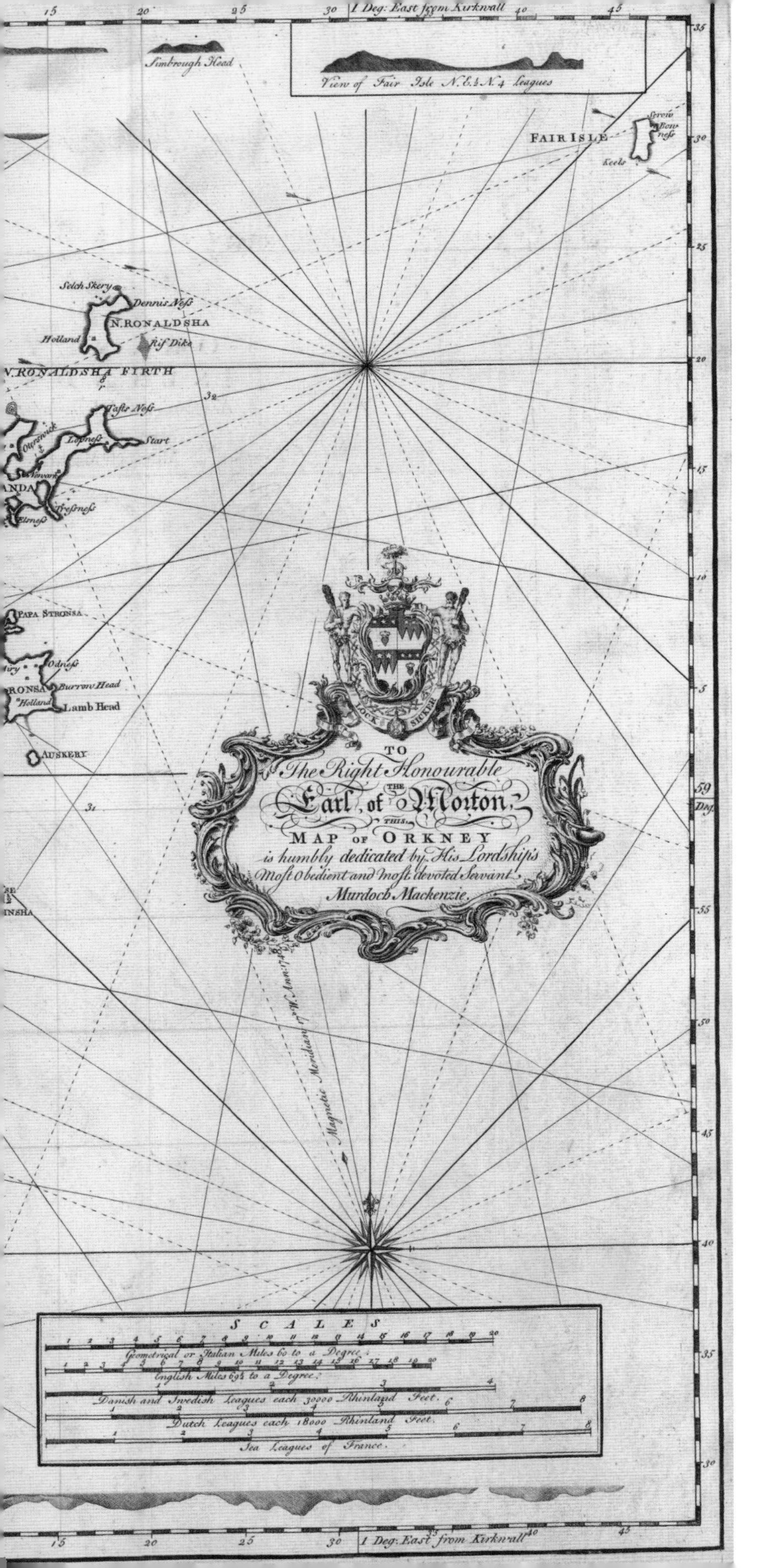

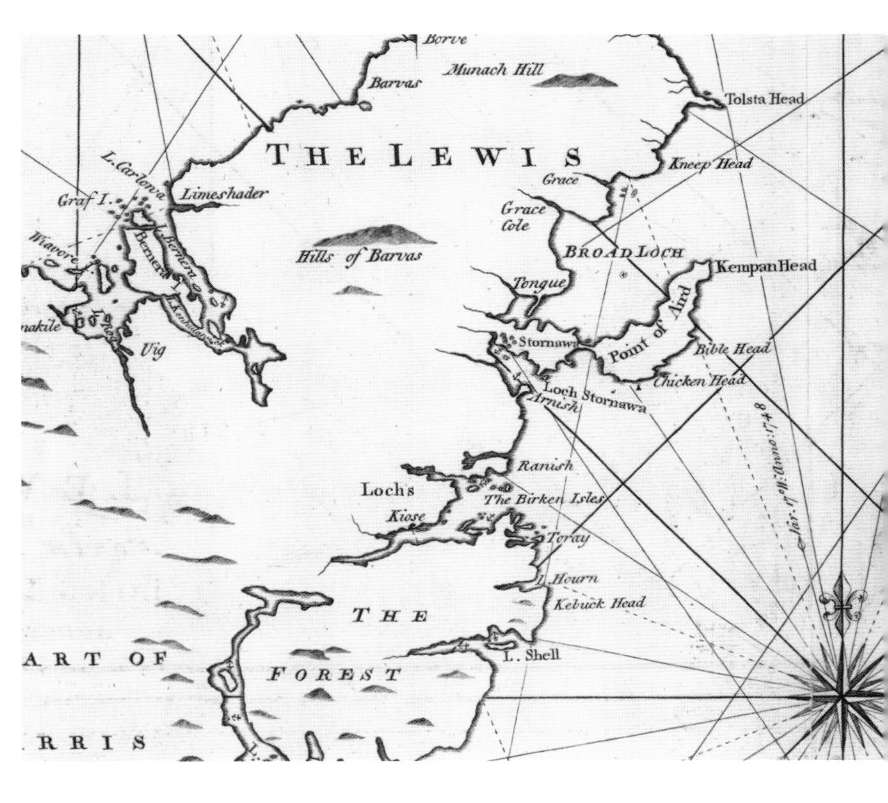

FIGURE 4.4

(*Opposite*) Orkney in the mid eighteenth century.
Source: Murdoch Mackenzie, 'The Orkney Islands', from *Orcades: or a Geographic and Hydrographic Survey of the Orkney and Lewis Islands in Eight Maps* (London: Mackenzie, 1750).

FIGURE 4.5

(*Above*) In this detail of the town and harbour of 'Stornawa' (Stornoway/Steòrnabhagh) taken from map 7 in his *Orcades*, Murdoch Mackenzie outlines the general position of the harbour and the shape of the Lewis/Leòdhas coastline in the immediate vicinity: note, too, the outline of a hill, in 'pop-up' style placed to the centre of the island.
Source: Murdoch Mackenzie, 'The North Part of the Lewis', from *Orcades: or a Geographic and Hydrographic Survey of the Orkney and Lewis Islands in Eight Maps* (London: Mackenzie, 1750).

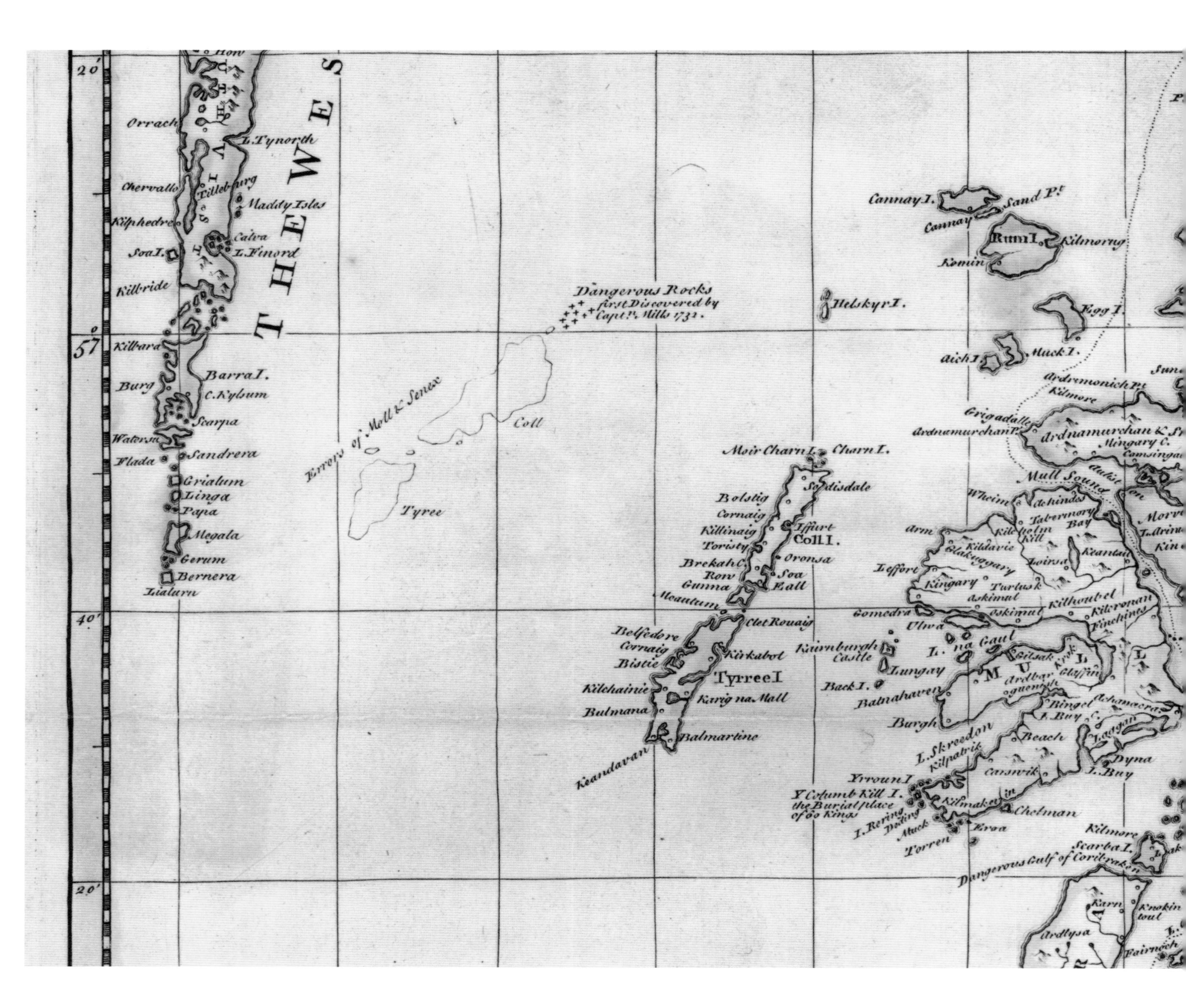

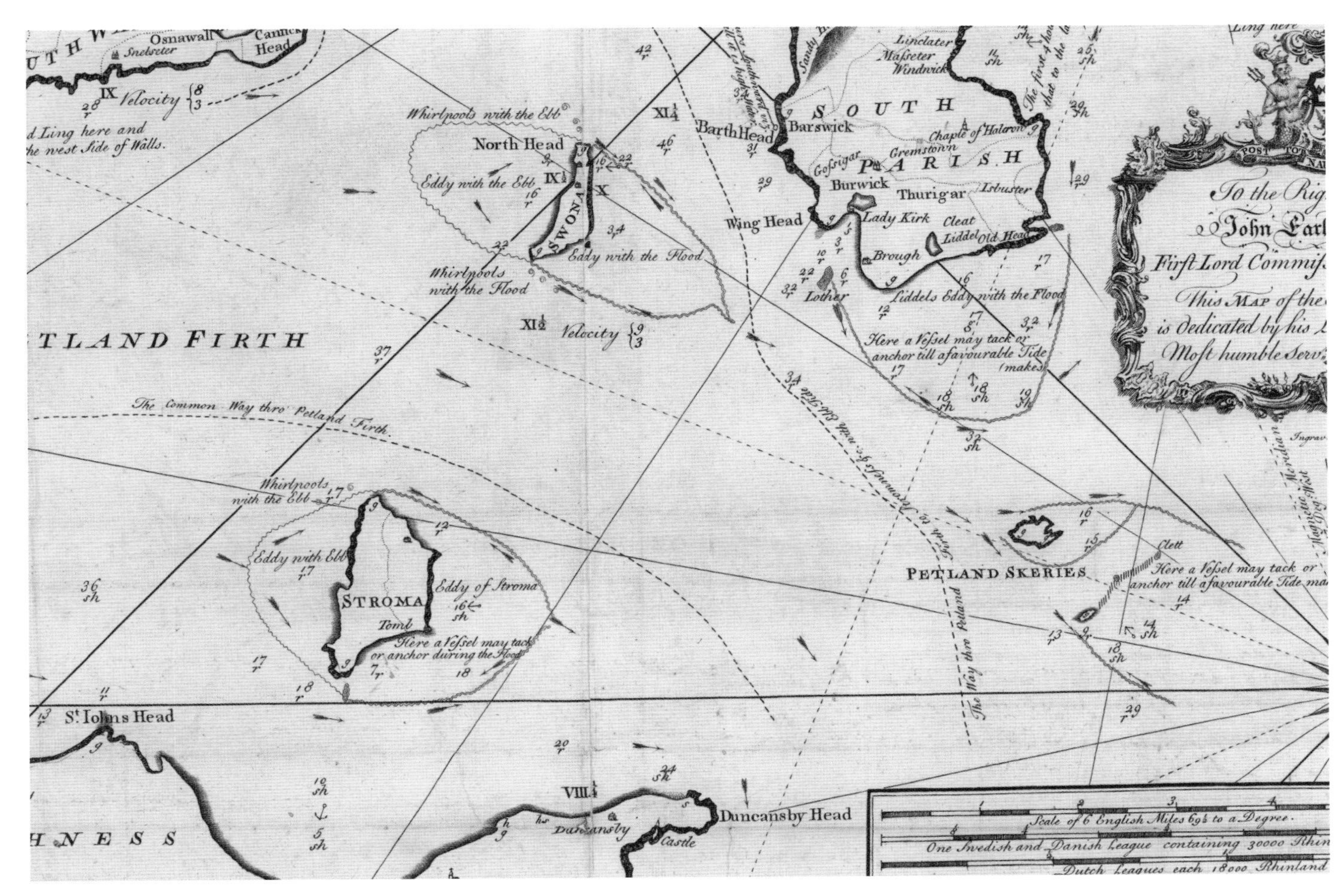

FIGURE 4.6

(*Opposite*) From this detail, we can see that part of the credibility of John Elphinstone's work depended upon revising the errors of earlier map makers. Here, in addition to depicting the islands of the west coast in his own terms (in a map whose title is suggestive of accuracy and, in being called 'North Britain', of political loyalty), he uses 'ghost outlines' of the islands of Coll and Tiree, taken from the work of the German map maker Herman Moll and the English map maker John Senex, to inform the reader of his map of their mistakes. The actual effectiveness in navigational terms of Elphinstone's corrections would depend, of course, on his map being used at sea, and on mariners being able to know exactly where they were – the 'proper' position of themselves and of the islands in question. What is also not clear from the map is whether the rocks 'first discovered' by Captain Mills in 1732 are in their correct position or if they refer to rocks off the north end of Coll.
Source: John Elphinstone, *A New & Correct Mercator's Map of North Britain* (1745).

FIGURE 4.7

(*Above*) Mackenzie's maps also included textual commentaries about the run, strength and direction of tides around islands and, as here, in the Pentland Firth, then as now a potentially dangerous stretch of water. Interestingly, and a feature of an age in which metrology for linear distances had not yet been standardised, he includes several countries' linear distances in the scale bar to the right-hand column, among them those of the Swedes, the Danish and the Dutch, with whom Orkney and Scotland traded.
Source: Murdoch Mackenzie, 'The South Isles of Orkney', from *Orcades: or a Geographic and Hydrographic Survey of the Orkney and Lewis Islands in Eight Maps* (London: Mackenzie, 1750).

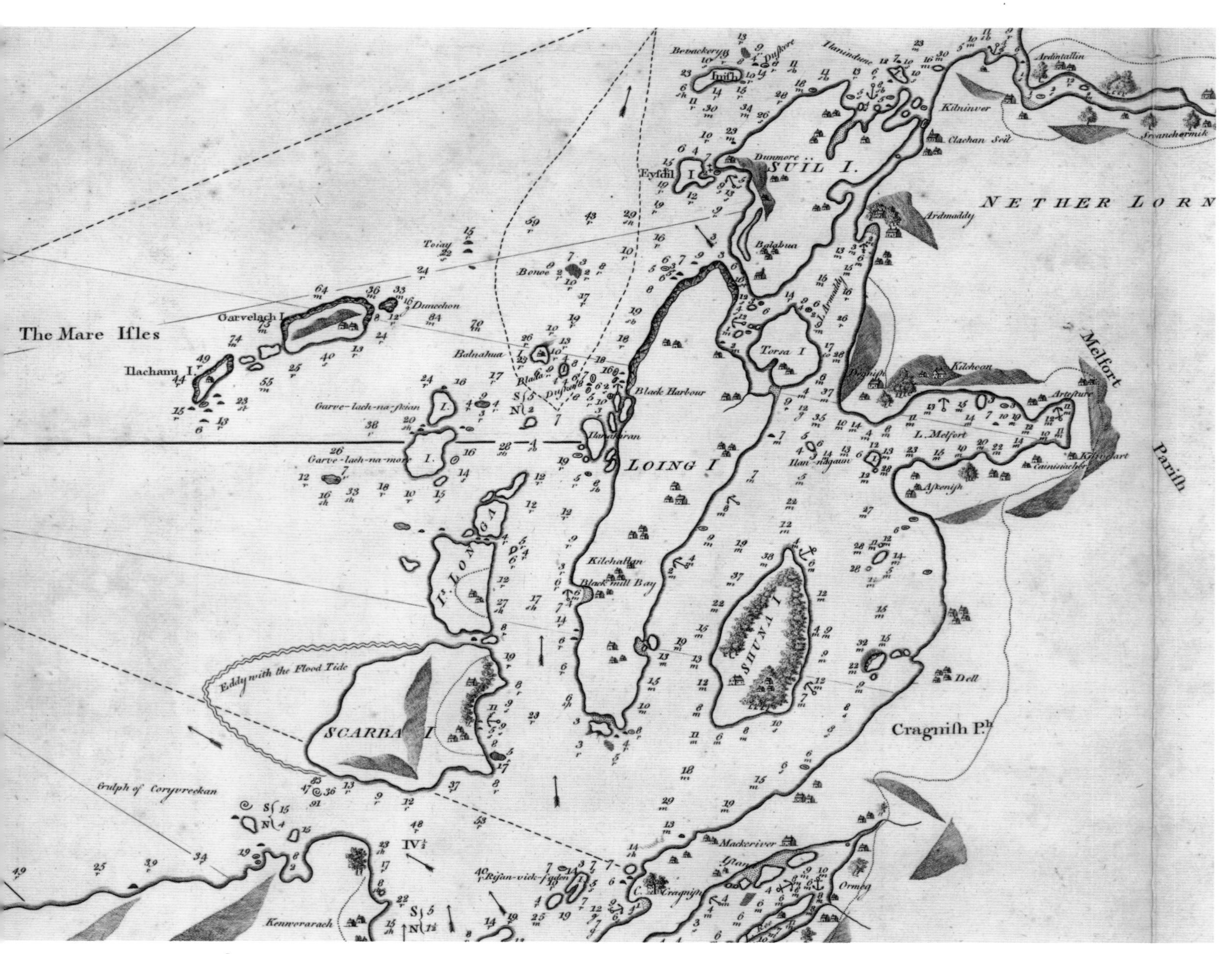

FIGURE 4.8

In this detail of Luing ('Loing' as Mackenzie has it) and other Argyllshire islands, Mackenzie includes both tidal information and soundings as navigational aids. Anchorages (with the relative depth of water) are also shown. His description of the Garvellachs as the 'Mare Isles' – 'the Isles of the Sea' – is a combination of Latin and a partial English transliteration of their Gaelic name; compare his representation with that a century or so later in the Ordnance Survey map of the Garvellachs (fig. 2.4). This chart is Chart 21 in Mackenzie's 1776 work.

Source: Murdoch Mackenzie, 'The West Coast of Scotland from Ila to Mull', from *A Maritim Survey of Ireland and the West Coast of Great Britain* (London: Mackenzie, 1776).

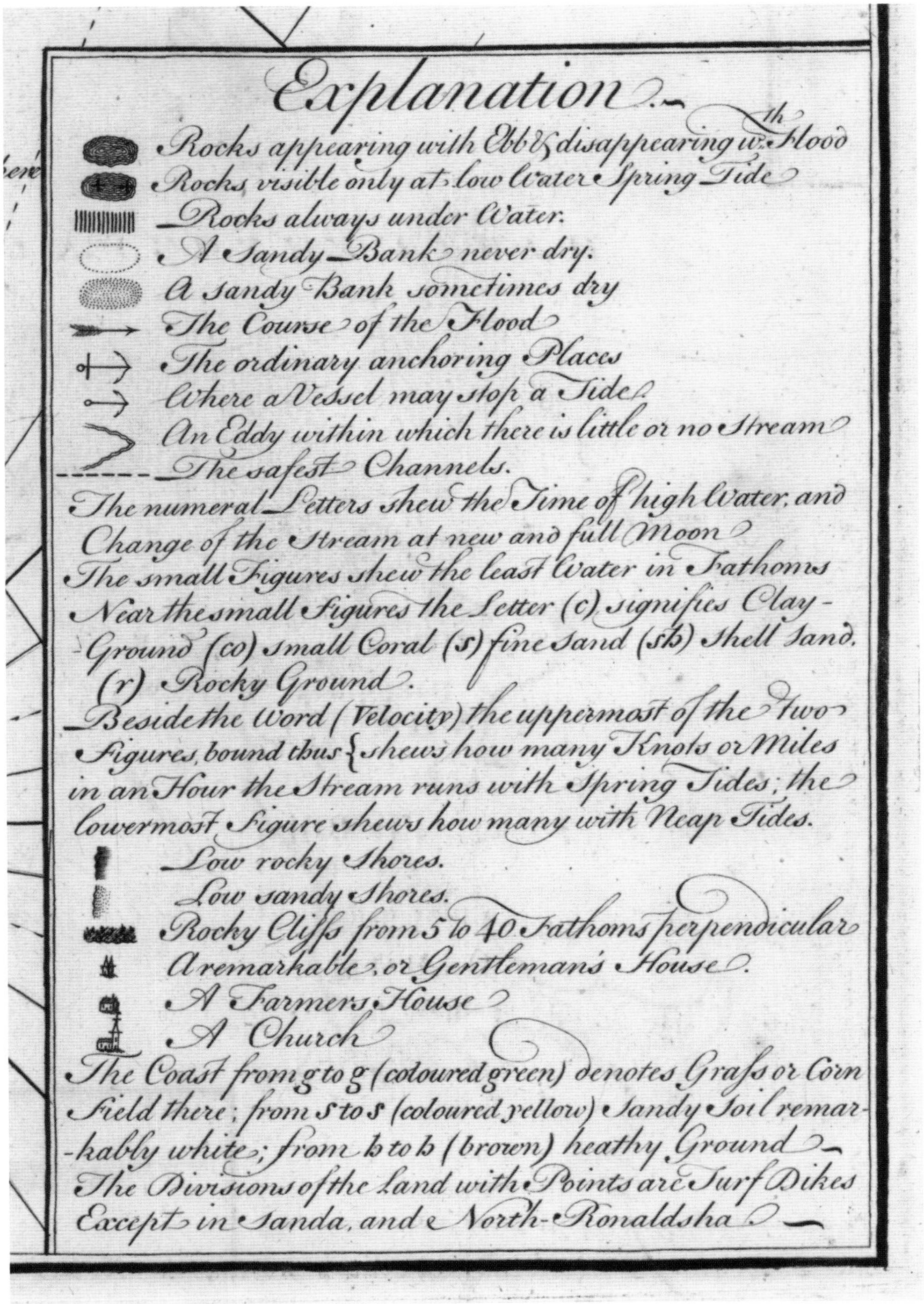

Explanation.
Rocks appearing with Ebb & disappearing wth Flood
Rocks visible only at low Water Spring Tide
Rocks always under Water.
A Sandy Bank never dry.
A Sandy Bank sometimes dry
The Course of the Flood
The ordinary anchoring Places
Where a Vessel may stop a Tide
An Eddy within which there is little or no Stream
The safest Channels.
The numeral Letters shew the Time of high Water, and Change of the Stream at new and full Moon
The small Figures shew the least Water in Fathoms
Near the small Figures the Letter (c) signifies Clay-Ground (co) small Coral (s) fine Sand (sh) Shell Sand. (r) Rocky Ground.
Beside the Word (Velocity) the uppermost of the Two Figures, bound thus { shews how many Knots or Miles in an Hour the Stream runs with Spring Tides; the lowermost Figure shews how many with Neap Tides.
Low rocky Shores.
Low sandy Shores.
Rocky Cliffs from 5 to 40 Fathoms perpendicular
A remarkable, or Gentleman's House.
A Farmers House
A Church
The Coast from g to g (coloured green) denotes Grass or Corn Field there; from s to s (coloured yellow) Sandy Soil remarkably white; from b to b (brown) heathy Ground.
The Divisions of the Land with Points are Turf Dikes Except in Sanda, and North-Ronaldsha.

FIGURE 4.9
The development of more standardised forms of hydrographic mapping which Murdoch Mackenzie did so much to pioneer depended upon the development of symbols which would make clear which feature in the landscape they referred to. These had to work for different places and over time. In his 'Explanation', Mackenzie provides a set of symbols and gives their meaning. But interpretation of them and thus navigational practice extending from them might differ from user to user. It is, after all, not easy to distinguish on the illustration and thus on the map between his first two pictorial symbols, those designed to show rocks in various tidal conditions. It would be harder still to distinguish the difference in reality.
Source: Murdoch Mackenzie, 'The North East Coast of Orkney', from *Orcades: or a Geographic and Hydrographic Survey of the Orkney and Lewis Islands in Eight Maps* (London: Mackenzie, 1750).

and from anxieties about foreign naval power. The fear of invasion by the French was very real during the Revolutionary Wars from 1791 and in the Napoleonic Wars between 1803 and 1815. Island mapping was an important means to state security.

Navigating the ship of state in times of war is one thing. Navigating a ship at any time depends in part upon knowing where not to steer. Scotland's first island lighthouse was built on the Isle of May in 1636. Its light, that is, the flames from the burning daily of up to four tons of Fife coal, was apparently much less effective as a navigational aid than was the resultant smoke. The construction of lighthouses as a more formal government matter followed the establishment of the Northern Lighthouse Board in 1782. On 27 June 1786, Parliament passed an Act 'for erecting Light-houses in the Northern Parts of Great Britain', its wording making it clear that this would 'conduce greatly to the security of navigation and the fisheries'. Lighthouses on Scotland's islands soon followed the Act: on North Ronaldsay in the Orkneys and on Scalpay, off Harris/Na Hearadh, in 1789; on Pladda, off Arran, in 1790; on the Pentland Skerries in the eastern reaches of the Pentland Firth in 1794 (fig. 4.10 and pp. 112–3). Many others would follow in the nineteenth century.

The construction of island lighthouses reflected general anxieties about navigating, safety, and the loss of life and commerce. Some, such as the Bell Rock – which was more reef than island, the rocks being exposed only at low tide – were of particular concern. Government commissioners charged with this problem were in no doubt about its significance: 'The extreme danger which arises from such a rock, situated in a tract pursued by an immense number of vessels, may be easily figured. It has long, accordingly, been an object of consideration, not only with traders, but with his Majesty's naval officers, who were cruising upon the coast; and who appear to have been fully aware, of the tremendous perils to which it exposed them.' The loss with all hands of the 64-gun HMS *York* in 1804, presumed to have hit 'the Cape, or Bell Rock' (also known as 'Inchcape'), finally prompted the government into action. The Bell Rock lighthouse was built between 1807 and 1810 in work managed by John Rennie and Robert Stevenson (fig. 4.11a and b).

The first Hydrographer to the Admiralty was a Scot, Alexander Dalrymple, of Newhailes near Edinburgh, who was appointed in 1795. For all that Mackenzie's work half a century earlier had established new standards, and for all that the military need for defence was heightened from the 1790s (see chapter 5), the mapping of Scotland's islands and coasts remained uneven. Charts and island maps were produced in the decades following Mackenzie's work, some by leading chart makers such as Joseph Huddart and William Heather, but not as the result of any systematic and coordinated central agency (fig. 4.12). The principal responsibility of the Hydrographic Office was to oversee and classify existing marine charts and, from that, to incorporate data from a variety of sources onto extant maps. Dalrymple's work essentially involved running a cartographic bureau, overseeing existing work rather than undertaking surveys to produce new maps. Even so, Dalrymple sought to establish new standards for coastal and island mapping, laying out the scheme in his *Essay on Nautical Surveying* in 1806. This involved a centrally anchored vessel surrounded by four smaller ships, also anchored, from which a series of 'sea triangulations' would position and plot the island and coasts in question.

Dalrymple's method was never used, being too expensive in ships, time and manpower, as well as too slow. His standards as a naval map maker were so exacting that many extant charts produced by others were never sanctioned by him and so remained unused, to the annoyance of many of his contemporaries. Dalrymple was dismissed from office in 1808. His replacement was Thomas Hurd. Under Hurd, and notably in the work of his successors, Rear-Admiral Sir William Parry between 1823 and 1829, and Rear-Admiral Sir Francis Beaufort from 1829 to 1855, the Hydrographic Office planned and realised what has become known as the 'Grand Survey of the British Isles'. Much of this work was undertaken in association with Ordnance Survey. Scotland's islands, Shetland and the Orkneys especially, were vital mapping points as the Grand Survey put coastal and maritime Britain into

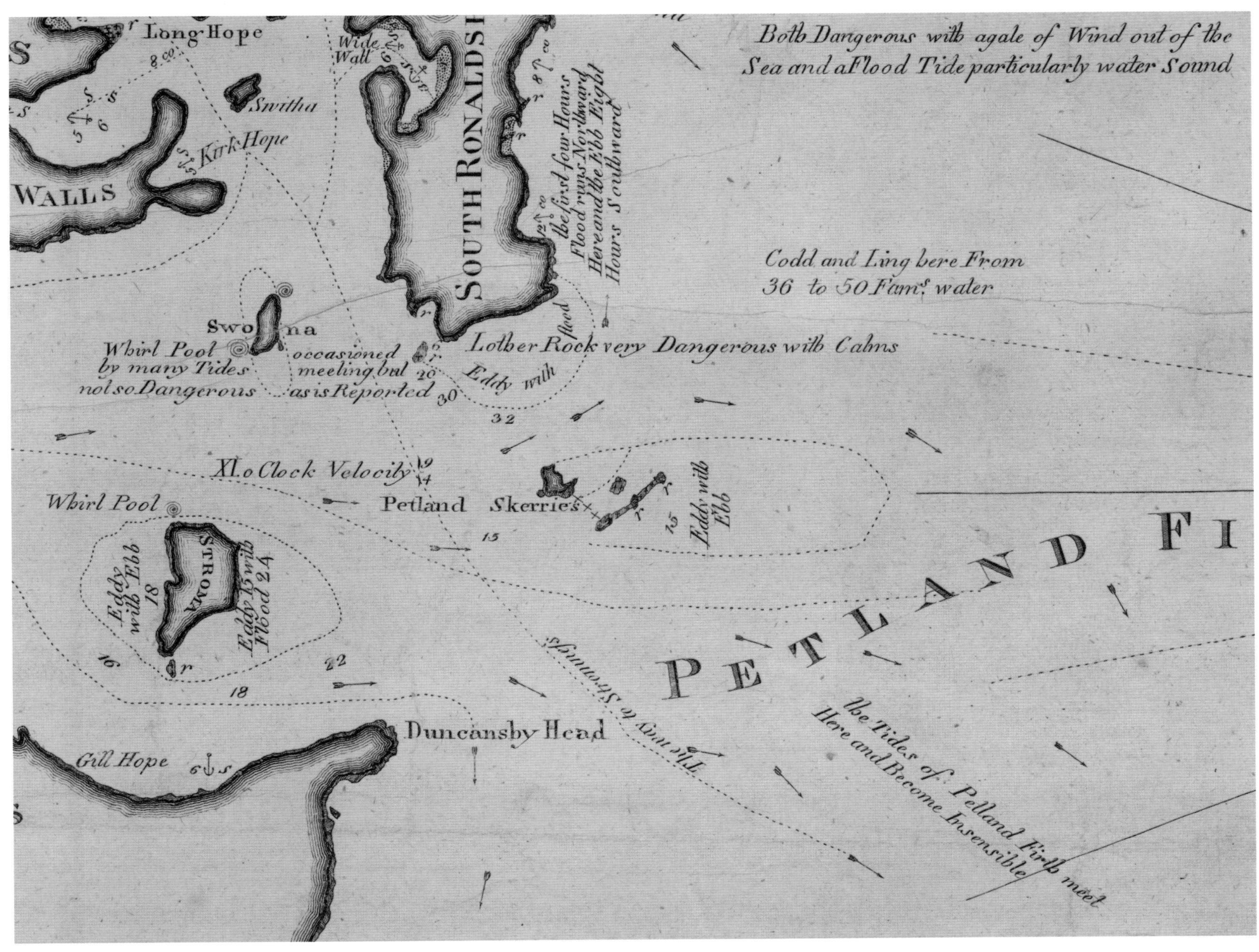

FIGURE 4.10

The written remarks on this inset of Eunson's map of 1795 were no doubt intended to allay the fears of mariners but might, perhaps, have had the opposite effect. Consider, for example, his remarks relative to the waters off the small island of Swona in the Pentland Firth: 'Whirl pool occasioned by many Tides meeting but not so Dangerous as is Reported'. Quite how dangerous would always have been a matter of interpretation based on direct encounter – and direct encounter was not something that Eunson was advocating.

Source: George Eunson, *A Chart of the Islands of Orkney, with the Adjacent Part of the Coast of Scotland* (1795).

FIGURE 4.11
(*Right and overleaf*). These two images disclose the detail, in terms of the names of the rocks themselves and in the scale of the operation to erect the lighthouses, that is belied by the use of the single name 'Bell Rock'. Detail (a) shows the names given to the many outcrops of rock. Detail (b) gives some indication of the scale of the task and the precision required to construct the lighthouse.
Source: (a) G. C. Scott, 'North Eastern Parts of the Bell Rock shewing the position of the light house, railways, wharfs, &c'. Plate VI from Robert Stevenson's *An Account of the Bell Rock Light-House, Including the Details of the Erection and Peculiar Structure of that Edifice* (Edinburgh: Printed for Archibald Constable & Co. Edinburgh; Hurst, Robinson & Co., 90 Cheapside; and Josiah Taylor, 59 High Holborn, London: 1824). (b) W. Miller (engraver), 'Progress of the Bell Rock Works', Plate IX from the above Robert Stevenson work.

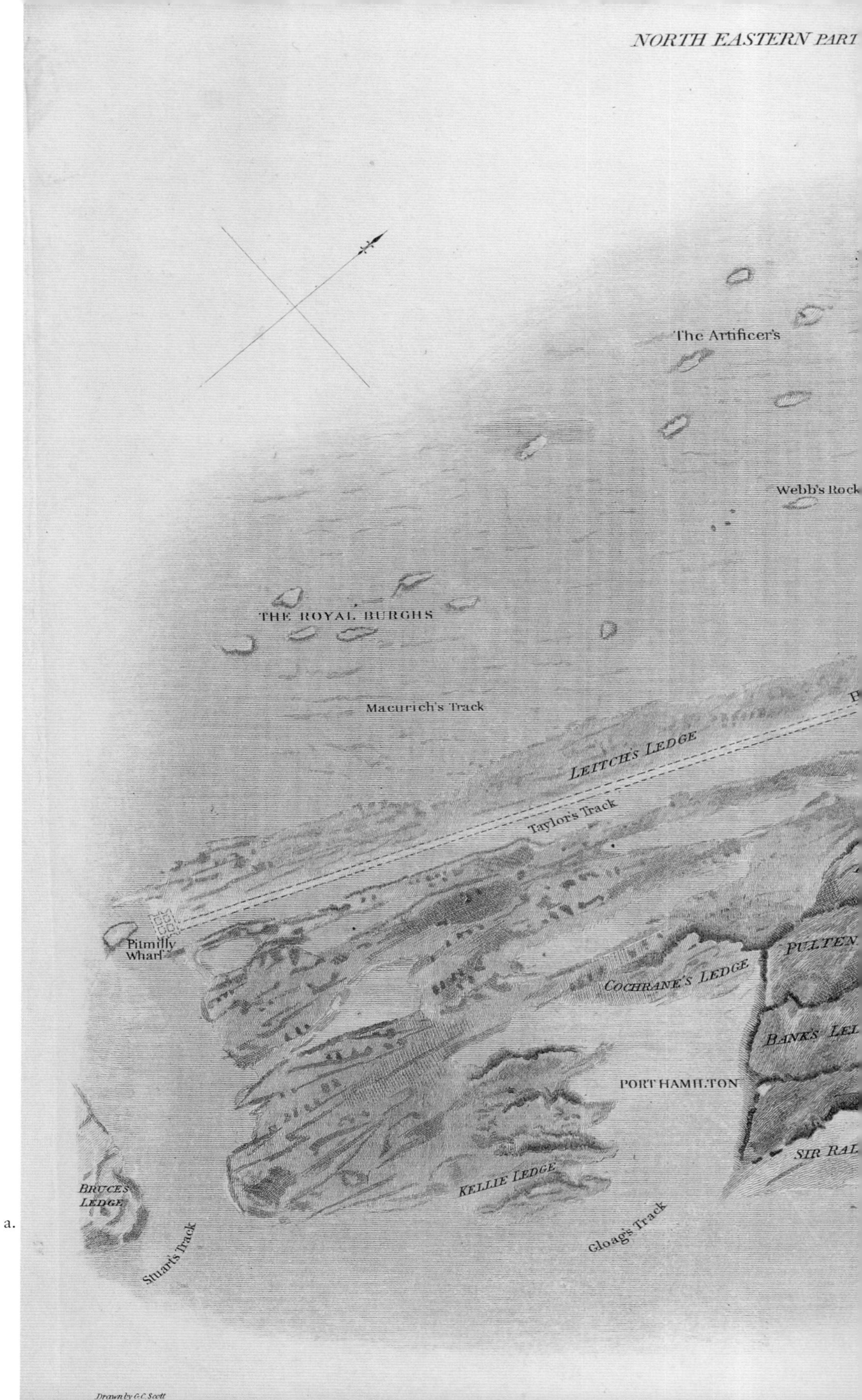

a.

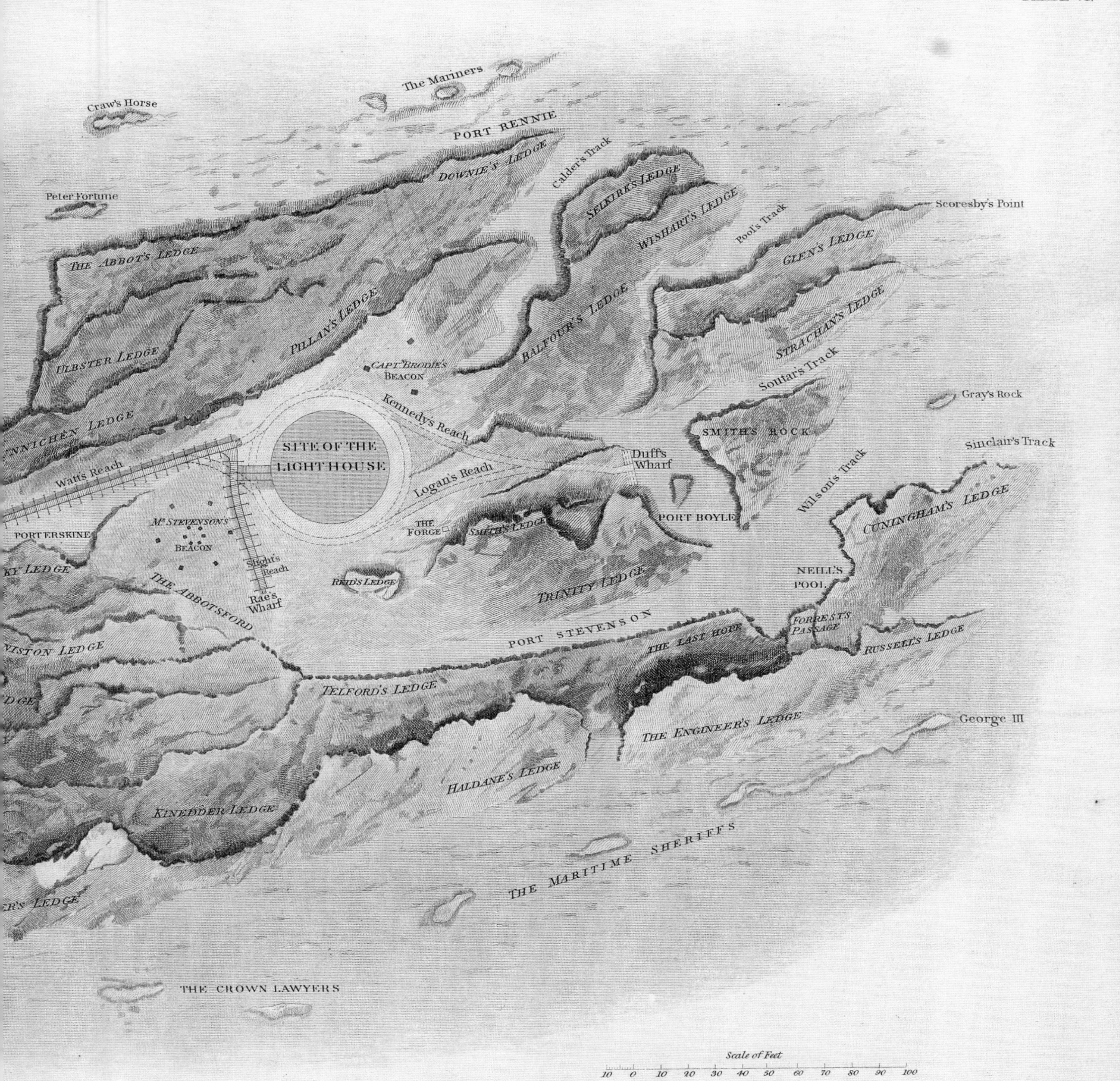
The Mariners
Craw's Horse
PORT RENNIE
DOWNIE'S LEDGE
Peter Fortune
Calder's Track
SELKIRK'S LEDGE
WISHART'S LEDGE
Pool's Track
Scoresby's Point
THE ABBOT'S LEDGE
GLEN'S LEDGE
PILLAN'S LEDGE
BALFOUR'S LEDGE
STRACHAN'S LEDGE
ULBSTER LEDGE
CAPT. BRODIE'S BEACON
Soutar's Track
Gray's Rock
Kennedy's Reach
NNICHEN LEDGE
SMITH'S ROCK
SITE OF THE LIGHT HOUSE
Duff's Wharf
Sinclair's Track
Watt's Reach
Logan's Reach
Wilson's Track
CUNINGHAM'S LEDGE
PORTERSKINE
Mr. STEVENSON'S BEACON
THE FORGE
SMITH'S LEDGE
PORT BOYLE
KY LEDGE
Slight's Reach
REID'S LEDGE
NEILL'S POOL
THE ABBOTSFORD
Rae's Wharf
TRINITY LEDGE
FORREST'S PASSAGE
PORT STEVENSON
THE LAST HOPE
RUSSELL'S LEDGE
NISTON LEDGE
TELFORD'S LEDGE
DGE
George III
THE ENGINEER'S LEDGE
HALDANE'S LEDGE
KINEDDER LEDGE
THE MARITIME SHERIFFS
R'S LEDGE
THE CROWN LAWYERS
Scale of Feet
10 0 10 20 30 40 50 60 70 80 90 100
Engd. by E. Mitchell

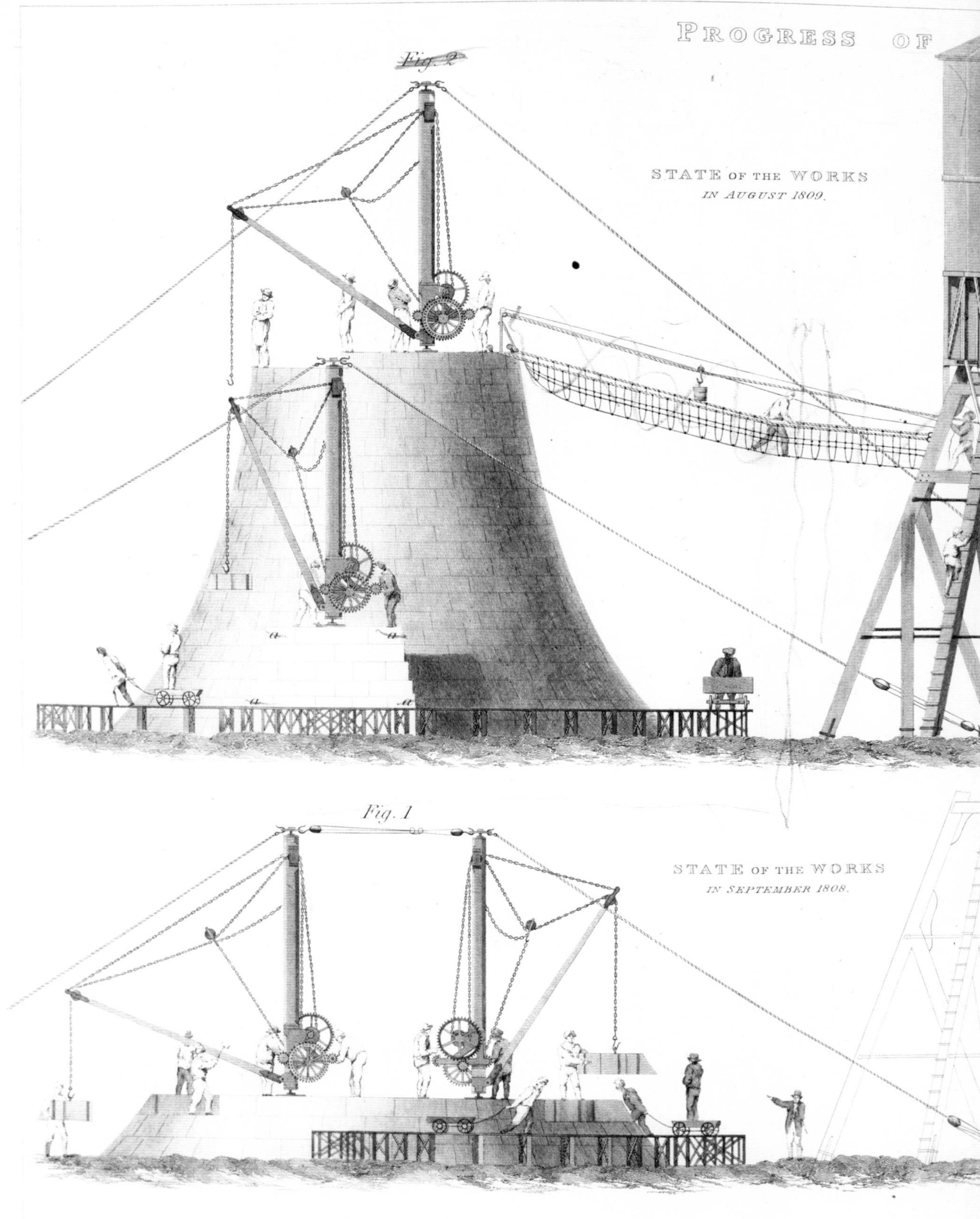
PROGRESS OF
Fig. 2
STATE OF THE WORKS
IN AUGUST 1809.
Fig. 1
STATE OF THE WORKS
IN SEPTEMBER 1808.
Scale of Feet.
10
5
0
10
20

b.

E BELL ROCK WORKS.
PLATE IX.
Fig. 3
STATE OF THE WORKS
IN JULY 1810

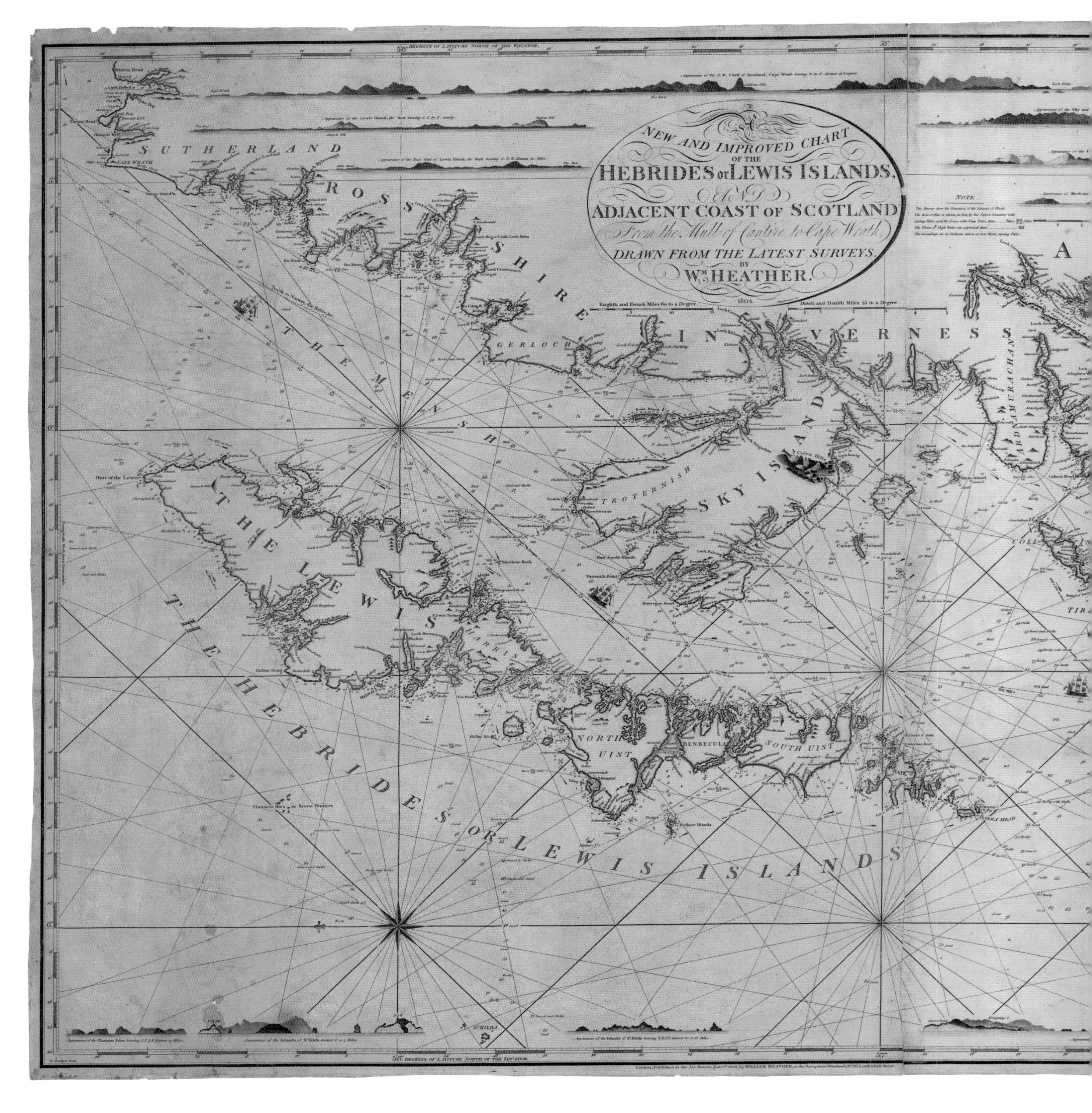

DEGREES OF LATITUDE NORTH OF THE EQUATOR.
NEW AND IMPROVED CHART
OF THE
HEBRIDES or LEWIS ISLANDS,
AND
ADJACENT COAST OF SCOTLAND
From the Mull of Cantire to Cape Wrath,
DRAWN FROM THE LATEST SURVEYS,
BY
Wm. HEATHER.
1804.
English and French Miles 60 to a Degree.
Dutch and Danish Miles 15 to a Degree.
SUTHERLAND
ROSS SHIRE
INVERNESS
CAPE WRATH
GERLOCH
THE MENSH
SKY ISLAND
TROTERNISH
THE LEWIS
HARRIS
NORTH UIST
BENBECULA
SOUTH UIST
ARDNAMURACHAN
Butt of the Lewis
Flannen Isles or Seven Hunters
St. KILDA
THE HEBRIDES OR LEWIS ISLANDS
NOTE
DEGREES OF LATITUDE NORTH OF THE EQUATOR.
London, Published as the Act directs, June 1st 1804, by WILLIAM HEATHER, at the Navigation Warehouse, No. 157, Leadenhall Street.

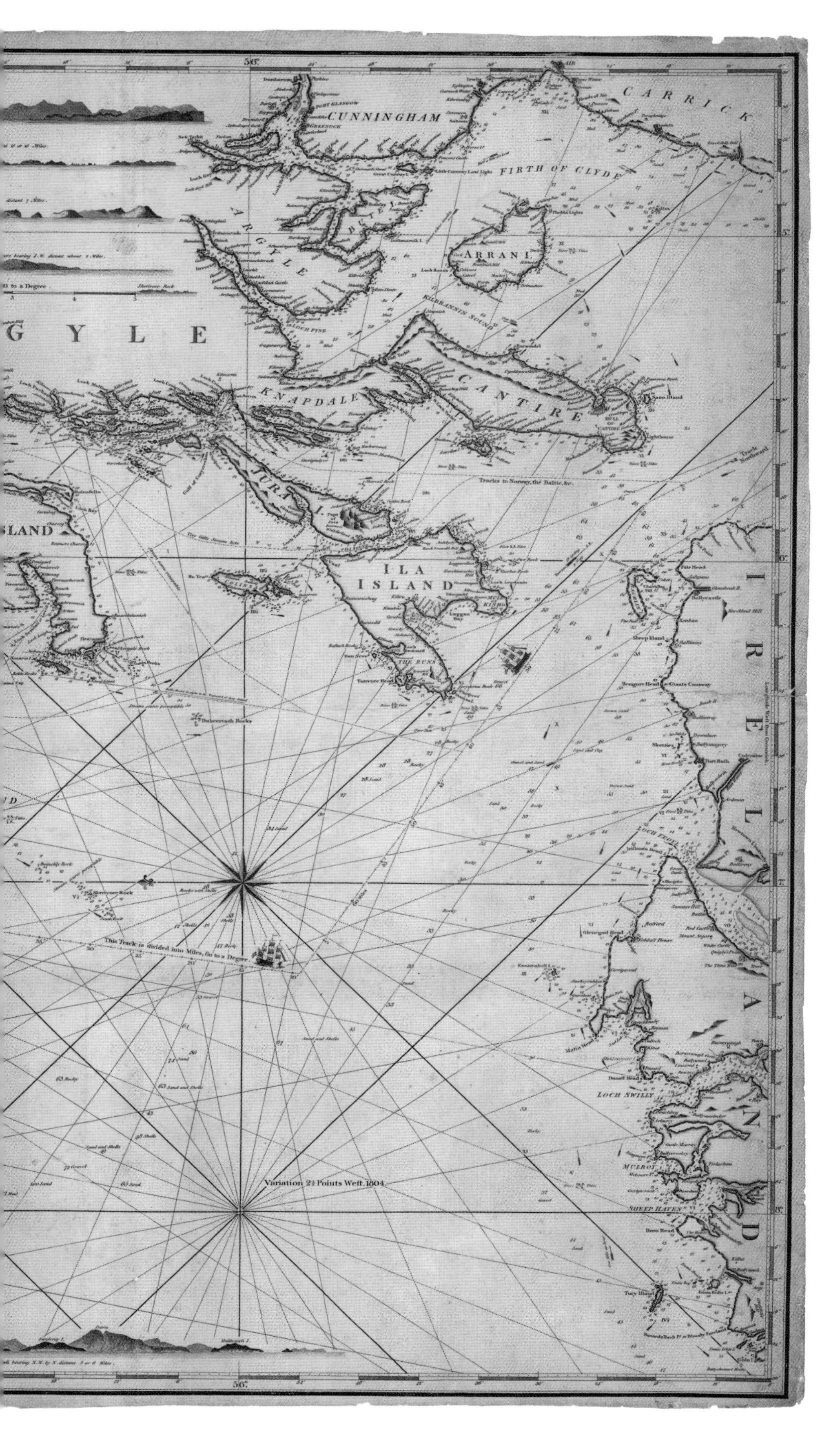

FIGURE 4.12
(*Left and overleaf*) William Heather's map is interesting given its combination of features. He has not represented the main islands shown with accuracy. He has attempted, much in the style of Murdoch Mackenzie, to incorporate the internal topography of some islands as an aid to the sailor and to the map's user: consider in this respect his depiction of part of the 'Culen [sic] Hills' on Skye. To the top and bottom of the map frame, Heather includes a number of sea-level views of Scotland's western islands together with the compass bearings of that view. In the bottom left-hand corner (shown enlarged in (b) on the next page), he includes two such stylised views for 'the islands of St Kilda', showing that island group as it appeared from 10–12 miles away and from 6–7 miles distant.
Source: William Heather, *A New and Improved Chart of the Hebrides, or Lewis Islands and Adjacent Coast of Scotland* (1804).

a.

b.

shape. One island group, Shetland, and one island in particular, Unst, the British Isles' most northerly point, would become the site of international discord over the methods and instruments to be used in doing so.

To understand why, we need to look at what was happening elsewhere in mapping at this time. From 1787 until his death in 1790, William Roy was involved with William Mudge (who continued Roy's work in what became Ordnance Survey), and with their French counterparts, in the mapping by triangulation of southern England and north-eastern France in order to fix the positions of the countries' national observatories, at Greenwich and in Paris. Between 1792 and 1798, under the direction of Pierre François André Mechain and Jean Baptiste Joseph Delambre, the French measured France and parts of Spain north and south of Paris using that revolutionary unit, the metre, as part of a new metric system for calculating national space and the earth's dimensions. Roy's cross-Channel cartography and the French metric schemes were both interrupted by war. By 1817, however, the French picked up where they had left off in mapping their nation, and the British and the French resumed their collaborative enquiries. One vital element in these joint schemes was the extension of national lines of mapping, the meridian arc, further north still, to Orkney and Shetland. For the British, this would have the benefit of fixing these islands more accurately and so be a boon to navigation. For the French, island mapping would assist in delineating the world according to the uniform standards of the metric system.

For mapping to work and for words like 'accuracy' and 'standards' to have any meaning, however, it is important that map makers collaborate and that they use the same devices to the same ends. On Shetland in the autumn of 1817, when British and French map makers resumed their interests in what was, simultaneously, local island mapping, national delineation and the shared international measurement of the earth, things went wrong almost from the off. Following William Mudge's ill health, the British party was led by Thomas Colby: as Director of Ordnance Survey, he would go on to oversee that body's mapping work in Ireland. The French were led by the distinguished astronomer and mathematician Jean-Baptiste Biot. The British used a theodolite made by the Halifax-born instrument maker Jesse Ramsden, as they had under Roy when triangulating between Greenwich and Paris. The French used a different instrument, Jean Charles de Borda's repeating circle, as they had in participating in what they regarded as the Paris–Greenwich work and in promoting the metre. Biot and Colby disliked one another intensely. Biot's work on Unst helped extend the metric meridional arc of the French. Colby's work, mainly on the smaller island of Balta, was important to the overall triangulation of Britain. Yet the different instruments were never properly tested against one another nor was the work shared between the two nations. What should have made Shetland in late 1817 a key cartographic site of international cooperation and instrumental comparison became instead a site of personalised cartographic bickering and squabbles over the methods and purpose of mapping (fig. 4.13).

The mapping as a whole of Scotland's islands and their positioning as base points in the 'Grand Survey of the British

FIGURE 4.13
(*Opposite*). Balta Sound and Balta Island as shown from this Admiralty chart. Note how the topography of the island is represented in graphic form rather than through use of contour lines, but only partly: the hills are indicated but almost truncated.
Source: Hydrographic Office, *Plan of Balta Sound in the Island of Unst, Shetland*. Admiralty Chart 116 (surveyed 1827, published 1829).

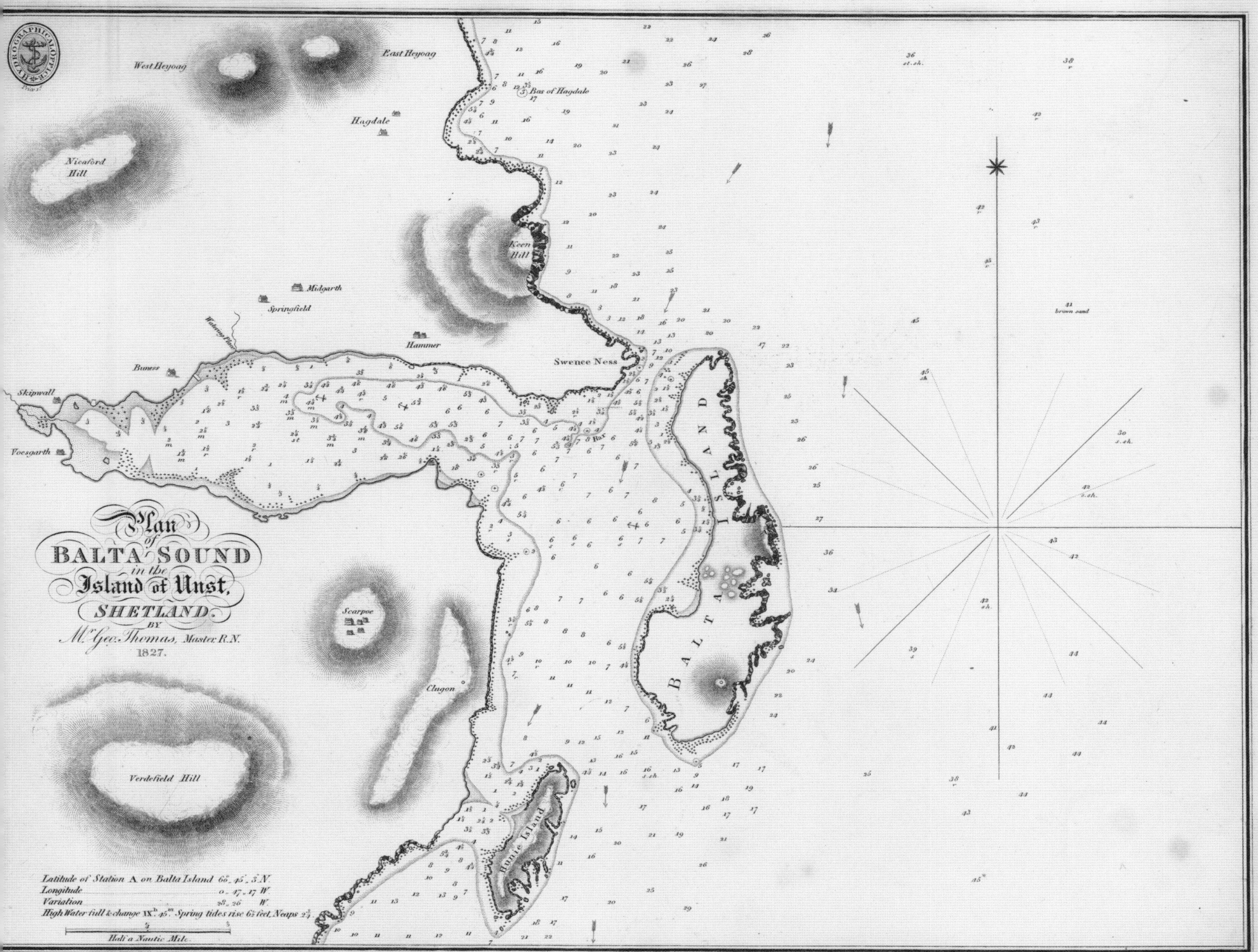
Plan of BALTA SOUND in the Island of Unst, SHETLAND, BY Mr. Geo. Thomas, Master R.N. 1827.
HYDROGRAPHICAL OFFICE
West Heyoag
East Heyoag
Hagdale
Baa of Hagdale
Nicaford Hill
Keen Hill
Midgarth
Springfield
Watering Place
Hammer
Buness
Skipwall
Voesgarth
Swence Ness
Bar
BALTA ISLAND
brown sand
Scarpoe
Clugon
Verdefield Hill
Hunie Island
Latitude of Station A on Balta Island 60°. 45'. 3" N.
Longitude 0. 47. 17 W.
Variation 28. 26 W.
High Water full & change IXh. 45m. Spring tides rise 6½ feet, Neaps 2¾.
Half a Nautic Mile.
London Published according to Act of Parliament, at the Hydrographical Office of the Admiralty, 19th March 1829.
J.&C. Walker Sculpt.

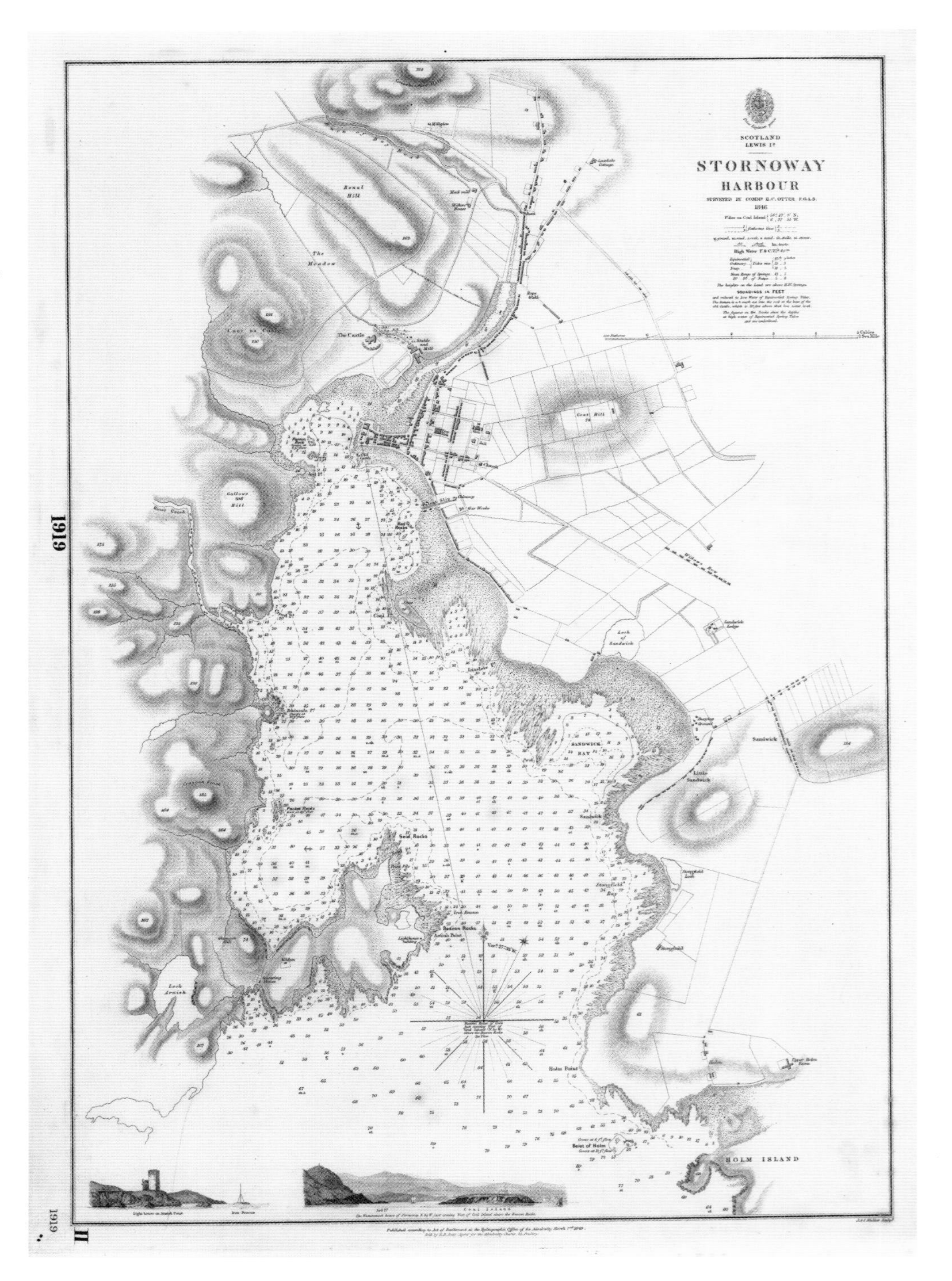

FIGURE 4.14
The rather aptly named Admiralty Hydrographer William Henry C. Otter cautioned about the navigational dangers of entering Stornoway/Steòrnabhagh harbour and the need to use the land as a guide to positioning at sea: 'The land in the vicinity of the east side of the harbour of Stornoway/Steòrnabhagh is comparatively low, and many accidents occurred before the erection of the lighthouse [in 1852] in consequence of vessels mistaking one or other of the small bays for the entrance to the harbour. In hazy weather, the Barvas range may probably be first made out, when it [the ship in question] should be brought to bear N.N.W., and kept on that bearing until the entrance is seen. In coming from the north, take care to bring the lighthouse in sight by day, and the light by night, before rounding Chicken Head, so as to clear the Hen and Chickens'. This image is interesting to compare with that produced by Murdoch Mackenzie in 1750: see fig. 4.5.
Source: Hydrographic Office, *Stornoway Harbour.* Admiralty Chart 1919 (surveyed 1846, published 1849).

Isles' during the first half of the nineteenth century was uneven in time and in geography. This was partly a consequence of getting two government bodies, the Hydrographic Office and Ordnance Survey, to work together. It was partly a reflection of the size of the task in mapping all of Scotland's islands to shared standards and scales. In part, too, it reflected the personnel available and their commitment to the task. Following Colby, George Thomas from the Hydrographic Office was at work in the Northern Isles for over a decade from 1825. His painstaking attention to detail brought islands slowly into view, but, rather like Dalrymple before him, was a source of frustration to his superiors.

Gradually, however, and especially after Thomas Colby's return to Scottish Ordnance mapping in 1838 after his Ireland work, islands appeared ever more accurately on maps, notably on the western seaboard. Respective mountain base points were used to fix islands' positions and, from those, the dimensions of the nation: Ben Hynish on Tiree and Ben More on Mull in 1822, Monach on Lewis/Leòdhas in 1839, Cleisham on Harris/Na Hearadh in 1839, Ben More on South Uist/Uibhist a Deas in 1840, from Jura in 1847, An Storr on Skye in 1848, Goat Fell on Arran in 1852. As Ordnance Survey triangulated on and between islands, so the Hydrographic Office plotted their dimensions and the features of the sea that were important to know – and to avoid. In accompanying texts, the Hydrographic Office published descriptions of each island and its distinctive features for the use of sailors – of the hazards of Stornoway/Steòrnabhagh harbour for example (fig. 4.14). Admiralty charts were continuously updated, and more frequently than Ordnance Survey maps: it was, and is, in everybody's navigational interests that this be done for island maps especially.

Flight paths and sea lanes

Scotland's island lighthouses were important points of observation from the late seventeenth century – not just warning of navigational hazards to be avoided, but as viewing platforms in their own right. In the late nineteenth century, however, and as part of plans developed by scientists meeting in the British Association for the Advancement of Science, Scotland's lighthouse keepers were employed not just to keep a navigational eye open but as 'citizen scientists' to collect meteorological information and, in pioneering wildlife studies, to act as ornithological observers.

Between 1880 and 1887, William Eagle Clarke, a member of the Natural History Department in what was then the Museum of Science and Art in Edinburgh, conducted a series of surveys of bird migratory paths. Over 100 of Britain's lighthouses were used as observational stations. It was, Eagle Clarke recounted, 'impossible to over-estimate the value of observations made at islands and rock stations, and other places removed from the usual haunts resorted to by the various species'. In 1882, the Isle of May topped the list of places for the numbers of species enumerated. In 1904, Clarke spent time on Eilean Mor, the largest of the Flannan Isles: 115 different species were recorded there and the lighthouse keepers maintained further records until 1910. In 1911, Clarke was on St Kilda, there observing 96 different migrant species – all part of the avian flight lines and observational sites which made up the migratory geographies of north-west Europe (fig. 4.15).

That other transient species, the island-hopping tourist, appeared in only limited numbers before about 1800, but was more and more common thereafter. Certain islands attracted more than others: Staffa given its several historical associations; St Kilda luring new visitors even as the indigenous inhabitants departed forever. Commercial steamer routes ferried more than tourists, of course, and the growth in number of these routes and the geography of the links speaks to the connectedness, rather than the separation, of Scotland's islands (fig. 4.16). Air travel has meant a different sort of navigating and so has produced maps of a different type (fig. 4.17). And, as tourist volumes increased and interests diversified, so maps of a yet different type have appeared. Some incorporate as visitor attractions the sites not just of bird species but of shipwrecks (fig. 4.18) – the perils of not navigating islands.

FIGURE 4.15
(*Above*) This map, 'Routes traversed by migratory birds', is taken from the Printing Records of the archives of Bartholomew, the world-leading cartographic firm based in Edinburgh. The handwritten notation to the foot – '1020 shts [sheets], 15/3/12' – indicates the size and the date of the print run. Writing of the ninety-six species which came to his notice on St Kilda in the autumn of 1911, Eagle Clarke noted that 'No less than forty-eight of these species were new to the avifauna of the St Kilda group; and not a few had not previously been observed in any of the isles of the Outer Hebrides.'
Source: William Eagle Clarke, 'Routes traversed by migratory birds', from *Studies in Bird Migration*, two volumes (London and Edinburgh: Gurney and Jackson and Oliver and Boyd, 1912), Volume 1, facing page 72. Published with kind permission of HarperCollins Publishers, *www.collinsbartholemew.com*

FIGURE 4.16
(*Opposite*) The history of David MacBrayne's steamers – whose name lives on today in Caledonian MacBrayne Ltd or 'Calmac' – can be traced to the early 1850s with the formation of David Hutcheson & Co., with David MacBrayne as a partner. Following the retirement of the Hutcheson founders in 1878, the company was renamed after David MacBrayne. By this time, as illustrated by this map, it had already become the main carrier to most of the islands off the west coast of Scotland, helped by the growing railway and coach connections on the mainland (indicated respectively by black continuous and dashed lines). This map was included in a promotional booklet produced by MacBrayne's for tourists, with descriptions of the main steamer routes.
Source: Maclure and MacDonald, 'Routes of David MacBrayne's Steamers 1886', from MacBrayne's *Summer Tours in Scotland* (Glasgow, 1886).

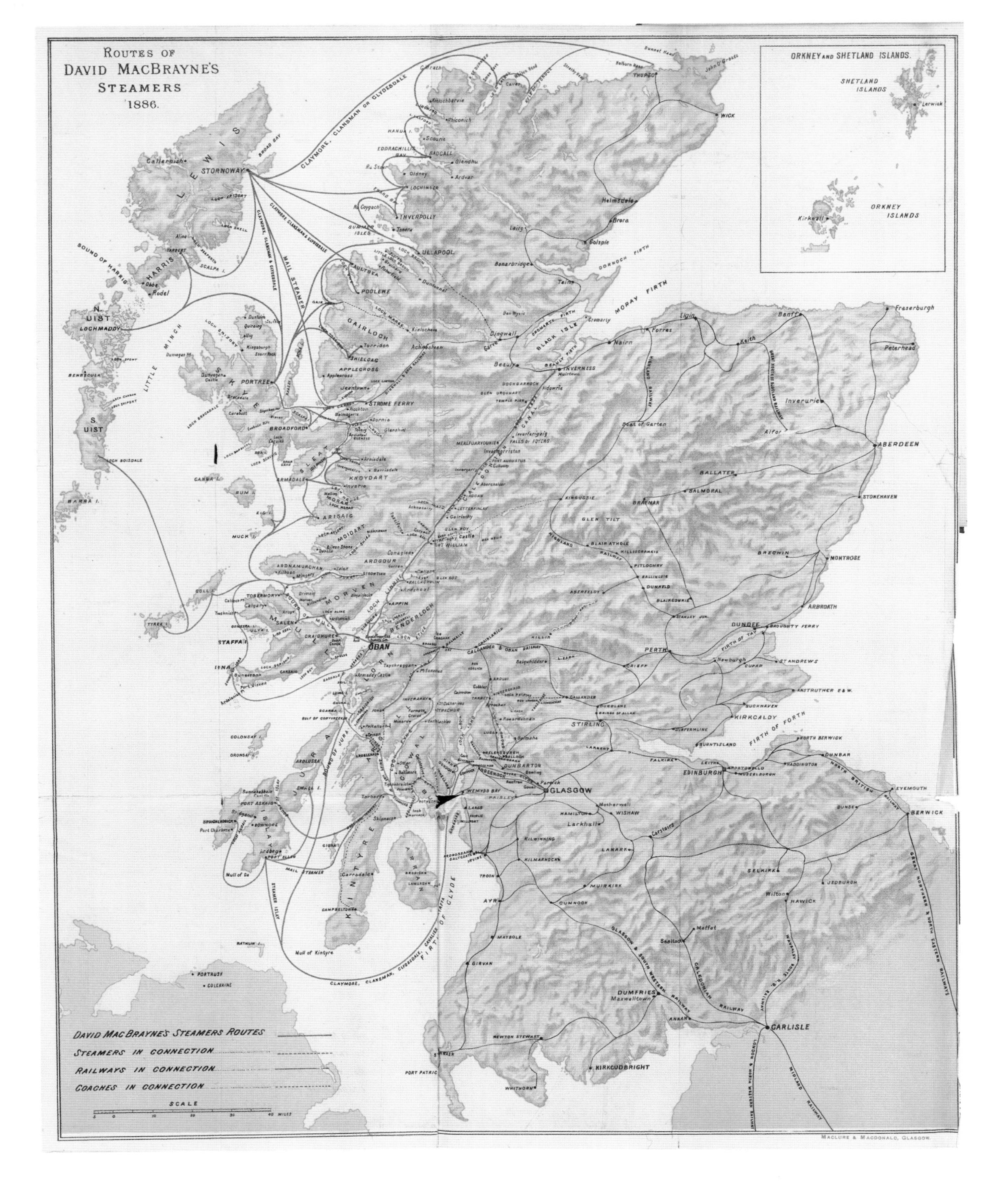

ROUTES OF
DAVID MACBRAYNE'S
STEAMERS
1886.
ORKNEY AND SHETLAND ISLANDS.
SHETLAND ISLANDS
ORKNEY ISLANDS
DAVID MACBRAYNE'S STEAMERS ROUTES
STEAMERS IN CONNECTION
RAILWAYS IN CONNECTION
COACHES IN CONNECTION
SCALE
MACLURE & MACDONALD, GLASGOW.

HP
JP
KP
HN
JN
KN
LN
OUTER HEBRIDES OR WESTERN ISLES
THE MINCH
THE LITTLE MINCH
SEA OF THE HEBRIDES
FLANNAN ISLES
Gp Fl (2) 30 sec lighthouse
Butt of Lewis lighthouse
Fl 20 sec
LEWIS
GREAT BERNERA
Callanish
Stornoway
STORNOWAY
STZ Chan 98
Barvas
Balallan
SCARP
GASKER
TARANSAY
Toe Head
NORTHTON
PABBAY
BERNERAY
Leverburgh
HARRIS
Tarbert
Renish Point
SHIANT ISLANDS
Gp Fl (3) 20 sec lighthouse
SOLLAS
NORTH UIST
Lochmaddy
HEISKER OR MONACH ISLANDS
SOUND OF MONACH
BENBECULA
BEN Chan 91
RONAY
WIAY
West Gerinish
SOUTH UIST
Lochboisdale
ERISKAY
BARRA
SOUND OF BARRA
VATERSAY
SANDRAY
MINGULAY
BERNERAY
Fl 15 sec
EG D701
0-30 000 AMSL
EG D701A
0-UNLTD
EG D701B
0-UNLTD
EG D701C
0-10 000 AMSL
Fl WR 20 sec
Vaternish Point
Dunvegan Head
LOCH SNIZORT
Gp Fl (2) 30 sec
Idrigill Point
ISLAND OF SKYE
Portree
SOUND OF RAASAY
RAASAY
INNER SOUND
RONA
SCALPAY
Kyleakin
ISLE OF SKYE
SOAY
Elgol
SOUND OF SLEAT
CANNA
SOUND OF CANNA
RHUM
SOUND OF RHUM
EIGG
ISLE OF EIGG
MUCK
COLL
TIREE
Gp Fl (3) 30 sec
Gp Fl (2) 30 sec lighthouse
Fl WRG 2 sec
Gp Fl (2) WRG 10 sec
Gp Fl (6) 30 sec
11½°W (1978)
HANDA ISLAND
EDDRACHILLIS BAY
Fl 15 sec lighthouse
ENARD BAY
Lochinver
SUMMER ISLES
Ullapool
LOCH EWE
LONGA ISLAND
Gairloch
LOCH GAIRLOCH
EG D710
0-1500 AMSL
LOCH TORRIDON
Torridon
Kinlochewe
Shieldaig
Strathcarron
LOCH CARRON
Achmore
Dornie
TACTICAL TRAINING AREA
Glassburn
Fort Augustus
Invergarry
Glenfinnan
Roybridge
Fort William
Salen
Acharacle
PMR 705
58° 30′
58°
57° 30′
57°
56° 30′
8°
7° 30′
7°
6° 30′
6°
5° 30′
5°
4° 30′

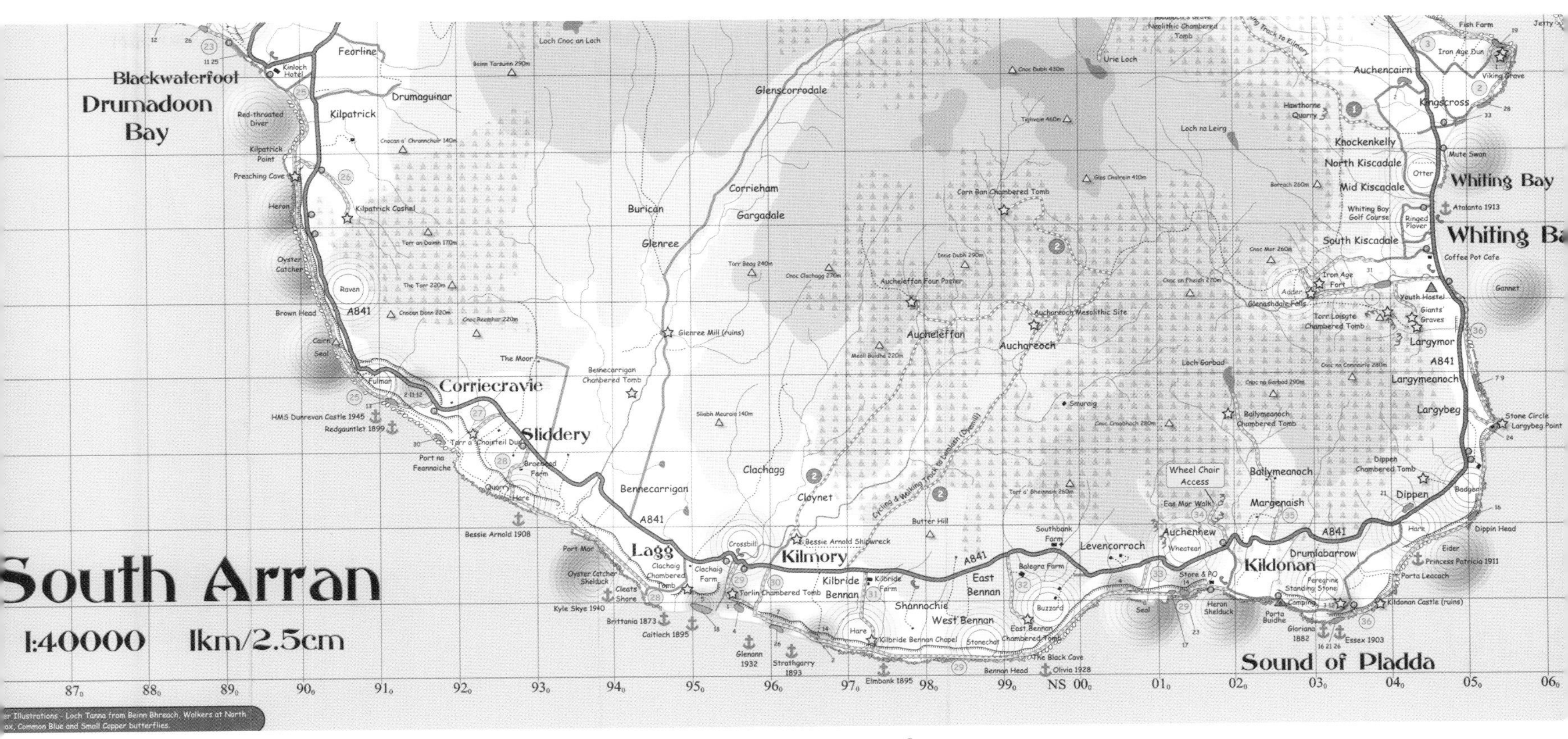

FIGURE 4.17

(*Opposite*) This detail shows the geography of different flight paths in and around the Western Isles/Na h-Eileanan an Iar in the early 1970s. The red lines are military flight paths, the blue lines are civil flight paths. The restrictions to do with military low flying over St Kilda are symbolised here by a duck in profile and the cautionary remark 'Bird Sanctuary 1 April to 30 September'. Because of its mapping of military flight paths, the map was until recently marked as 'NATO Restricted (U.S. Official use only)'.
Source: Ministry of Defence, *Low Flying Chart (Second Series). Scale 1:500,000. Sheet 3 United Kingdom, North* (1978). © Crown Copyright 1978. © UK MOD Crown Copyright, 2016.

FIGURE 4.18

(*Above*) This extract, of the southern section of the island, is taken from *Touring Arran* (2003). The map offers an innovative use of symbols (compare these with those of Murdoch Mackenzie, for example, in 1750: see fig. 4.9) to encourage tourism of different sorts: flower species, birdlife, historical monuments and shipwrecks. The site of the *Bessie Arnold*, to the south-west of the island, marks the location of one of Arran's most disastrous shipwrecks when the schooner of that name ran aground in heavy seas in 1908, four of the five-man crew losing their lives.
Source: Hugh McKerrell, *Isle of Arran: Map, Touring Arran, Walks, Wildlife, Historic Sites.* (Lochranza, 2003). Reproduced with kind thanks to the late Hugh McKerrell.

George Eunson, *A chart of the islands of Orkney... with the coast of Scotland* (1795).

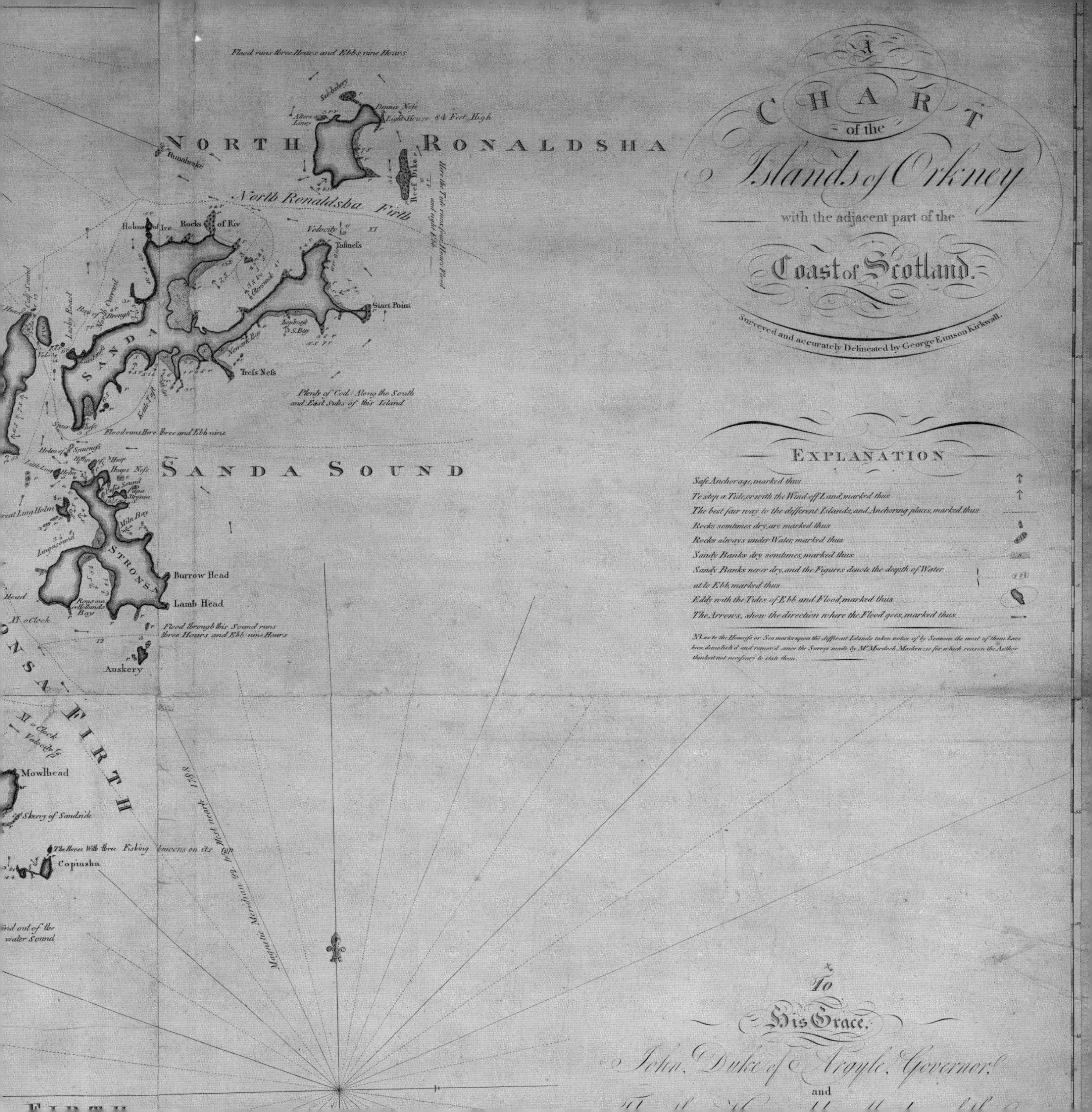

A CHART of the Islands of Orkney
with the adjacent part of the
Coast of Scotland.
Surveyed and accurately Delineated by George Eunson Kirkwall.
Flood runs three Hours and Ebbs nine Hours
Dennis Ness
Light House 84 Feet High
NORTH RONALDSHA
Runabrake
Reef Dike
Here the Tide runs four Hours Flood and eight Ebb
North Ronaldsha Firth
Holms of Ire
Rocks of Riv
Velocity
Tafiness
Start Point
SANDA
Newark Bay
Tress Ness
Plenty of Cod Along the South and East Sides of this Island
Kettle Toft
Flood runs Here three and Ebb nine
SANDA SOUND
Holm of Spurness
Huip
Huips Ness
Papa Sound
Papa Stronsa
Miln Bay
Lingasound
STRONSA
Burrow Head
Lamb Head
XI o Clock
Flood through this Sound runs three Hours and Ebb nine Hours
Auskery
FIRTH
Velocity
Mowlhead
Skerry of Sandside
Copinsha
EXPLANATION
Safe Anchorage, marked thus
To stop a Tide, or with the Wind off Land, marked thus
The best fair way to the different Islands, and Anchoring places, marked thus
Rocks somtimes dry, are marked thus
Rocks always under Water, marked thus
Sandy Banks dry somtimes, marked thus
Sandy Banks never dry, and the Figures denote the depth of Water at lo Ebb, marked thus
Eddy with the Tides of Ebb and Flood, marked thus
The Arrows, show the direction where the Flood goes, marked thus
NB as to the Houses or Sea marks upon the different Islands taken notice of by Seamen the most of them have been demolish'd and remov'd since the Survey made by Mr. Murdoch Mackenzie for which reason the Author thinks it not necessary to state them.
To
His Grace,
John, Duke of Argyle, Governor
and

CHAPTER FIVE

DEFENDING

Scotland's islands are naturally defensible – surrounded by the sea or loch, and often cliffs with strong tides – and some have had a defensive role for thousands of years. The much shorter history of mapping the islands, primarily over the last five hundred years, provides only glimpses of earlier geographies and histories. True, these glimpses become more meaningful when maps are combined with other sources such as historical records, archaeological excavation, and aerial photography, but even then we are often left with substantial gaps in understanding. Put simply, the most active periods of what we might term the 'militarisation' of Scotland's islands are not depicted on maps. On the one hand, this is because the main periods of conflict in the islands – of initial colonisation, national political struggles over territory and instances of internecine feuding – all pre-date the main period of detailed map production. On the other hand, as a result of security measures over official maps in the last century, and the large volume of classified military mapping in more recent times, many modern military sites are hardly shown at all. Fortunately, however, numerous maps survive to illustrate military and defensive themes and the importance of these themes in the life of islands.

At a broad level, it is useful to distinguish 'military maps' – that is, maps created with a specific military purpose in mind – from more general maps that depict defensible military structures (such as castles and forts) as part of broader concerns. Military maps form a distinctive genre given their specific draughtsmen (such as military engineers), readership (primarily army or navy personnel) and purpose (such as reconnaissance, attack, defence or battle). We can see their work in the Scottish islands primarily through maps from the Board of Ordnance, the War Office, and the Ministry of Defence. Military structures – usually significant features in the landscape – were often depicted by civilian surveyors. In the past two centuries, the influence of military map making on bodies such as Ordnance Survey, the Hydrographic Office of the Admiralty, together with the extension of military draughtsmanship into fields such as landscape art, topographic views and civil engineering, has blurred the distinction in terms of maps' aesthetic form and content.

Opposite. Detail from John Slezer, *The Prospect of ye Bass from ye South Shore* (1693).

Early defensive structures on islands

The earliest detailed survey of Scottish islands by Timothy Pont in the late sixteenth century provides the first graphic evidence for many defensive structures, some built centuries before Pont. Crannogs are a type of defensive freshwater loch dwelling found throughout Scotland and Ireland dating from around 2,500 years ago. Pont's manuscript maps provide evidence for the names, locations and (sometimes) the usage of over fifty crannogs in Scotland (fig. 5.1). When we add to this evidence the printed maps of Blaeu based on Pont's survey, and Pont's written notes, the number rises to 114.

Crannogs, some used as recently as the seventeenth century, served defensive and many other roles: as status symbols, homesteads, hunting and fishing stations, for example. Similarly, Iron Age brochs – drystone hollow-walled defensive structures, unique to Scotland, thought to date mainly from the first centuries BC and AD and which are particularly concentrated in the Northern and Western Isles/Na h-Eileanan an Iar – may be identified on maps. Pont recorded the more significant and well-known sites, for example picking out the broch on Sumburgh Head as 'The ancient fort of Swenbrugh' on the Pont/Blaeu map of Shetland (1654). There are thought to be nearly 600 brochs today. Antiquarians and others took a growing interest in these structures in the eighteenth and nineteenth centuries, but the features themselves often only appeared on maps in later periods, from the later nineteenth and even early twentieth centuries (see also chapter 2). Captain Fred Thomas, the Admiralty hydrographer who joined his father George Thomas in 1827 at the age of 10 or 11 to survey the Northern Isles for the British Admiralty, was a respected antiquarian in his later years. In papers for the Society of Antiquaries of Scotland, Fred Thomas commented on the 'Pictish Broughs', distinguishing them from burial mounds with which they had often been incorrectly grouped in the past. Like crannogs, there are continuing debates over the roles of brochs, but it seems likely that with disputes over land and resources often erupting into conflict, they served as defensive sites.

Migrations and invasions, *c.*500–1500

From around the fourth to the fourteenth centuries AD, the Western Isles/Na h-Eileanan an Iar were colonised or invaded from all directions: initially by Gaels from Ireland, followed in the eighth and ninth centuries by Vikings from Norway, and, latterly, by mainland Scots. The Viking settlement extended Viking control through the 'Norðreyjar', or Northern Isles, to the 'Suðreyjar', or Inner Hebrides, and beyond, as far south as the Isle of Man. Many castles in the Western Isles/Na h-Eileanan an Iar – Kisimul/Chiosmuil on Barra/Barraigh (fig. 5.2), Cairnburgh Castle on the Treshnish Isles (fig. 5.5) and Dunvegan on Skye (fig. 6.14) – date from this time. Only following the inconclusive Battle of Largs (1263) and the Treaty of Perth (1266) was Scottish sovereignty officially recognised over the Western Isles. The Norwegians held on to Orkney and Shetland until 1468, when Margaret, daughter of King Christian of Denmark, was betrothed to James III, King of Scots.

The subsequent Lordship of the Isles – the aristocratic clan-based maritime administration that encompassed the entire seaboard of Highland Scotland and Antrim and which stretched as far inland as Dingwall on the Moray Firth – lasted until 1493, when the MacDonald lords forfeited their estates and titles to King James IV of Scotland. The Lordship is often regarded as a period of relative stability and peace, but the lords still practised 'piracy', raided cattle, attacked castles and forts, and traded mercenaries and slaves. They wielded power particularly through birlinns, or galleys, developed from the Viking longships. After 1493, various sporadic rebellions took place to try to re-establish the Lordship, and hopes that it might again rise persisted until the Jacobite Rebellion of 1745. In practice, the kings of Scotland gradually brought the Hebrides under control through colonisation, military force and the steadily expanding Campbell hegemony as earls of Argyll (fig. 5.3).

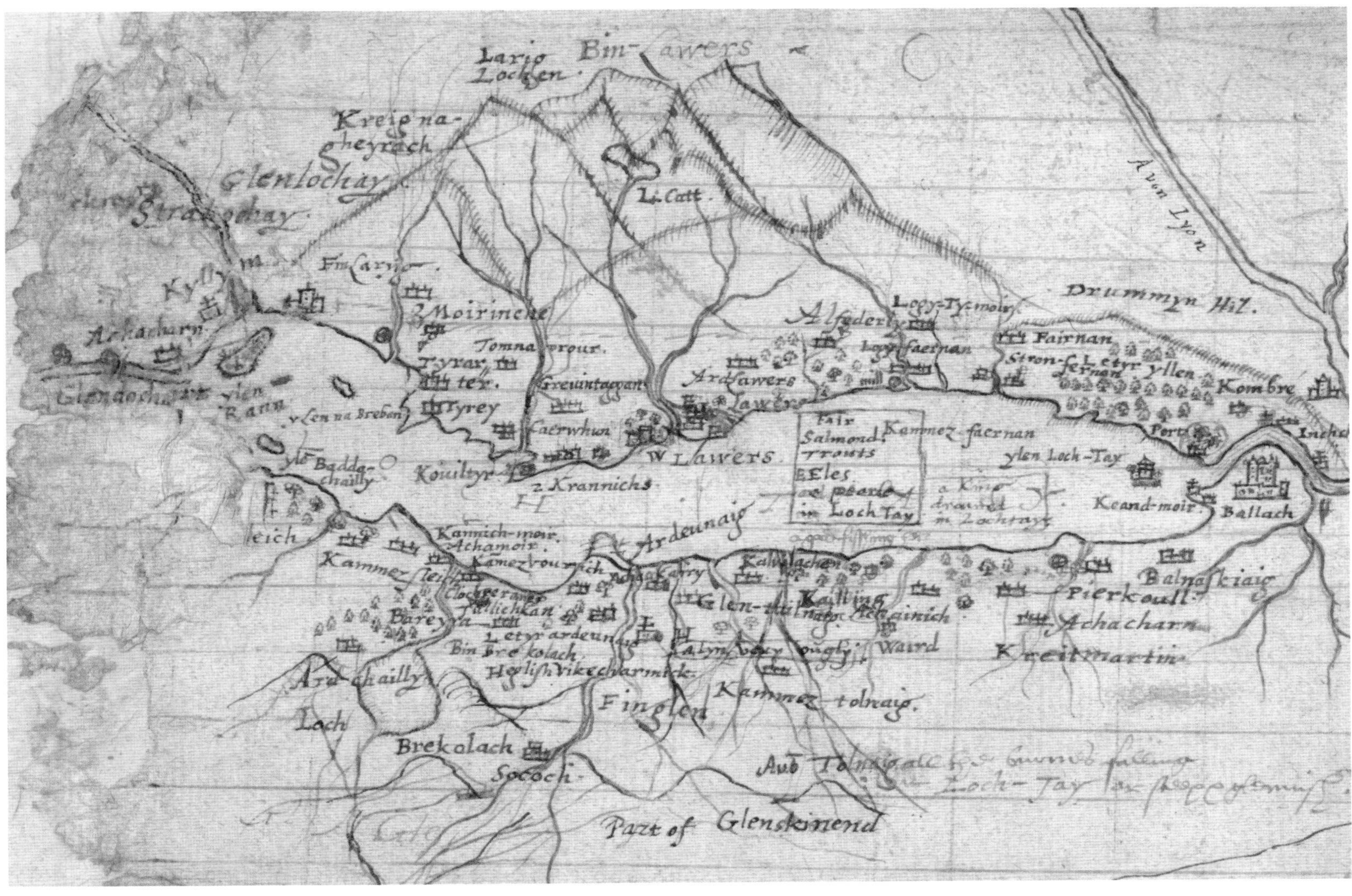

FIGURE 5.1

Although the term 'crannog' has been popularised in the last two centuries, and it was not used by Pont – the '2 Krannichs' near Lawers are thought to refer to nearby fermtouns – islands on lochs were significant as places of defence, refuge, as religious centres, for hunting, even for the pasturing of livestock. Pont clearly depicts and names them as an integral part of the landscape. An underwater archaeological survey of Loch Tay in 1979 confirmed the location of eighteen crannogs. Pont provides confirmation of several of them and their usage four hundred years earlier. Eilean nam Breaban (or yLen na Breban on the Pont map) is near the north shore at the western end: sixteenth-century charters confirm that its rights were held then by James Campbell of Lawers. The western end of Loch Tay is shallow and has extensive silting. It is possible that the large Eilean Sputachan (ylen Rann on the Pont map) is today the peninsula approximately 100 yards east of Killin. In a charter of 1568 Sir Colin Campbell of Glenorchy allowed Patrick Campbell to set nets around this island, and build a stable and a residence on it. Priory Island (or ylen Loch Tay), near Kenmore at the eastern end, is the largest island in Loch Tay, and man-made despite its size. In a twelfth-century charter, Alexander I granted the island to the monks of Scone Abbey, shortly after its foundation. The island was later fortified by the Campbells of Glenorchy: the ruins of a building, probably constructed by Duncan, 2nd Earl of Glenorchy, and damaged by a fire in 1509, can be seen on Pont's map and today.

Source: Timothy Pont, Loch Tay, Pont 18 (*c.*1583–1614).

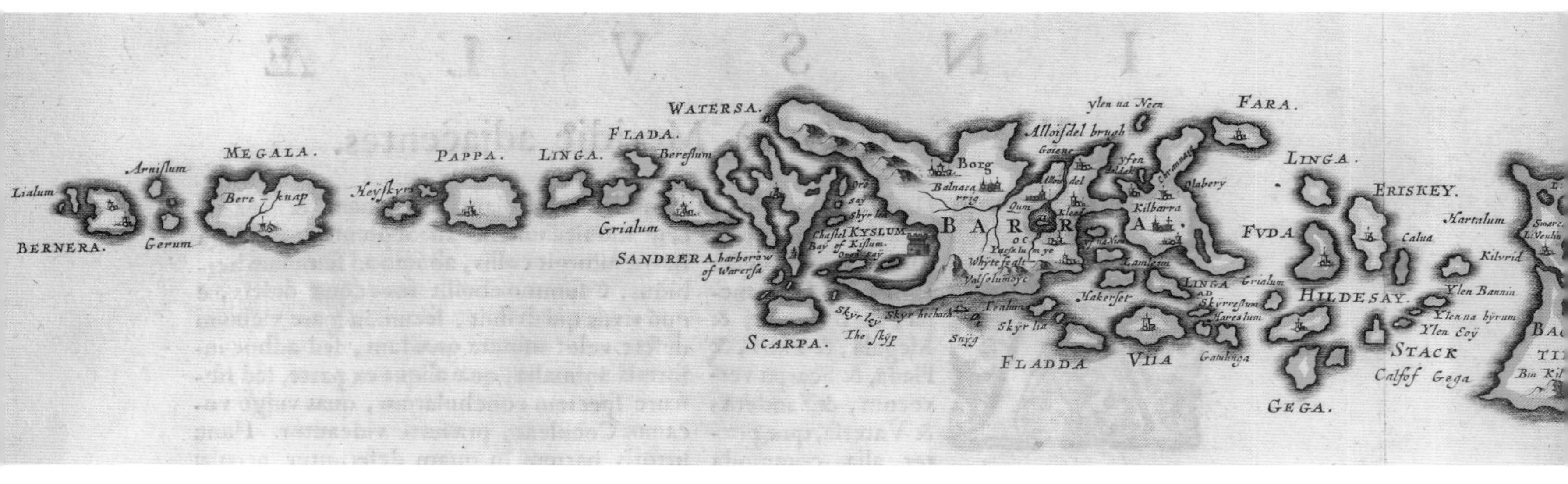

FIGURE 5.2

(*Above*) For most of the Western Isles/Na h-Eileanan an Iar, the only surviving detailed record of Timothy Pont's original survey is through the printed maps of Joan Blaeu, published in 1654. This map of North and South Uist/Uibhist a Tuath, Uibhist a Deas and Benbecula/Beinn na Faoghla is oriented with west at the top and includes detail of Barra/Barraigh and adjacent islands. As its place names testify, Barra/Barraigh had been under Norse domination for centuries, but Clan MacNeil, who trace their descent from Ireland, claim to have built a stronghold here in 1030. Blaeu's descriptive text, based on the writings of George Buchanan, notes that 'Two miles from Vatersay is Barra, seven miles long, stretching from south west to north east: not infertile in crops; especially known for haddock fishing. Into this pours a gulf of the sea with narrow jaws: inside it rounds out more widely. It has one island, and on it a fortified castle'. Chastel KYSLUM or Kisimul Castle/Caisteal Chiosmuil has a stunning defensive position on a rocky islet off Castlebay/Bàgh a' Caisteal. Blaeu's stylised symbol only hints at what is likely, originally, to have been one of Pont's sketches of this impressive fortified tower. While parts of the Castle may date from the thirteenth century, it was mostly constructed in the fifteenth century by Gilleonan MacNeil, the first lord. It was besieged many times in its turbulent history, following clan rivalry and resulting from the MacNeil's support for Charles II at the Battle of Worcester (1651), and for the Jacobites in 1689 and 1715. Following the MacNeil repurchase of Barra/Barraigh in 1937 after a century of its ownership by Colonel Gordon of Cluny, Kisimul/Chiosmuil was restored. In 2001 the Castle was leased by the chief of Clan MacNeil to Historic Scotland for 1,000 years for the annual sum of £1 and a bottle of whisky.

Source: Timothy Pont/Joan Blaeu, *Vistus Insula, vulgo Viist, cum aliis minoribus ex Aebudarum numero ei ad meridiem adjacentibus* (1654).

FIGURE 5.3

(*Opposite*) Innis Chonnell Castle on Loch Awe was the main stronghold of the Campbells of Loch Awe from the fourteenth century. The Castle stands on a small, rocky island, just off the south-east shore of Loch Awe near Ardchonnell. Probably fortified from a much earlier period, its earliest record in history dates from 1308 when it was held by the MacDougalls of Lorn. Following their defeat by Robert the Bruce, the Castle and surrounding lands passed to Clan Campbell by 1315, and it became their principal residence until the time of Colin Campbell, 1st Earl of Argyll (1453–93), who moved his main seat to Inveraray.

By the time of Pont's survey, the Campbells had entrusted Innis Chonnell to their hereditary captains, the MacArthurs, with the Castle housing a series of political and criminal prisoners. The subsequent captains from 1615, the MacLachlans, built their main residence on shore in the early 1700s, and the Castle was marked as ruinous on an estate plan of 1806. Further north, Pont records the island and church of Inishail, a place of worship from the twelfth to the eighteenth centuries. In 1257, the church of 'St. Findoc of Inchealt' and its surrounding lands were granted to the Augustinian canons of Inchaffray Abbey, who occupied it until the Reformation. Church services were still held until 1736 when a new church was built at Cladach.

Source: Timothy Pont, Mid-Argyll; from Dunoon to Inveraray and Loch Awe, Pont 14 (*c.*1583–1614).

Cromwellian and Hanoverian outposts, *c.*1639–1745

Scotland's islands escaped the worst of the Wars of the Three Kingdoms (England, Ireland and Scotland) between 1639 and 1651, and the subsequent Cromwellian invasion of Scotland, although the Bass Rock in the Forth (fig. 5.4) was used to harass English supply ships during the Cromwellian period, as well as being used as a prison for Covenanters. Some island castles, Tioram for example (fig. 5.8), were briefly besieged by Cromwellian forces, and new defences were constructed at Goat Island/Eilean na Gothail by Stornoway/Steòrnabhagh. The fortification of Fort Charlotte in Shetland dates largely from the Second Anglo-Dutch War in 1665 (fig. 5.9a and b).

Following the Jacobite Rebellion of 1689 and the construction of Fort William, Hanoverian troops were garrisoned in several former Jacobite castles, including Eilean Donan, Duart on Mull, Tioram and Invergarry. It is likely these sites and garrisons had a more symbolic than effective function, what one historian has described as 'sacrificial Hanoverian lambs surrounded by Jacobite wolves'. Following the Act of Union in 1707, the British Board of Ordnance was, albeit reluctantly, given expanded responsibilities for Scottish castles and defences. The best surviving early military plans of them are a direct result of this (figs 5.5 and 5.6).

At this time, French military engineers were the recognised experts in the field, and one such, Lewis Petit, was chosen by the Board for a special reconnaissance mission to north-west Scotland from September to November 1714. As a French Huguenot, Petit had joined the British Board of Ordnance in the later 1680s after the revocation of the Edict of Nantes, and he distinguished himself in the Wars of the Spanish Succession. In 1714, while trying to shore up the dilapidated defences at Fort William, he also surveyed several of the garrisoned castles nearby, including Eilean Donan (fig. 5.7) and Tioram (fig. 5.8 and pp. 138–9).

During the eighteenth century, it is possible to discern a developing military mapping aesthetic, as the result particularly of standardised training of military engineers in the Board of Ordnance Drawing Room in the Tower of London. The choice of features for inclusion, the use of standardised perspectives and conventions for colour and symbols were all promoted. We can see this in the draughtsmanship of the Roy Military Survey, and the depiction of the re-fortification of Fort Charlotte in the 1780s (fig. 5.9). The Military Survey of Scotland (1747–55), which deserves a special place in the mapping of Scotland, was a direct reaction to the 1745

FIGURE 5.4
Just over a mile offshore, to the north-east of North Berwick on the East Lothian coast, the Bass Rock was fortified from the early fifteenth century. The Bass was in the possession of the Lauder family from 1318. In the 1650s, it was used as a base to harass English supply ships during the Cromwellian invasion of Scotland. From 1671, it acted as a prison for Covenanters. With the abdication of James VII in 1688, the Covenanters were released, but its owners remained faithful to King James, and Jacobites held the Bass Castle until 1694, when they were granted amnesty. The Castle was partially dismantled in 1701, when the island was sold to Sir Hew Dalrymple (whose descendants still own it), but it was repaired when the lighthouse was built in 1902 on the site of the Governor's House.

John Slezer, whose 1693 view this is, was a talented artist and military draughtsman. He held the post of Chief Engineer in Scotland from 1671, a position which involved surveying the country's major fortifications and castles. His idea for an illustrated volume of views including castles, towns, abbeys and country houses, was eventually realised as the *Theatrum Scotiae* in 1693. The work included texts by the Geographer Royal, Sir Robert Sibbald, which highlight the value of the Bass as a garrison, as well as its bird life as a source of food: 'Upon the Top of this Island there is a Spring, which sufficiently furnishes the Garrison with Water; and there is a Pasturage for Twenty or Thirty Sheep. 'Tis also famous for the great Flocks of Fowls, which resort thither in the Months of May and June, the Surface of it being almost covered with their Nests, Eggs and young Birds. The most delicious amongst these different Sorts of wild Fowl, is the Soaling Goose, and the Kittie Waicke.'

Although the *Theatrum Scotiae* sold poorly, with Slezer ending his days in the debtors' sanctuary in Holyrood, his views have grown in appeal over time, capturing as they do the essence of places three centuries ago with the eye of a military engineer. Slezer provides a rare view of the Bass Castle fortifications, with battlemented parapets, the curtain wall running at right-angles to the sea close to the landing-place, and three buildings within the walls. Further up the slopes, St Baldred's Chapel can be clearly seen, rebuilt several times by the Lauder family, and commemorating the seventh-century hermit reputed to have spent time here. Sea birds were clearly in evidence at this time (a fact confirmed by visitors' written accounts), their numbers were then relatively small. Today, an estimated 150,000 gannets breed on the rock, making it the world's largest gannet colony – and perhaps both the noisiest and smelliest island in Scotland.
Source: John Slezer, *The Prospect of ye Bass from ye South Shore* (1693).

aris Australi. The Prospect of yᵉ BASS from yᵉ South shore.

A Plan of the Two Carrinburghs drawen

Perpendicular Rocks 6 ffathom

B. O

18

35 ft

Old Chappell

The Well

56 ft

21

Barrwick

House for fireing

Guard house

Port

The Island of Great Carrinburgh

5 10

7

FIGURE 5.5
This is a unique and significant detailed plan of Cairnburgh Castle on the Treshnish Isles, north-west of Mull. The location was of strategic importance, guarding the main southern approach into the Inner Hebrides. It was fortified from the thirteenth century, or possibly earlier, by Ewen, Lord of Lorn. The Castle, which changed hands numerous times, was held by the MacLeans of Duart from the fourteenth to the eighteenth centuries. Dean Munro's description in 1549 hints at the difficulties, apparent even today, of landing given the strong currents: 'the ane callit *Kerniborg moir*, the uther callit *Kerniborg beg*; baith strenthie craigis be nature biggit in the sea, and fortifeit about be the devise of man, lyand in the middis of it great stark streams of the sea, bruikit by Mcgillane of Doward, very perilous for shippis . . .' Following the Castle's surrender to Archibald Campbell, 10th Earl of Argyll, in 1692, it was garrisoned by Government forces, particularly in the build-up to the 1715 and 1745 rebellions.

Robert Johnson – whose map this is – was the overseer of Fort William from 1708 to 1715. This roughly drawn sketch map of Cairnburgh is accompanied by another of the island of Mull, picking out the major castles (fig. 5.6). The plan shown is invaluable in being the earliest surviving depiction of the location, extent and purposes of the Castle at this time: the former chapel, barracks, guard houses, wells and other buildings are noted, including the ports for both islands. What Johnson may have lacked in terms of the professional military draughtsmanship and standardised colouring observable on later Board of Ordnance maps, he made up for with an evocative and distinctive style, and informative notes such as the 'Perpendicular Rocks 6 ffathoms in heighth all around these Islands'. Cairnburgh was not taken by the Jacobites, and, perhaps because it ceased to have military significance from the later eighteenth century, the structures shown by Johnson can mostly still be seen today.
Source: Robert Johnson, *A Plan of the Two Carrinburghs Drawen on the Place* (*c*.1714?).

FIGURE 5.6

Robert Johnson's map of Mull complements his more detailed map showing the location of Cairnburgh Castle on the Treshnish Islands (fig. 5.5), and was probably drafted at a similar time, around 1714. The Treshnish Isles appear to the upper left, and this map allows them to be visualised in a broader context. Its outline clearly departs from the more distorted Blaeu form of the previous century and, in spite of its rough appearance, picks out many features of military interest as well as significant places.

Source: Robert Johnson, *A Plan of the Island of Mull with the Adjacent Islands Drawen on the Place* (*c.*1714?).

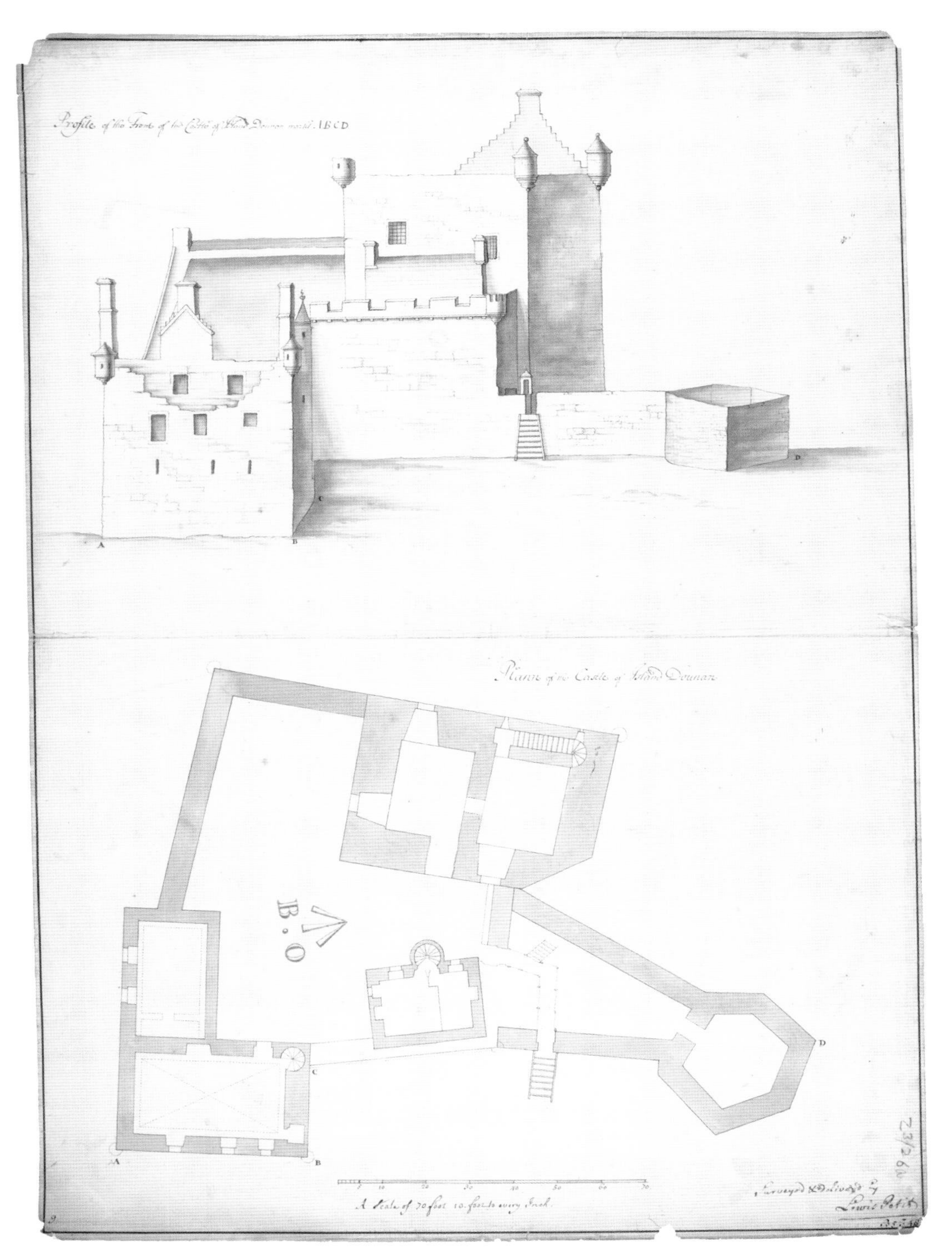

FIGURE 5.7
Eilean Donan is today one of the most photographed castles in Scotland. In its present form it is very much a modern creation, reconstructed between 1912 and 1932 by the then owner, Lt Col. MacRae-Gilstrap. For two centuries before then, it had lain in a ruinous state, following heavy bombardment by Hanoverian forces shortly after the Battle of Glenshiel in 1719. This uniquely important plan and profile was made five years earlier by the French military engineer Lewis Petit, and shows us how the original castle looked, as well as its detailed layout. In 1714 there was an east entrance, as well as wing walls connecting with a hexagonal bastion (to the right) which dates from the sixteenth century. It is an island site – the three-arched bridge connecting castle and shore is another twentieth-century feature.

Historically, Eilean Donan was the main stronghold of the Mackenzie Earls of Seaforth. It has had a turbulent history given its strategic location guarding the Kyle of Lochalsh and Sound of Sleat. In 1504 the Castle was captured by the Earls of Huntly, and in 1539 it was besieged by Donald MacDonald, a claimant to the Lordship of the Isles. Although the Castle saw little action in the 1715 Jacobite rebellion, it was recaptured by William MacKenzie, Earl of Seaforth, who held it with Spanish troops preceding the Battle of Glenshiel (1719). Following its bombardment by three English men-of-war, the Spanish troops withdrew, allowing a party to lay mines and blast Eilean Donan to ruins. *Source*: Lewis Petit, *Plann of the Castle of Island Dounan* (1714).

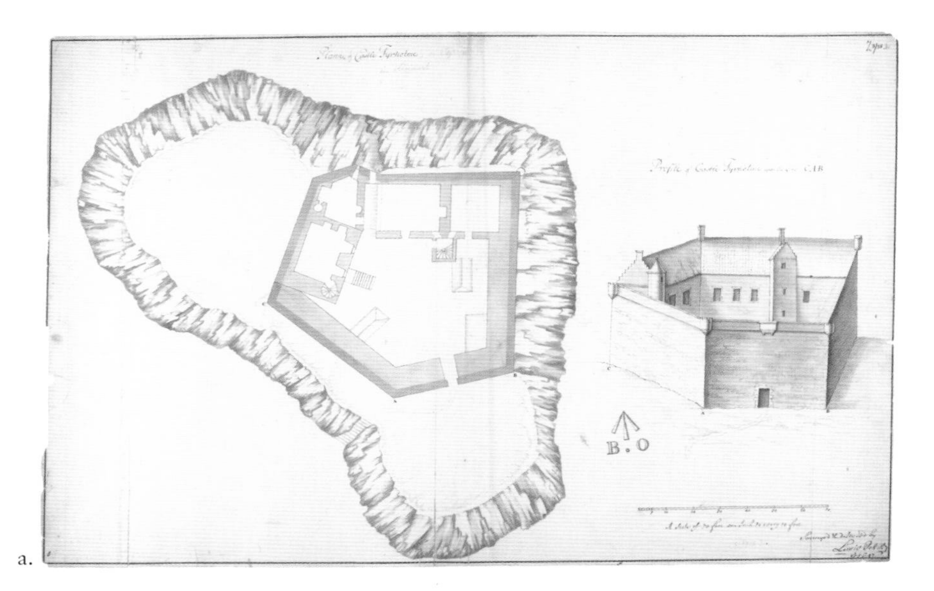

a.

b.

FIGURE 5.8

Castle Tioram stands in a spectacular location on a tidal island near the mouth of Loch Moidart. Tioram was the ancestral home of the MacDonalds of Clanranald from the fourteenth century, and the massive curtain wall depicted here – occupying the whole summit of the rock, surrounding an irregularly shaped courtyard – dates from that time. The Castle was briefly held by Cromwell's forces in 1651 following a siege, but was re-taken by the MacDonalds, who held it until the early eighteenth century, the date of this view.

The French military engineer Lewis Petit surveyed Tioram in the autumn of 1714 as part of a rapid reconnaissance survey for the Board of Ordnance (a). The Board had received repeated requests for money and men from Fort William, given mounting (and well-founded) concerns that the clans were 'all ready to rise, and they expect the Pretender to land every day'. Much of Fort William's wooden defences were rotten, its gates broken, and the platforms for men and guns were out of order. The financially hard-pressed Board of Ordnance had no money or resources available; the best it could do was send Petit to survey the situation and nearby strongholds, as part of a report for the Board. His plan is valuable for us today, as the Castle was badly damaged by fire in 1715 (deliberately torched by Allan, the 14th Clan Chief), in an attempt to prevent its being used as a Hanoverian garrison. Allan led his men to battle at Sheriffmuir in November 1715, and was killed in action. Tioram remained unoccupied thereafter, and appears as a roofless (albeit picturesque) ruin in Paul Sandby's view of the 1740s (b).

Source: (a) Lewis Petit, *Plann of Castle Tyrholme* (1714). (b) Paul Sandby, *Plan of Castle Tyrim in Muydart* (1748).

Rebellion, and involved the first significant use of triangulation and measured traverses in Scottish topographical mapping. Significantly, the Scottish islands were largely excluded from the Roy Military Survey, as they were not considered to be the likely main theatre of future military action, but some smaller islands near the mainland were included (fig. 6.14).

Fortress Britannia: the islands and global warfare, *c.*1789–

Following the defeat of the Jacobites at Culloden in 1746, the main threats to the Scottish islands were from a new direction: overseas, particularly continental Europe. Following the French Revolution, and subsequent Napoleonic Wars (1803–15), growing concerns in Britain over a possible French invasion led to a substantial construction programme of coastal defences. The martello towers off Leith, and at Hackness and Crockness in Orkney, date from this time. Several islands along the east coast of Scotland, including Inchcolm (fig. 5.10) were also fortified. Fortunately, the French never invaded, but the Napoleonic Wars did much to help islands' economies through, among other things, acting to encourage the kelp industry (see chapter 6).

During the First and Second World Wars, the Northern Isles, Orkney in particular, were heavily fortified, notably in the use of Scapa Flow as the base for the British Grand Fleet. Tens of thousands of sailors were based in Orkney during the First World War, and over sixty block-ships were sunk in the many channels into Scapa Flow to facilitate the use of anti-submarine nets and booms. Following the German defeat in 1918, the German fleet was interned in Scapa Flow, and seventy-four ships were scuttled there in 1919 to prevent them falling into British hands. Scapa Flow's role as a major naval base was repeated during the Second World War, not least given its distance from German airfields. The catastrophic sinking of HMS *Royal Oak* in October 1939 was one of the first major enemy strikes against Britain during the Second World War, as well as one of the most daring submarine attacks of all time (fig. 5.11).

In the early 1940s, concerns over German intentions to develop biological weapons led to research into the feasibility of retaliatory germ warfare by British military scientists at Porton Down. An uninhabited and remote island was required at short notice for the testing of an anthrax bomb attack, and Gruinard Island, half a mile off the coast of Wester Ross between Gairloch and Ullapool, was requisitioned by the government from its owners in 1942 (fig. 5.12a). Around eighty sheep were taken to the island and exposed to anthrax bombs; all became infected and died within days. In this sense the experiment was a success, but at the cost of long-term contamination of the area, and as it was felt the weapon was almost as much risk to any army using it as to the enemy, fortunately the anthrax bomb was never used in practice. Gruinard Island remained contaminated and quarantined for four decades after the War with notices prohibiting landing also appearing on relevant Admiralty charts (fig. 5.12b). It was only following a major decontamination project in the 1980s, involving spraying the island with formaldehyde and seawater, and removing contaminated topsoil, that it was finally declared 'safe' (at least officially) and returned to its previous owners.

The post-war era saw yet further militarisation of Scottish islands, although security restrictions often reduce the visibility of these sites on maps. The active testing of Britain's first guided nuclear missile took place from the late 1950s off Benbecula/Beinn na Faoghla, with an extensive military zone extending out to the west including St Kilda (fig. 5.13). Several other sites were selected for radar stations – Aird Uig on the north-west coast of Lewis/Leòdhas, Saxa Vord on the north coast of Unst – as part of the British early warning system for missile attack. Other sites, such as Rona between Skye and the Scottish mainland, were used in testing the acoustic signatures of surface ships and submarines. From the 1950s, these sites employed hundreds of people, although changing military technology reduced these numbers in recent years. During the Cold War, the main threat to Britain was thought to be the Soviet Union. Their declassified military maps, available in the West from the 1990s, present a stark and powerful reminder of the ongoing value of mapping for military activity (fig. 5.14).

a.

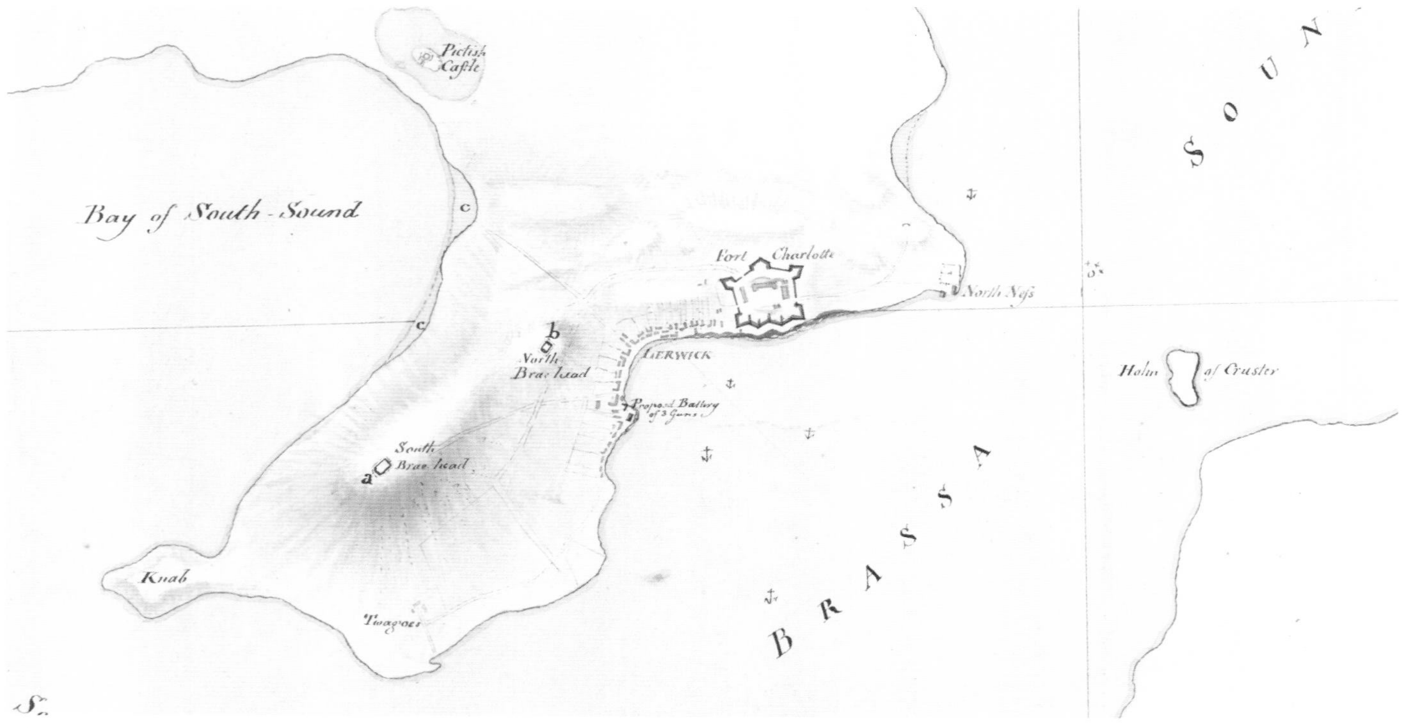

FIGURE 5.9

Fort Charlotte, which dominates the north side of the port of Lerwick in Shetland, was originally planned as a rough pentagon with corner bastions by King Charles II's master mason Robert Mylne, during the Cromwellian invasion of Scotland. With the re-opening of hostilities against the Dutch in 1665, and concerns over the Shetlanders' greater loyalties to this new enemy, it was assumed that the fort could also protect the King's navy while it lay within Bressay Sound. Building the fort proved difficult – lack of suitable stone, lime and workmen delayed the construction of the walls – and when the Second Anglo-Dutch War ended in 1667, further work on the fort was called off and the garrison was disbanded. Unfortunately, this decision proved premature and with the resumption of warfare against the Dutch during the Third Dutch War (1672–77), the Dutch landed and burnt the barrack block, along with several other buildings in Lerwick.

The fort lay abandoned and incomplete for over a century until concerns over American privateers in the 1780s resulted in a decision to rebuild it along similar lines, naming it Fort Charlotte in honour of George III's Queen. At this time, the skilled draughtsman and Board of Ordnance engineer Andrew Frazer created a set of detailed plans to show the planned reconstruction of the fort, barracks and environs (a). His plan of the fort (b), dated January 1783, is a work of careful survey. Oriented with north to the right, it allows us to visualise the detail. Compared with the Cromwellian fort which had barracks for 100 men, the new fort could accommodate up to 270 men in the large West Barracks (D) and North Barracks (F). The Barracks' internal layout is specified with the officers' mess and kitchen, numbers of rooms and even numbers of beds per room. To the south was the artillery storehouse (C), to the west a coal yard, to the north-east the gunpowder magazine (A), and a central reservoir (H) gathered rainwater from the roofs.

The *Explanation* further makes it clear, however, that all was not well: 'As neither Earth nor Sod could be got to make the Parapets in the Curtain, Stage or Plank for Musketing in laid upon the Butressess.' The main east wall overlooking Bressay Sound was vulnerable to enfilading fire (from the side), and Fraser had suggested the construction of the traverses (see (a)) to protect it. Artillery was also yet to arrive: 'The Platforms coloured yellow, are mounted with eight 18 Pounders and two heavy 9 Pounders which is all the heavy Guns yet sent to this Place - There is yet no Artillery in the Bastions towards the land'. A redoubt planned on North Brae Head to the south was not finished, and encroaching buildings on the south side would have given easy cover for any attacking force. Fortunately, this never came in spite of concerns over French attack during the Napoleonic Wars. By 1797, the garrison was being reduced. When Walter Scott visited Shetland in 1814, there were just two companies of Invalids left, both about to be withdrawn. In spite of significant investment on two separate occasions a century apart, the fort's short active life each time has meant it survives today as the best preserved Cromwellian fort in Scotland.

Source: (a) Andrew Fraser, *Plan of Brassa-Sound in the Southern Part of the Isles of Shetland . . .* (1783). (b) Andrew Fraser, *Plan of Fort Charlotte at Lerwick in Shetland* (1783).

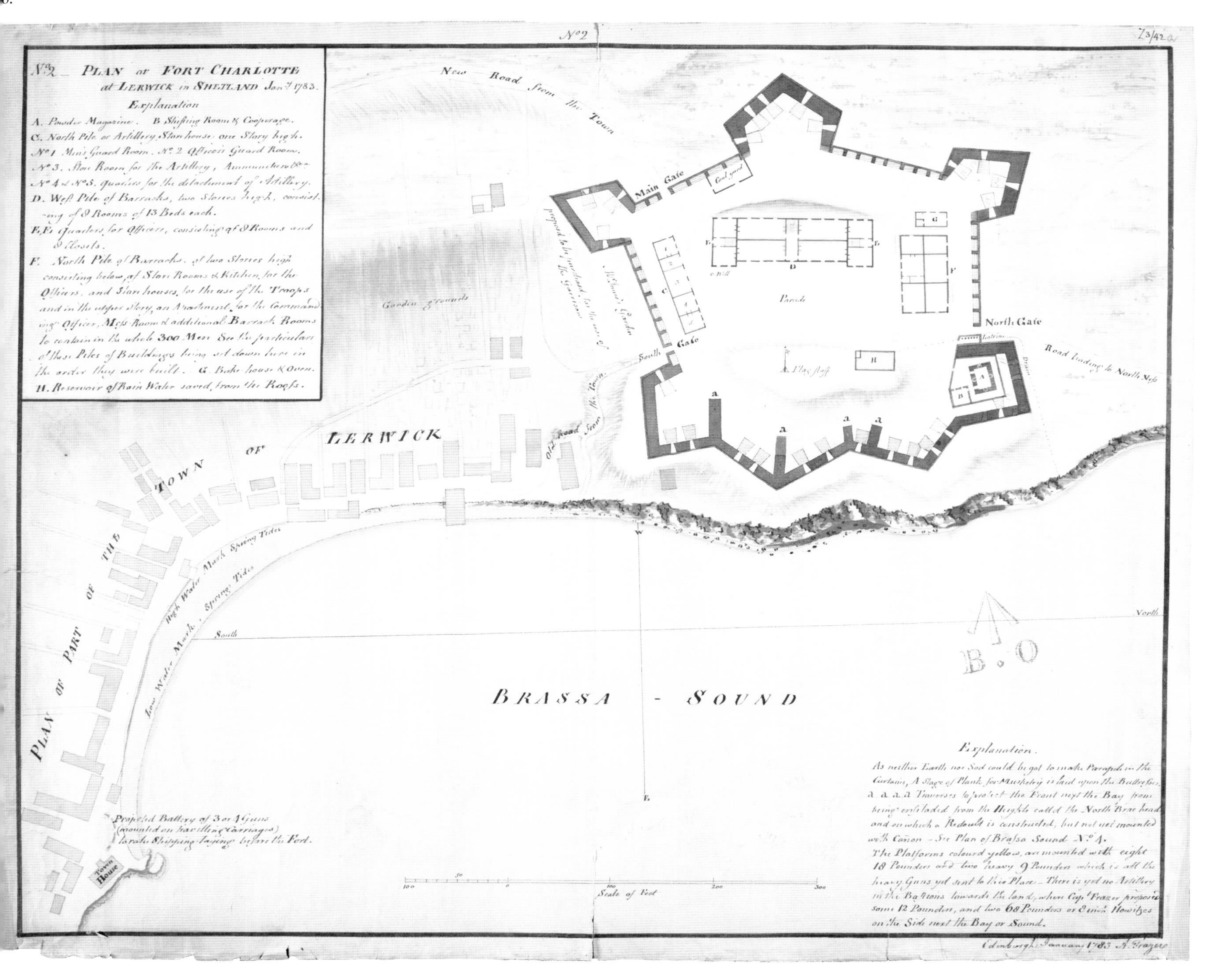
N.º 2
N.º 2 – PLAN OF FORT CHARLOTTE at LERWICK in SHETLAND Jan.y 1783.
Explanation
A. Powder Magazine. B Shifting Room & Cooperage.
C. North Pile or Artillery Store house one Story high.
N.º 1 Men's Guard Room. N.º 2 Officer's Guard Room.
N.º 3. Store Room for the Artillery, Ammunition &c.
N.º 4 & N.º 5. quarters for the detachment of Artillery.
D. West Pile of Barracks, two Stories high, consisting of 8 Rooms of 13 Beds each.
E, E: Quarters for Officers, consisting of 8 Rooms and 8 Closets.
F. North Pile of Barracks, of two Stories high consisting below, of Store Rooms & Kitchen for the Officers, and Store houses, for the use of the Troops and in the upper story, an Apartment for the Commanding Officer, Mess Room & additional Barrack Rooms to contain in the whole 300 Men See the particulars of these Piles of Buildings being set down here in the order they were built. G. Bake-house & Oven.
H. Reservoir of Rain Water saved from the Roofs.
New Road from the Town
Main Gate
Coal yard
North Gate
South Gate
Parade
Flag Staff
Latrine
Road leading to North Ness
Garden Grounds
Old Road from the Town
LERWICK
PLAN OF PART OF THE TOWN OF LERWICK
High Water Mark Spring Tides
Low Water Mark Spring Tides
North
South
BRASSA - SOUND
B. O
Propsed Battery of 3 or 4 Guns (mounted on travelling Carriages) to rake Shipping laying before the Fort.
Town House
Scale of Feet
Explanation.
As neither Earth nor Sod could be got to make Parapets in the Curtains, A Stage of Plank for Musketry is laid upon the Buttresses, a a a a Traverses to protect the Front next the Bay from being enfiladed from the Heights called the North Brae head and on which a Redoubt is constructed, but not yet mounted with Canon – See Plan of Brassa Sound N.º 4.
The Platforms coloured yellow, are mounted with eight 18 Pounders and two heavy 9 Pounders which is all the heavy Guns yet sent to this Place – There is yet no Artillery in the Bastions towards the land, where Capt Frazer proposes some 12 Pounders, and two 68 Pounders or 8 inch Howitzes on the Side next the Bay or Sound.
Edinburgh January 1783 A. Frazer

FIGURE 5.10
This attractive plan of the island of Inchcolm in the Firth of Forth is the work of a former military engineer who developed an interest in antiquities following his years of active service. George Henry Hutton was the only surviving son of Charles Hutton (1737–1823), chair of mathematics at the Royal Military Academy at Sandhurst. George also sought a military career, and was appointed second lieutenant in the Royal Artillery in 1777. His early years of service were in the West Indies fighting the French, but in a later action, a musket ball resulted in the loss of his right eye, and he was held prisoner of war before being discharged in 1796. Hutton thereafter cultivated a passion for architectural and antiquarian research, travelling widely from his main home in Aberdeen. He amassed over 500 documents and drawings of Scottish ecclesiastical antiquities and was elected a Fellow of the Society of Antiquaries.

Hutton was probably attracted to Inchcolm on account of its Augustinian monastery, first established as a priory in the twelfth century, and elevated to abbey status in 1235. Its most famous abbot was perhaps Walter Bower, who wrote the *Scotichronicon* there in 1441. Inchcolm was the subject of a number of attacks by English forces who plundered the monastery, and following the Reformation and the demise of the abbey, the island's strategic location in the Forth resulted in it being fortified against new enemies. In 1795 a gun battery was built due to concerns over a French invasion, and this can be seen as an Upper and Lower Gun Battery (E and F) on the plan with a nearby Guard House (D) and Store Shed (C). Hutton's key also notes that 'two Rooms have been fitted up as Quarters for Troops' in the monastery (A). The invasion never materialised. By 1822, these military needs were over, and Hutton was as interested in depicting the tombstones, inscriptions and burial grounds as in the fortifications.

Source: George Hutton, *Sketch of the Island of Inch Colm in the Firth of Forth* (1822).

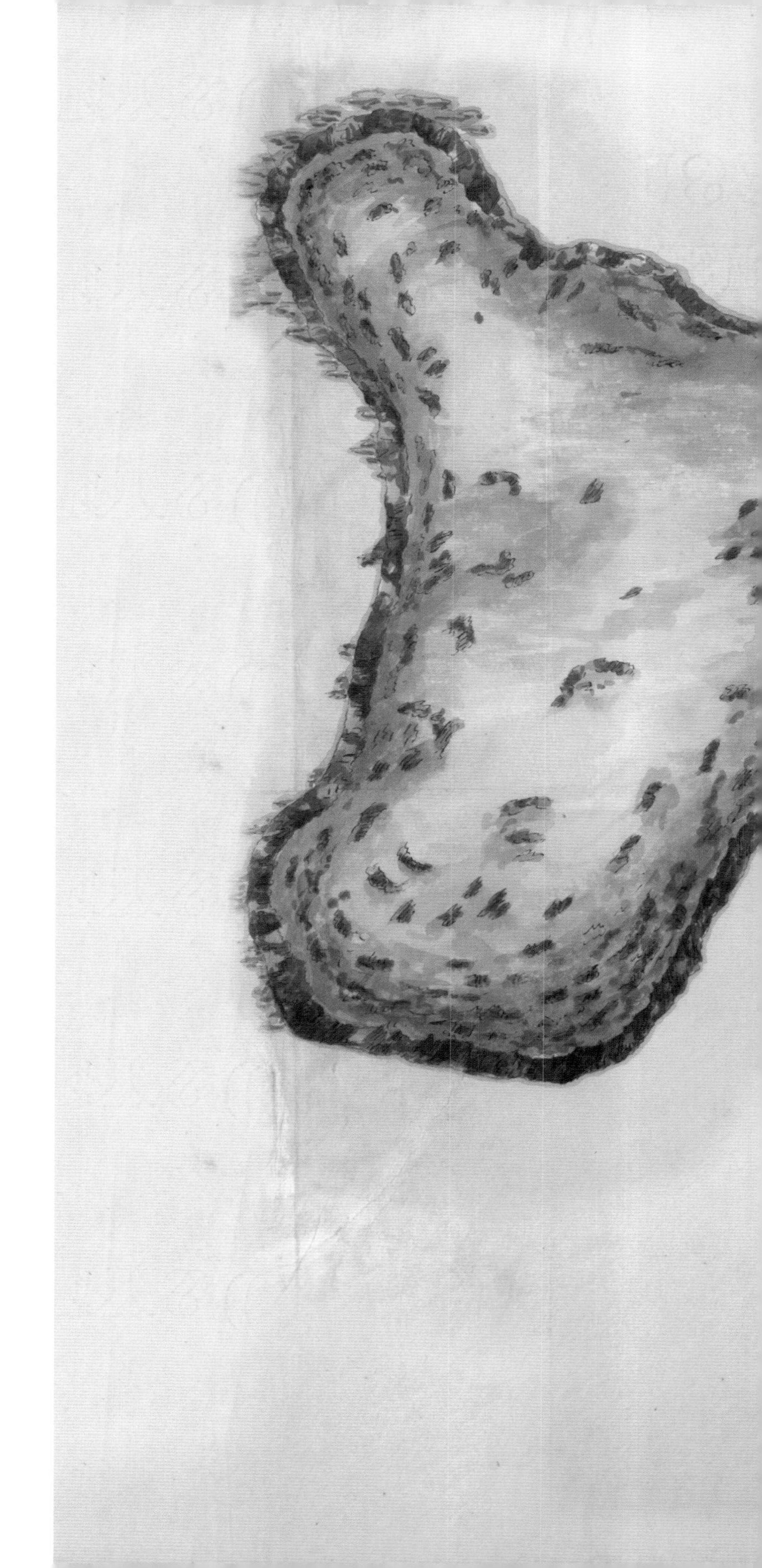

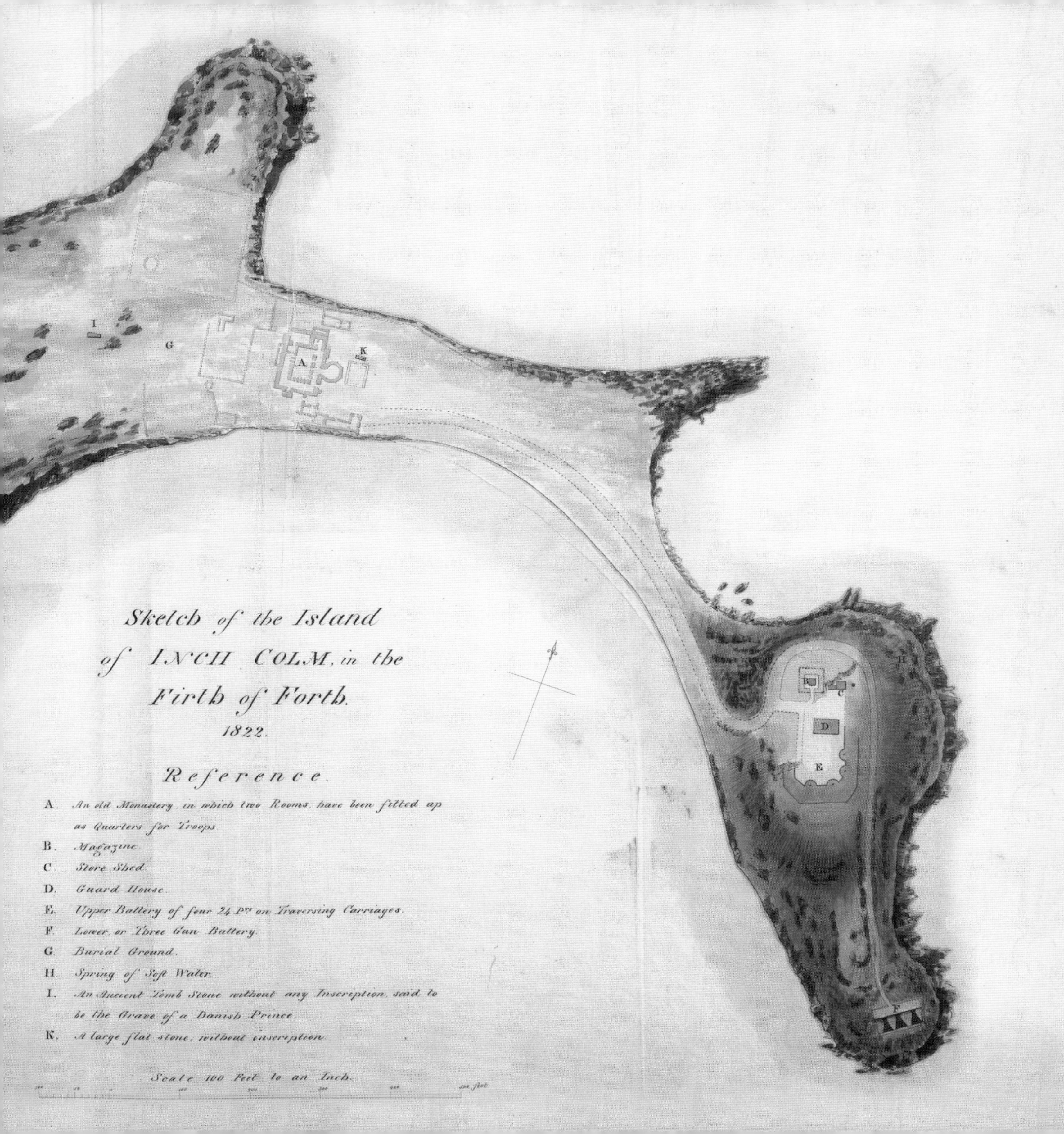
Sketch of the Island
of INCH COLM, in the
Firth of Forth.
1822.
Reference.
A. An old Monastery in which two Rooms have been fitted up as Quarters for Troops.
B. Magazine.
C. Store Shed.
D. Guard House.
E. Upper Battery of four 24 Prs on Traversing Carriages.
F. Lower, or Three Gun Battery.
G. Burial Ground.
H. Spring of Soft Water.
I. An Ancient Tomb Stone without any Inscription, said to be the Grave of a Danish Prince.
K. A large flat stone, without inscription.
Scale 100 Feet to an Inch.
A
B
C
D
E
F
G
H
I
K

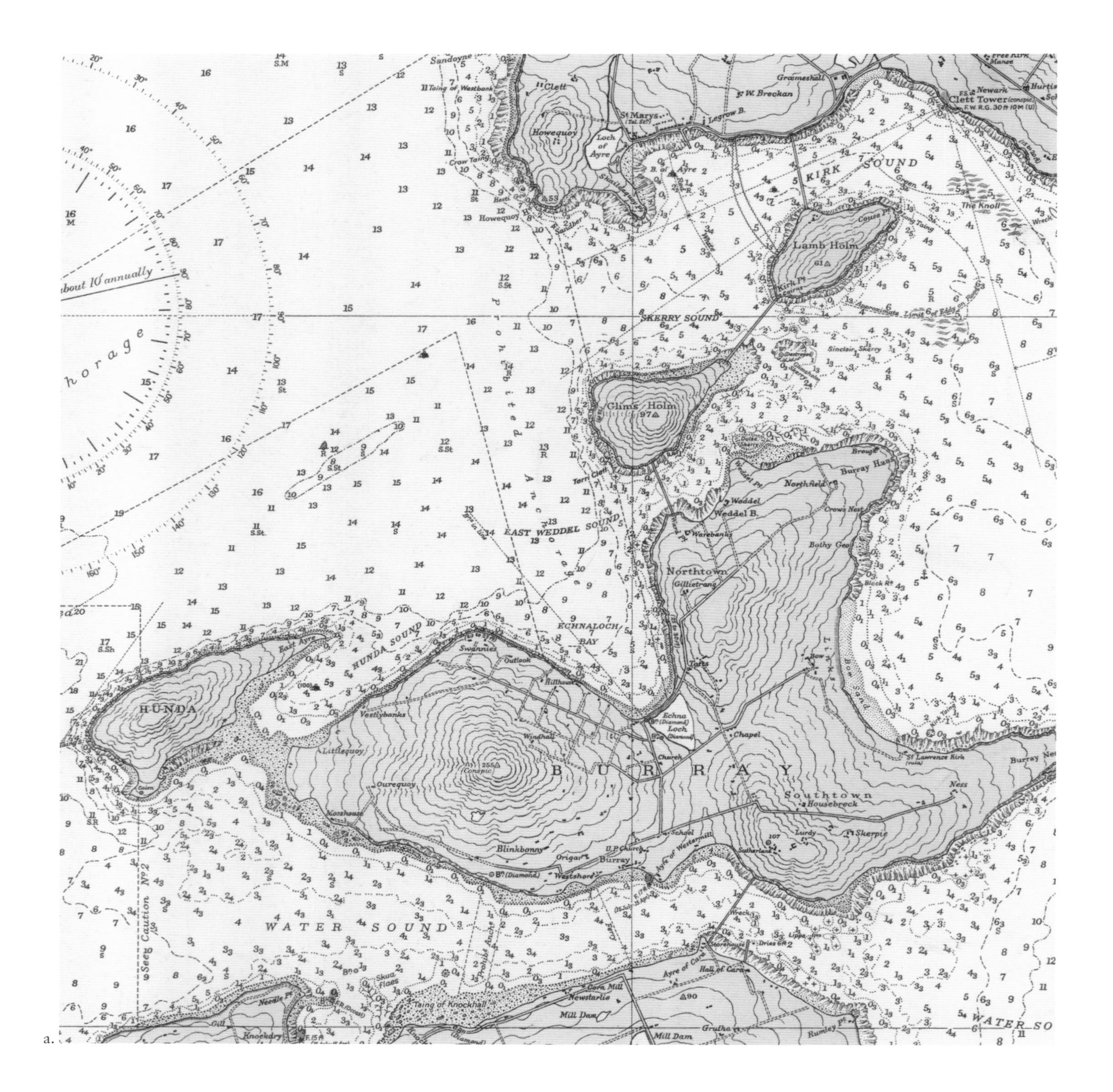

KIRK SOUND
Lamb Holm
SKERRY SOUND
Glims Holm
EAST WEDDEL SOUND
ECHNALOCH BAY
HUNDA SOUND
HUNDA
B U R R A Y
WATER SOUND
Clett Tower
St Marys
Howequoy
Northtown
Southtown
Weddel B.
Burray Ha
Taing of Knockhall
Mill Dam

a.

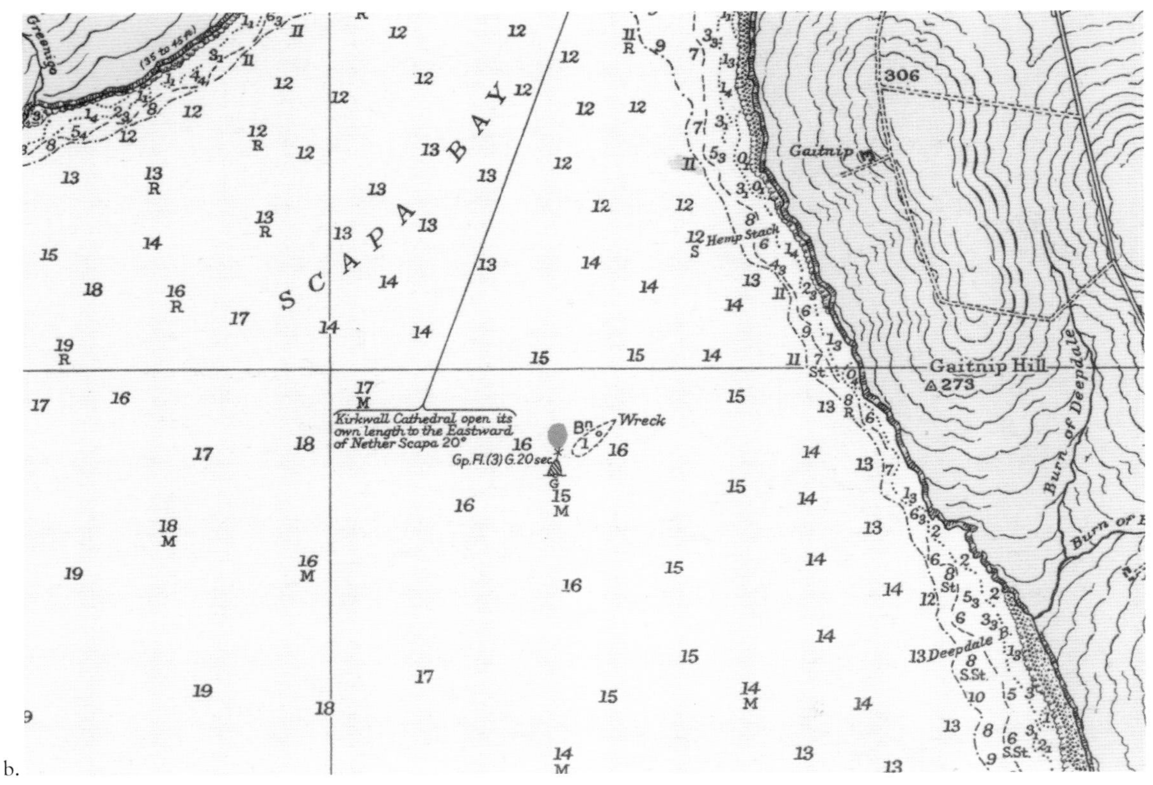

b.

FIGURE 5.11

(a) The fine natural harbour of Scapa Flow in Orkney was justly prized by mariners for centuries. After successfully protecting the British Grand Fleet during the First World War, it was chosen as the main British naval base at the outset of the Second World War. By this time, however, the defences guarding the main channels into the Flow had fallen into disrepair, with block ships and anti-submarine nets both rotting, and there were few patrol craft. On 14 October 1939, only one month after the British declaration of hostilities, a German U-boat penetrated the Flow through Kirk Sound (between Lamb Holm and Orkney Mainland) and torpedoed the Royal Navy battleship HMS *Royal Oak*. She sank within fifteen minutes, and of her 1,200-man crew 834 were lost. This loss prompted the construction of permanent barriers linking Orkney Mainland in the north to the island of South Ronaldsay in the south via Lamb Holm, Glimps Holm and Burray, to prevent the repetition of such events. Winston Churchill, then First Lord of the Admiralty, officially ordered the work, which began in May 1940 and was completed by late 1944. Rubble from overhead cableways was sunk between the islands, a task which proved challenging, given the speed of the tide in places and also depths: as indicated here, Kirk Sound is over 7 fathoms (43 feet) deep. After this, huge concrete blocks were cast in St Mary's Holm, each weighing 5–10 tons and laid on top of each other, eventually forming a wave barrier above sea level and a base for the roads along them. Some of the labour was done by Orcadians and Scots, but work accelerated following the arrival of Italian prisoners-of-war, based at Camp 60 on Lamb Holm, as well as two camps on Burray. Initially, the Italians reasonably claimed their work was prohibited under the Geneva Convention, but, after negotiation, the works were redefined as 'improvements to communications' to the southern Orkney Islands and work resumed. Those at Camp 60 also built the ornate Italian Chapel, which survives today.

This Admiralty chart shows the Churchill Barriers in 1944, only just complete, and with none of the connecting roads between them. The chart was based on an original survey by Captains Purey-Cust (1905) and Pudsey-Dawson (1906–09), but with 'small corrections' in 1944. During the twentieth century, most Hydrographic Office charts were photo-lithographed: this explains the coarse linework relative to nineteenth-century engraved charts. Note, too, the site of the wreck of HMS *Royal Oak* off Gaitnip Hill with its warning beacon (b). Although the Barriers served a limited military purpose and were not officially opened until 12 May 1945, four days after VE Day, they significantly improved post-war communications from Orkney Mainland to these former islands and (via the ferry in St Margaret's Hope) to mainland Scotland.

(b) Detail of the wreck site of HMS *Royal Oak* off Gaitnip Hill.

Source: Hydrographic Office, *Scapa Flow North*. Admiralty Chart 35 (1944).

a.

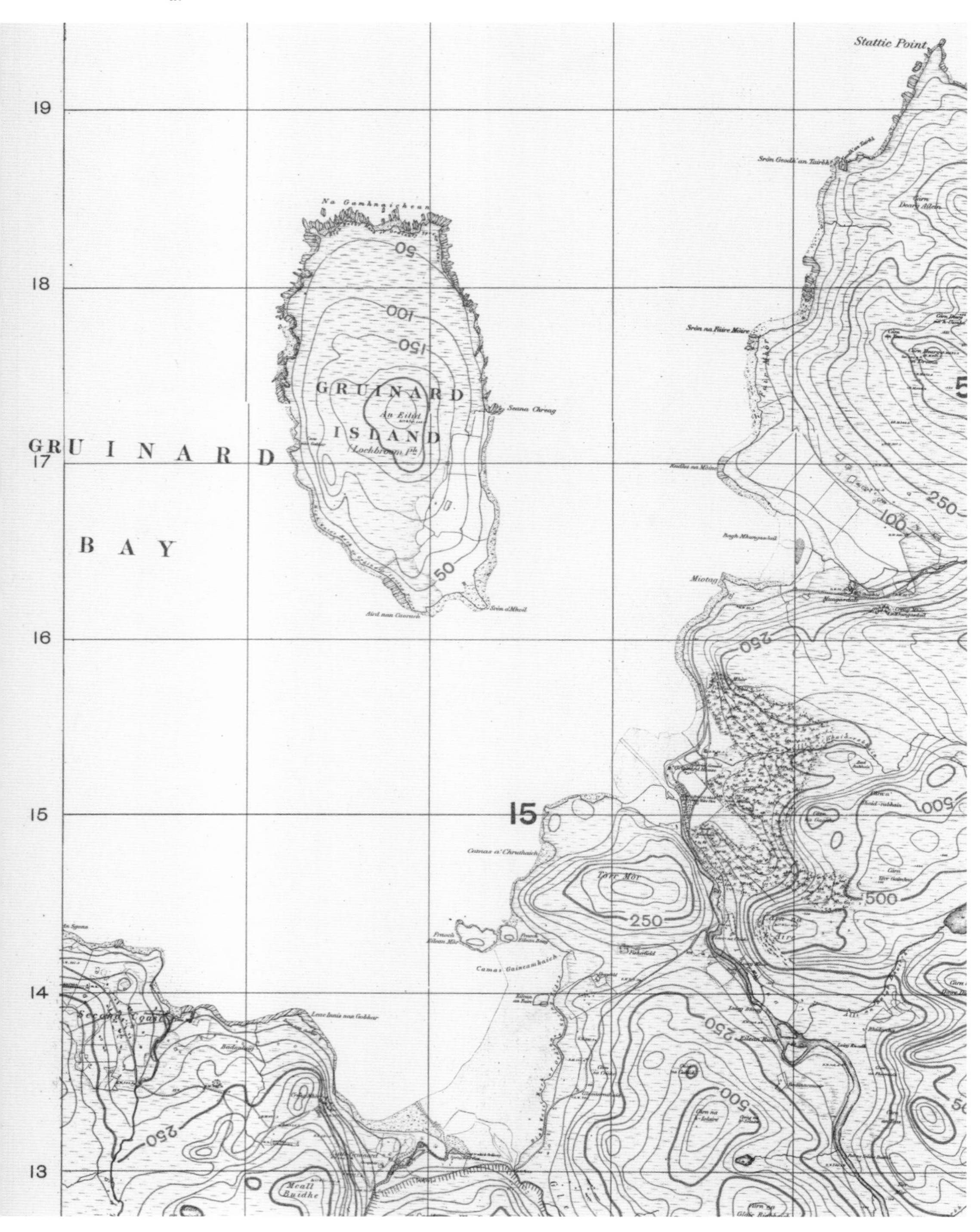

b.

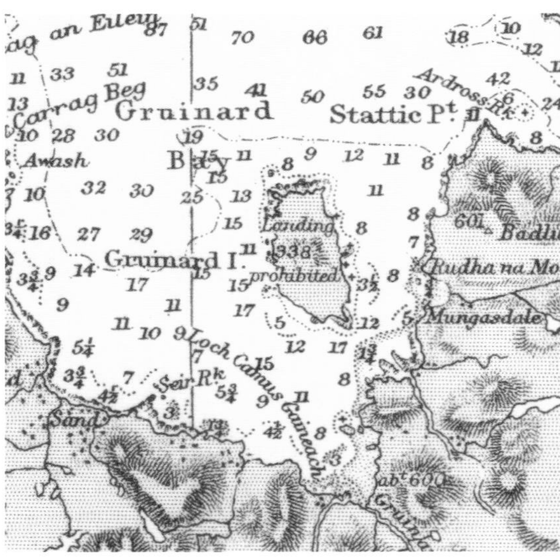

FIGURE 5.12
(a) In the 1930s, Ordnance Survey produced various series at a scale of 1:25,000, reduced from more detailed six-inch to the mile (1:10,560) mapping, and G.S.G.S. 3906 (the sequential War Office or Geographical Section General Staff reference for this series) was rapidly completed following the outbreak of war in 1939. Its distinctive style is due partly to the very small black linework of the reduced six-inch base mapping, and the very bold overprinted brown contours, photographically enlarged from OS one-inch to the mile (1:63,360) sources. Declassified official documents in The National Archives from 1943 confirm that this particular map sheet of Gruinard Island was used and annotated to plan the anthrax bomb testing.

(b) From the 1940s to the 1980s, most detailed Admiralty charts carried a 'Landing Prohibited' notice on them for Gruinard Island. *Source*: (a) Ordnance Survey, *Great Britain, 1:25,000 G.S.G.S. 3906, Sheet 23/90. N.E*, 2nd Provisional edn (1941). (b) Hydrographic Office, *Scotland – West Coast – Sheet 4 – Ardnamurchan to the Summer Isles – Including the Inner Channel and the Minch*. Admiralty Chart 2475 (1952).

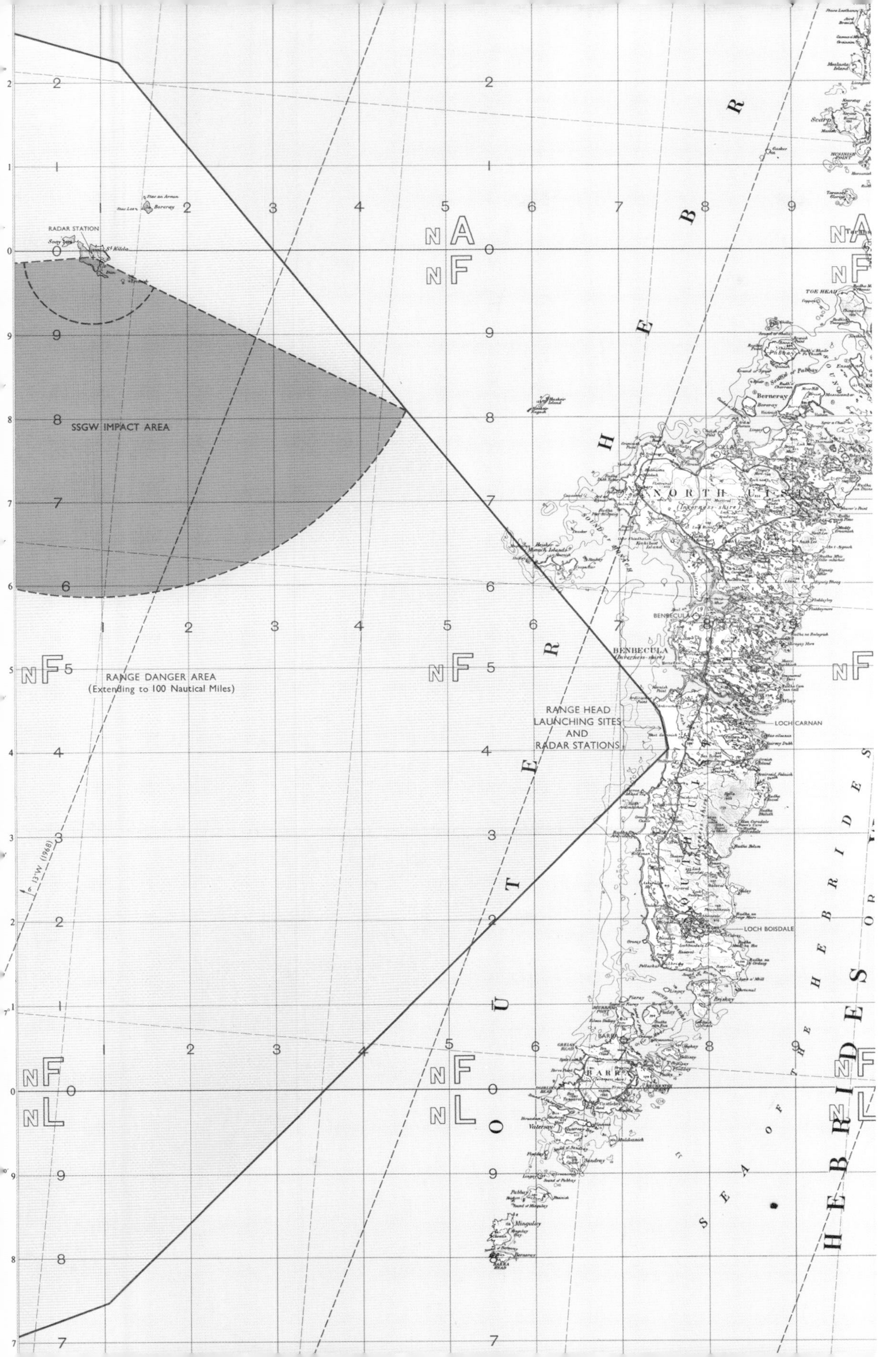

FIGURE 5.13
This declassified military map by the cartographic unit within the Ministry of Defence shows the danger area of the new rocket range sited in the late 1950s on Benbecula/Beinn na Faoghla. Although a military airfield with supporting infrastructure was constructed here during the Second World War, the main missile testing range shown was built by the Royal Air Force between 1957 and 1958 to test the Corporal missile, Britain and America's first guided nuclear weapon. Although the Corporal missile was quickly superseded by other missiles by the 1960s, including the longer-range Sergeant and Lance tactical nuclear missiles, over 200 launches of rockets took place between 1962 and 1982, some reaching altitudes of nearly 200 kilometres. In the early 2000s, the airfield underwent an upgrade, allowing it to participate in the Eurofighter Typhoon project, test firing advanced air-to-air missiles.

The Ministry of Defence decision to site the new rocket range here in the 1950s was based on a number of essential geographical requirements which this map reveals: a level stretch of beach over 3 miles long and a mile deep for the range launch area; an area of sea extending around 250 miles by 100 miles where the rockets would fall, relatively free from shipping and containing an uninhabited island in the line of fire to monitor the trajectory of the guided missile; high land on both flanks of the range launch area and in its rear for the siting of radar stations; and two airfields of 6,000 feet, or extendible to that size, within reasonable distance. Benbecula/Beinn na Faoghla was also a long way from the main centres of British population, and the recently depopulated island group of St Kilda was perfectly positioned for a radar tracking station, constructed at the same time. The siting of the range met with significant local and national objection given the loss of valuable crofting land, forced evictions of locals, as well as objections from fishermen and others who regularly used a far from 'empty' sea. Others supported the base for the benefits it brought to the local economy, the influx of new RAF servicemen, and the role of the island in supporting national defence needs.

This map reflects how military concern over Britain's vulnerability to attack, as well as the need to demonstrate flagging military prowess in the new Cold War world, dominated by superpowers, encouraged significant interest and investment in Britain's outer isles. As the range of these missiles increased, it also encouraged a new interest in attempting to assert Britain's ownership of the island of Rockall, some 200 miles to the west of the Hebrides, in an attempt to reduce the likelihood of enemy shipping or submarine observation of the range.

Source: Mapping and Charting Establishment, Ministry of Defence, *Hebrides Range Area. Scotland Scale: 1:253,440. G.S.G.S. 4982* (1966).

Арос
Лохалайн
г. Глаш-Венн
Эйгнейг
Иннинбег
м. Ардторниш
Сейлен
849
Бейл-Гевра
Пеннигоун
бух. Фишниш
Бейлменах
Фишниш
Килбег
пр. Линн-оф-М
Фракерсей
м. Руа-ан-Ридире
о. Бернера
замок
Тарранлохань
Гармойс
разв. Коррахад
морской паром
Грулайн-Хаус
Алтерих
Скалласл
Джава-Лодж
г. Беннь-Веон
Роэйл
Крейнгньюр
пирс
Фиарт
Томслейб
Далнароу
оз. Лох-Ба
вдп.
Торосей
Дуарт
Аппер-Ахнакройш
А Й Л Е Н Д - О Ф - М А Л Л
г. Беннь-Тала
(МАЛЛ)
Гортенбуи
г. Скурр-Дярг
Лохдон
разв. Торнесс
Гортен
Хейзелбанк
г. Норра-Венн
бух. Лох-Дон
Ишрифф
Страткойл
Ардера
Охнакрейг
пещ.
Коладойр
Крейг
Киллен
Слейтрих
Гортенакру
Средняя величина прилива 2,6 м
оз. Лох-Арлеглаи
Барнабак
Феллонмор
Ардмор
бух. Лох-Спелв
м. Рува-на-Фейлиннь
Кроган
Балур
г. Бен-Буи
Гайлен-Парк
Кинлохспелв
Далхана
Портфилд
м. Рува-Шенах
Барахандроман
Ан-Лярг
Лохбуи
оз. Лох-Ускь
Лагган-Лодж
Гленбайр
бух. Лох-Буй
г. Друм-Фада
пещ.
прист.
м. Рува-нам-Фер
о. Инш
пр. Инш
Арденкейпл-Хаус
Дуахи
гост.
Обан-Сил
Клахан-
Средняя величина прилива 2,6 м
Исдейл
Барнаярри
Балвикар
24

FIGURE 5.14
During the post-war period, the Soviet Army invested in a massive programme of global military geographic reconnaissance. Perhaps as many as 36,000 staff were employed in military map-making. In these Soviet military map series, the sheets follow those of the International Map of the World, allowing the globe to be completely organised, classified and brought together through four-degree-wide latitude bands, and six-degree-wide longitude zones. North of the equator, the bands are coded from (N)A to (N)V, while the zones are coded from 0 to 60, working eastward from 180 degrees longitude, i.e., the International Date Line. Grid square (N)O-30 here therefore lies between 52 and 56 degrees North and between 0 and 6 degrees west, and is further subdivided into 36 1:200,000 sheets per grid square, with suffix I to XXXVI. This 1:200,000 sheet is centred on Oban and the surrounding islands and Scottish mainland. Assembled from a variety of published mapping of different dates, the maps also include information gathered on the spot, including information about width, clearance and the carrying capacity of bridges along major roads. On the back of this 1:200,000 sheet is a full written description of the specific area under the headings 'Population Centres', 'Transport Network', 'Relief', 'Hydrography', 'Vegetation' and 'Climate', with a geological diagram. *Source*: ГУГК, Обан = GUGK, Oban *(O-30-XXXI)* (1984).

Paul Sandby, *Plan of Castle Tyrim in Muydart. Plan of Castle Duirt in the Island of Mull* (c. 1748).

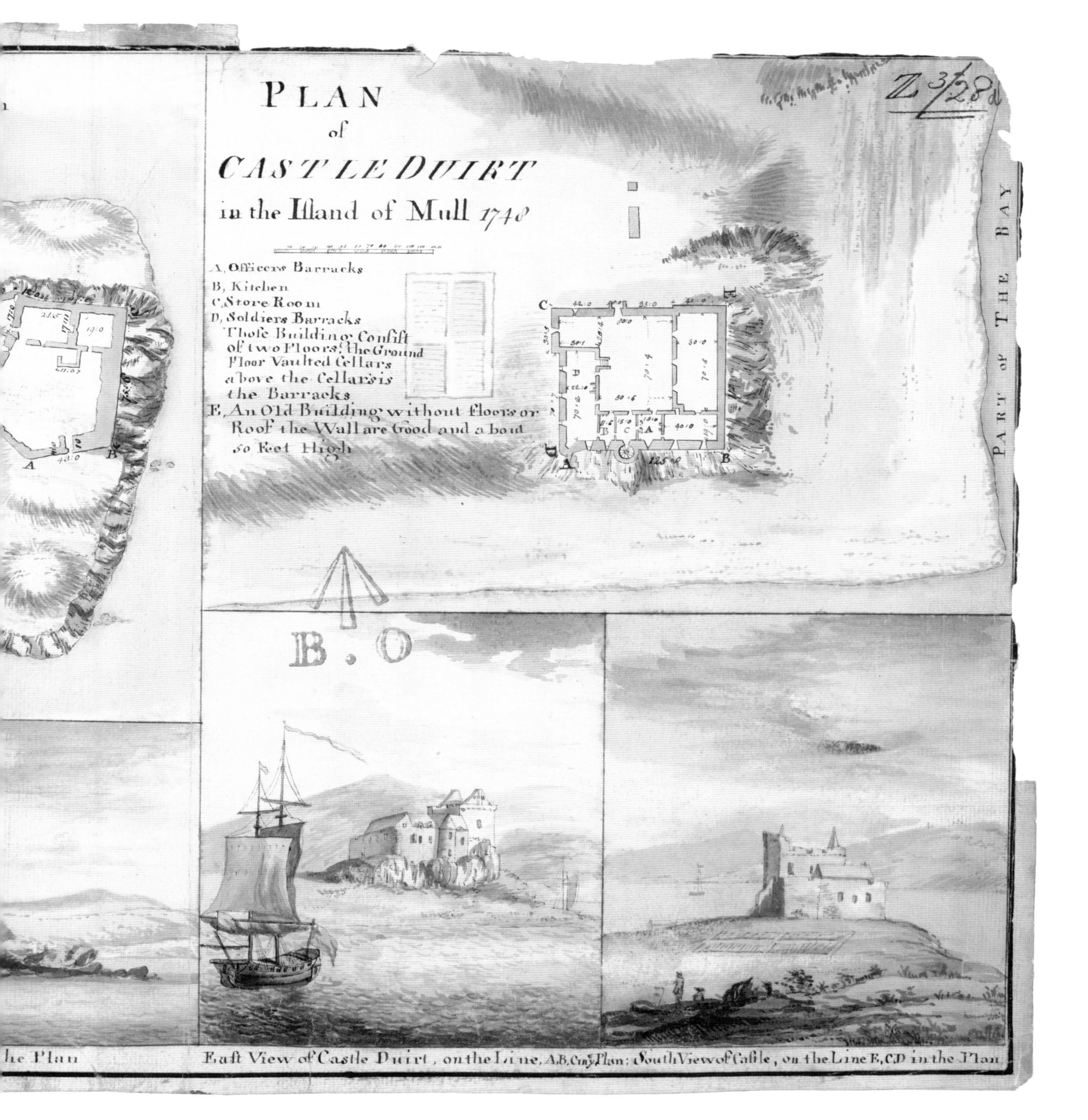
Z 3/28 d
PLAN
of
CASTLE DUIRT
in the Island of Mull 1748
A, Officers Barracks
B, Kitchen
C, Store Room
D, Soldiers Barracks
Those Building Consist of two Floors; the Ground Floor Vaulted Cellars above the Cellars is the Barracks
E, An Old Building without floors or Roof the Wall are Good and a bout 50 Feet High
PART of THE BAY
B . O
he Plan
East View of Castle Duirt, on the Line, A.B. in y Plan; South View of Castle, on the Line E, C.D in the Plan

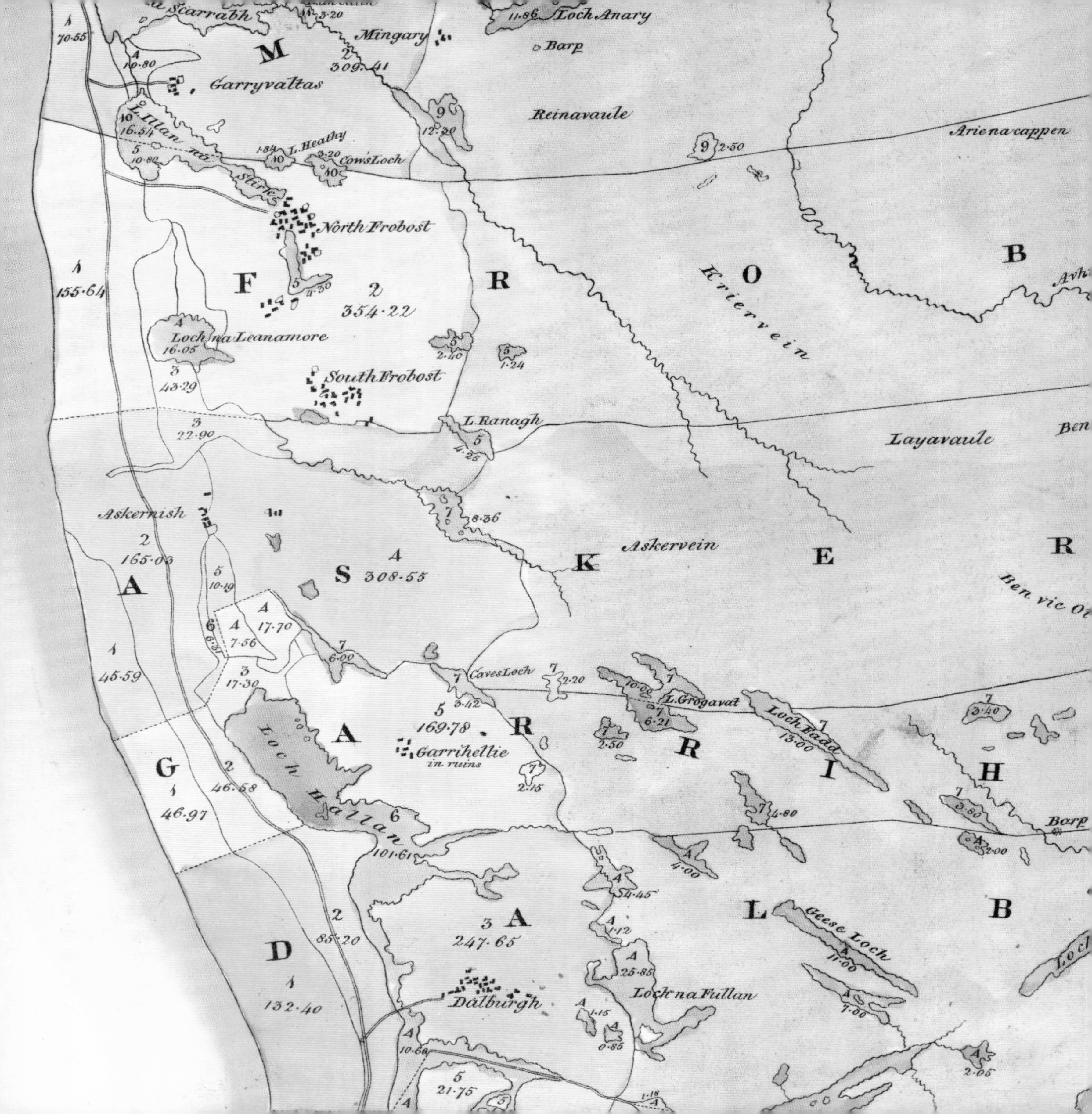

Scarrabh
Loch Anary
Mingary
Barp
M
Garryvaltas
Reinavaule
L. Illan na Stirie
L. Heathy
Cows Loch
Arienacappen
North Frobost
B
F
R
O
Kriervein
Loch na Leanamore
South Frobost
L. Ranagh
Layavaule
Ben
Askernish
Askervein
A
S
K
E
R
Ben vic Oi
Caves Loch
L. Grogavat
Loch Eadd
G
A
R
R
I
H
Loch Hallan
Garrihellie in ruins
Barp
D
A
L
B
Geese Loch
Loch na Fullan
Dalburgh

CHAPTER SIX

IMPROVING

Thinkers and writers in the Scottish Enlightenment combined a belief in the importance of human reason with an optimism over humanity's ability to 'improve' society and nature. These intellectual commitments had practical outcomes; if we think of maps as documents of improvement, we can see their direct influence on the landscape. Enlightenment ideas saw one practical expression in the Agricultural Revolution in Scotland with widespread innovations including drainage, enclosure, new crops, new rotations and new techniques. These both resulted in and reflected fundamental social and economic changes, and the creation of new and different landscapes. Improvement ideals also found expression in harnessing natural resources towards productive ends – in forestry and in fishing for example – and, in urban planning, the creation of planned villages and Georgian 'new' towns along rational lines. Techniques of map making, map styles and the purposes of maps also changed profoundly during this period. There was a greater emphasis on the importance of trigonometry, accuracy and quantification. Maps became plainer, with fewer artistic embellishments and with a greater use of standard scales, grids and graticules. There was a growing societal assumption that maps improved in accuracy and detail, and, in their purposes, that they both drove and mirrored the creation of new 'improved' landscapes.

Broadly speaking, we can trace the spread of ideas on improvement northwards and westwards from south-east Scotland and from urban centres in the early eighteenth century, with their practical application in the Western Isles by the early nineteenth century, and in the Northern Isles by the mid nineteenth century. In local contexts, specific factors within particular islands – microclimate, soil, location and matters of land management – all affected the timing and nature of improvement. National and international economic factors were also important: the integration of islands' economies into global markets made them always part of the wider world yet always vulnerable to circumstances beyond local control. Significant investment in improvements often did not occur until the inflation of agricultural commodity prices from the later eighteenth century, with prices heightened even further during the Napoleonic Wars. For example, when

Opposite. Detail from William Bald, *Plan of the Island of South Uist* (surveyed 1805, published *c*.1825).

the supply of cheap alkali from continental Europe was hindered with the outbreak of hostilities against France, many landlords in the Highlands and Islands specifically encouraged kelp production, the burning of seaweed to produce a calcinated ash useful in a variety of chemical and industrial processes. But after 1815, the renewed import of cheaper Mediterranean barilla alkali led to a collapse in kelp prices, causing poverty throughout the islands, social dislocation and emigration.

This chapter looks at the linked but different types of improvement evident in the islands – in agriculture, fisheries, urban development, woodlands and gardens – through maps, recognising the close connections between 'improvement' and 'exploitation', the theme of the next chapter. Quite often, improvement involved the exploitation of natural resources. This chapter focuses more on those activities associated with the idea of making islands' resources more valuable: the next chapter addresses the idea of making use of, or taking advantage of, those resources.

Agriculture

> The north-west isles are of all other most capable of Improvement by Sea and Land; yet by reason of their distance from Trading Towns, and because of their Language which is Irish, the Inhabitants have never had any opportunity to Trade at Home or Abroad . . . They have not yet arriv'd to a competent knowledge in Agriculture, for which cause many Tracts of rich Ground lie neglected, or at least but meanly improv'd, in proportion to what they might be . . . If two or more Persons skill'd in Agriculture were sent from the Low-lands, to each Parish in the Isles, they would soon enable the Natives to furnish themselves with such plenty of Corn, as would maintain all their Poor and Idle People.

Martin Martin's points in his *Description of the Western Isles of Scotland* (1703) share many similarities with later writers on improvement in the centuries that followed: a conviction that things needed to change, unrealistic optimism and an assumption that the locals need only learn from outsiders for everything to be better. Many of the early island improvers lacked practical know-how, however, and assumed that practices which worked elsewhere, in the Scottish Borders say, could easily be transferred to the islands. The Honourable Society of Improvers in the Knowledge of Agriculture in Scotland, founded in Edinburgh in 1723, had some 300 members, but not one who described themselves as a farmer, land-steward, factor or surveyor. This slowly began to change by the late eighteenth century: the formation of the Highland and Agricultural Society in 1784 and of the British Board of Agriculture in 1793 encouraged the up-take of new ideas. Landlords dominated agricultural activities and led the funding of county and estate mapping. The county map of Argyllshire by Langlands (fig. 6.1) is an example of a county map from this period that includes at least some islands.

The most detailed early topographic maps of the Western Isles/Na h-Eileanan an Iar were made directly for agricultural improvement in the early nineteenth century, with surveyors mapping islands' estates and often facilitating accompanying reports and advice recommending improvements to be made. Robert Brown, the commissioner or factor for numerous estates in the Outer Hebrides between 1807 and 1830, and a tenant of Carracroy in Lewis/Leòdhas, encouraged agricultural improvement there and the creation of new roads. Rental income derived from the manufacture of kelp allowed for the major investment required to commission estate mapping at this time.

One influential early estate surveyor was William Bald, born in Burntisland, Fife, in 1789. After schooling in Burntisland and Edinburgh, Bald was apprenticed in 1803 to John Ainslie, Scotland's leading surveyor and map maker at that time. Within only two years, aged just 17, Bald was sent to survey the islands of Harris/Na Hearadh, Benbecula/Beinn na Faoghla (fig. 6.2) and South Uist/Uibhist a Deas. His maps are

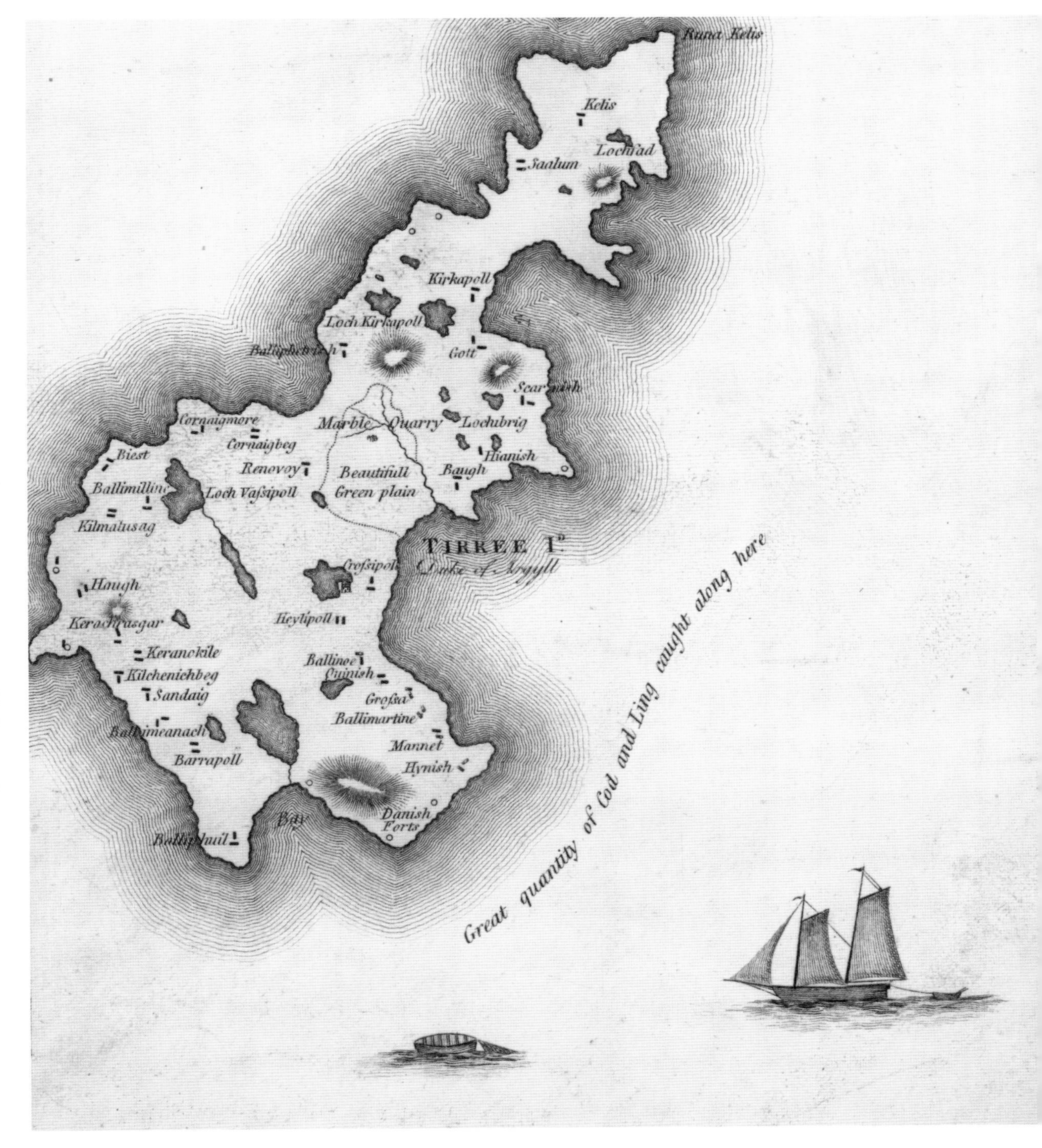

FIGURE 6.1

John Campbell, 5th Duke of Argyll (1723–1806), had an active military and political career before inheriting his father's estates in Argyll in 1770. He became a keen improver of his own estates, was a Governor of the British Fisheries Society, had a seat on the Board of Agriculture, and was influential in the construction of the Crinan Canal. His tenants were encouraged to follow habits of industry, sobriety and thrift. New agricultural techniques were encouraged, exports of black cattle and rents increased, and new textile-based industries were promoted. Land surveyors, such as George Langlands and his son Alexander, who were employed by the duke to work across his vast estates, were key instigators of change, drafting detailed estate maps, and producing smaller-scale maps of the county.

This is the main county map of Argyll – John Cowley's map of the heritable dukedom of Argyll in 1734 covers a different area than the administrative county depicted on the whole map – and as is hinted by the title cartouche and view and inset of Inveraray Castle, it was partly funded by the duke. The map promotes an optimistic view of Argyll bustling with potential. This is particularly so for Tiree, with the marble quarry at Balephetrish, and the note (in the sea to the south-east) 'Great quantity of Cod and Ling caught along here'. In reality, this claim was debatable: indeed, neither industry was a major source of revenue. The marble quarry was worked in the early 1790s by 'The Tiree Marble Company', and although a good deal of marble was extracted initially, the costs of transportation proved prohibitive, and the work was largely abandoned by the time this map was drawn. As with elsewhere on the west coast, the optimism expressed over fisheries at the end of the eighteenth century proved unfounded. By the time this map was published, income from the burning of kelp provided greater returns.

Source: George Langlands and Sons, *This Map of Argyllshire* . . . (1801).

invaluable to us today as the earliest surviving surveys to show the land use, farms and agricultural potential of these islands (fig. 6.3). In general, only a small fraction of the available land was arable, but there were important variations which Bald clearly shows: for South Uist/Uibhist a Deas, about 21 per cent was arable, with fertile machair lands in the west; on Harris/Na Hearadh, less than 7 per cent was arable.

On Lewis/Leòdhas, Francis Mackenzie, 1st Baron Seaforth, appointed James Chapman as Chamberlain between the late 1790s and about 1810. Chapman also worked as an estate surveyor on Uig in Skye in about 1803, and, later, practised as a land surveyor in Inverness, in partnership with Alexander Gibbs. Chapman's ideas for improvement are interesting to compare with the work in Uist/Uibhist and Benbecula/Beinn na Faoghla. There, crofting was being promoted as the best way of maximising rents through combining kelp manufacturing with agricultural smallholding. Chapman oversaw the creation of thirty-four crofts on Bernera/ Bernaraigh by 1807, probably the earliest crofts created in the Outer Hebrides (fig. 6.4). He was one of the main promoters of sheep in Lewis/Leòdhas, managing major sheep runs in the north-west of the island including the farms of Kaunlochroay, Clelachoag and Saclisvoe to the east of Little Loch Roag (taking these names from this map), as well as the island of Pabbay/Pabaigh, and on Valtos/Bhaltos to the north.

In Orkney, agricultural improvement came relatively late. The pre-improvement landscape that had existed since at least medieval times, illustrated in Murdoch Mackenzie's charts, survived in many cases until the nineteenth century (fig. 6.5). Inside the hill-dyke, closer to the townships, land use was a complex mixture of arable rigs, grassland and meadow, with oats and barley (bere) as the main crops. The kelp boom significantly delayed agricultural improvement until 1830, but, thereafter, new steamships allowed the export of black cattle, and the Public Money Drainage Act (1846) facilitated the improvement of much of the former runrig landscape (fig. 6.6 and pp. 164–5). Many Orkney commons were divided in the first half of the nineteenth century, although changes in ownership did not always result in alterations in the use of common land. What has been described by some as an Agricultural Revolution in Orkney, as the changes were so fast, would prove to be short-lived (fig. 6.7). Further improvements to communications, especially transatlantic steamships and American railroads, led to a collapse of the market by the 1880s. The problems caused for crofters by rising rents with declining agricultural prices were widely recorded by the Napier Commission in its reports of 1884 and 1885.

In the twentieth century, government-funded institutes encouraged improvement through mapping. The systematic survey of Scottish soils was initiated in the 1940s by the Macaulay Institute for Soil Research near Aberdeen. The initial priorities behind the work were connected with afforestation, but, after 1939, the focus switched to agriculture in order to increase food production. The work involved a technique known as 'free survey', digging a series of representative pits to describe and analyse the soil under laboratory conditions. Defining the boundaries between different soil types involved skill and judgement, usually done with the help of stereoscopes and aerial photographs to allow the terrain to be visualised in three dimensions. The colour coding was standardised, with detailed soil complexes grouped into soil associations, allowing broad groupings of soils – brown earths, podzols, gleys or alluvial soils – to be visualised (fig. 6.8). Soil mapping before the 1970s focused on better-quality agricultural lands, publishing maps at the one-inch to the mile (1:63,360) scale as shown in fig. 6.8. In the late 1970s, the soil survey was completed at the 1:250,000 scale, leading to the first complete and published coverage of Scotland's soils in 1982.

Fisheries

Compared with the systematic modernisation and state-sponsored improvement of the Dutch fisheries from the sixteenth century (see chapter 7), the development of Scottish fisheries was localised, piecemeal and lacking in focus. Various schemes to promote fisheries through joint stock companies or export bounties were considered during the sixteenth and

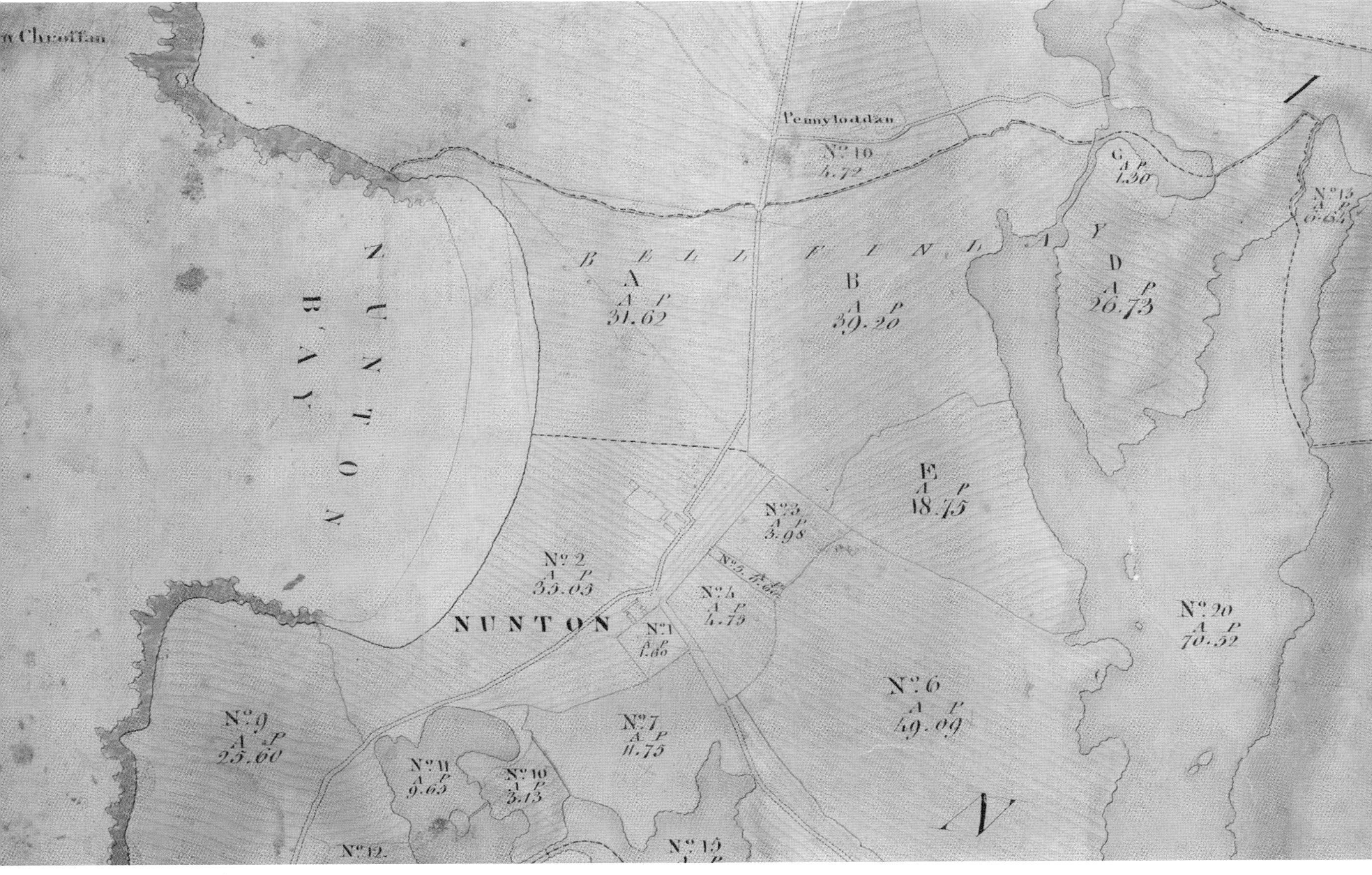

FIGURE 6.2

This original hand-drawn estate map of Benbecula/Beinn na Faoghla gives a useful and informative insight into the high quality of William Bald's original surveying work in the Western Isles. His original estate plans do not always survive, but, where they do, they are not only larger in scale than the lithographed reductions made from them in the 1820s (cf. figs 6.3 and 7.5), but also more informative, with details of farm buildings and their adjacent yards, as well as distinguishing ploughed land (with straight line shading) from lazy-beds (curved line shading). Comparison of the original estate map with the later lithographed maps also confirms significant changes to land use and ownership in this period. In 1805 in Benbecula/Beinn na Faoghla, the majority of farms were owned by tacksmen and leased to sub-tenants, but by 1829, many new crofts had been created, and the quantity of arable land had more than doubled through the reclamation of hill pasture.

Source: William Bald, *Plan of the Island of Benbecula, the Property of Ranald George McDonald Esq of Clanranald* (1805). Original deposited at the National Records of Scotland, RHP3028.

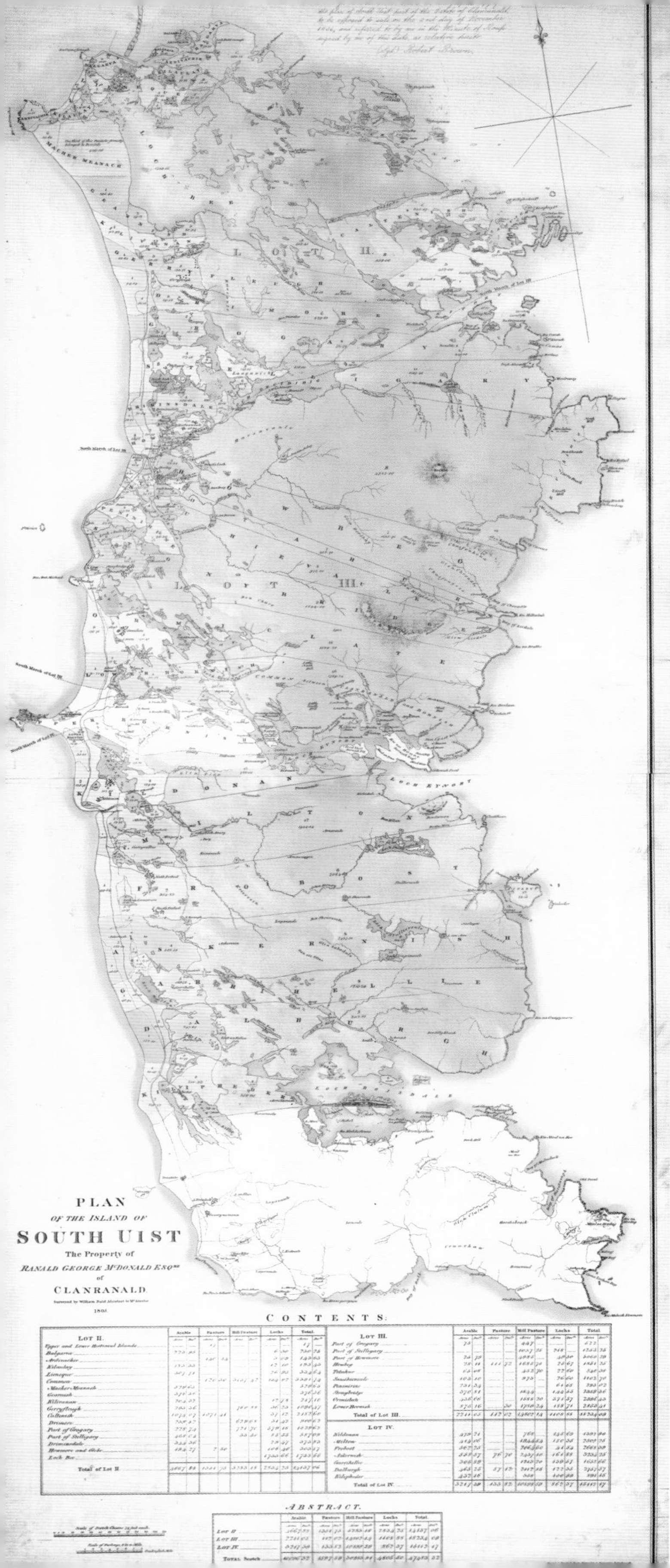

FIGURE 6.3

(*Left*). Bald's mapping of the Clanranald Estates (owned by Ranald George McDonald), covering most of South Uist/Uibhist a Deas, confirms that the majority of the farms extended from west to east. Most inhabitants lived near the inner edge of the fertile machair soils in the west. The main arable strip of land was on the western side, and improvement focused on maximising income from kelp and encouraging the reclamation of pasture land for arable purposes. Although most farms at this time were leased by tacksmen rather than tenants – a fact perceived as an important brake on agricultural improvement – there was a significant expansion in arable land through the creation of new crofts between 1805 and 1829. The plan shown here is a smaller-scale lithographed version of the original hand-drawn estate map, and made in the 1820s when extra copies were needed for prospective purchasers of the estate.

From the later 1820s as the kelp boom subsided and leases ran out, tacksmen and their sub-tenants were cleared, and extensive farms based on cattle and sheep rearing were created, paying much higher rents. Many townships lost parts of grazings and machairs to add to these already large farms. Although this was a time of significant emigration, few of the crofters were removed as most of their farms or crofting townships were located on poorer land, and the population continued to grow until a peak in the 1841 census. The estate of South Uist/Uibhist a Deas was finally sold in 1838 to Colonel Gordon of Cluny, Aberdeenshire, who was ruthless in evicting the remaining sub-tenants in the early 1840s.

Bald had a varied and successful career, particularly as a civil engineer and surveyor in Ireland. He became Director of the Trigonometrical Survey of Co. Mayo in the 1820s, and, in the 1830s, designed the Antrim Coast Road, an impressive engineering achievement today regarded as one of Ireland's most scenic routes. He worked briefly in France and in Scotland, for the Clyde River Trust and in improving Troon harbour. Geodetic comparison of his South Uist plan with modern Ordnance Survey maps shows a 0.003 percentage difference in acreages – an astonishing level of accuracy.

Source: William Bald, *Plan of the Island of South Uist* (surveyed 1805, published *c.*1825).

FIGURE 6.4

(*Opposite*). Although Chapman's original survey of Lewis/Leòdhas in 1807–09, his *Book of the Plans of Lewis*, is now lost, two later reductions survive. One of these, in Stornoway Public Library, dates from 1817. This is the other, drawn by William Johnson, and lithographed by Forrester and Ruthven in 1821. Johnson is better known as the draughtsman who assisted John Thomson in drawing the county maps for the latter's *Atlas of Scotland* (1832). The map shows the agricultural potential of Lewis/Leòdhas, with remarks on the quality of the pasture and moor as well as the ownership of the pasture rights. The boundaries between farms, and between the arable and better farmland closer to the shore, and the inland moors, are clearly shown. General topography is indicated by hachures, and parish boundaries are delimited in colour. Financial difficulties in 1819, which raised the possibility of selling the Lewis/Leòdhas estate, perhaps prompted multiple copies to be made of this lithographed plan, although the island remained in Mackenzie ownership until its sale to James Matheson in 1844.

Source: James Chapman/William Johnson, *Plan of the Island of Lewis, Reduced from Mr Chapman's Survey* (surveyed 1807–09, published 1821).

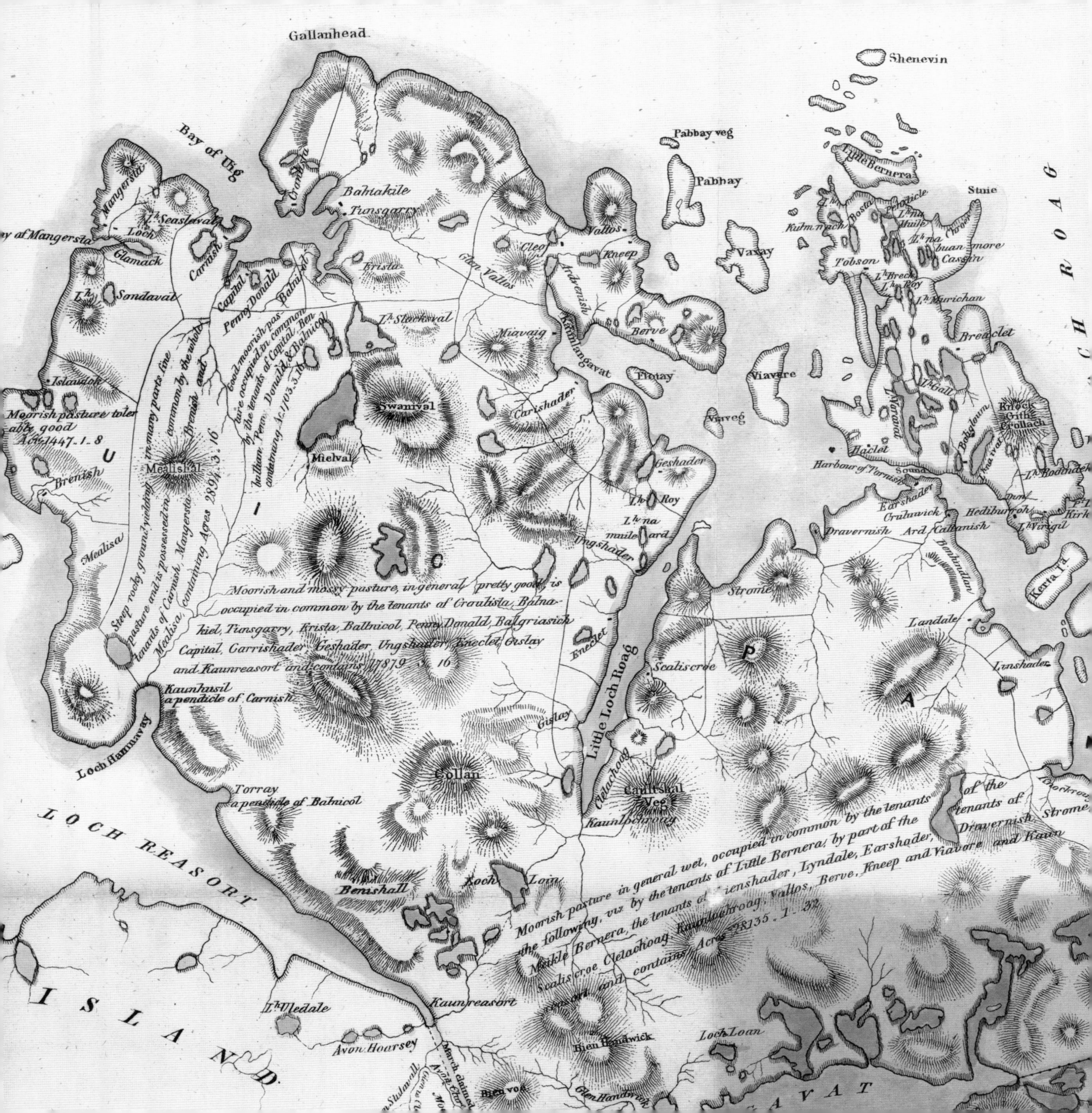

Gallanhead
Shenevin
Bay of Uig
Pabbay veg
Pabbay
Little Bernera
Stuie
Mangersta
Lh. Seaslevat
Loch
Bay of Mangersta
Glamack
Carnish
Balnakile
Tunsgarry
Vallos
Kneep
Cleop
Krista
Glen Vallos
Vaxay
Kulmnach
Bosta
Tobson
Lh. Breck
Lh. Roy
Lh. Murichan
Lh. Sondavat
Capital
Penny Donald
Balnicol
Lh. Stacksval
Ardvenish
Berve
Miavaig
Kirkinlangavat
Flotay
Viavere
Breaclet
Islaidok
Moorish pasture toler able good Acs. 1447. 1. 8
Swanival
Carishader
Viaveg
Lh. Gall
Knock Gith Crollach
Steep rocky ground, yielding in many parts fine pasture and is possessed in common by the whole tenants of Carnish, Mangersta, Brenish and Mealisa, containing Acres 2894. 3. 16
Good moorish pasture occupied in common by the tenants of Capital, Ben buillan, Penny Donald & Balnicol containing Ac. 1103. 3. 16
U
Mealishal
Brenish
Mielval
Geshader
Hacklet
Harbour of Tornish
Sound
Lh. Boondek
Lh. Roy
Lh. na mulle
Ungshader
I
Earshader
Crulawick
Hediburroh
Lh. Virigil
Dravernish
Ard Callanish
Mealisa
C
Kerta Id.
Moorish and mossy pasture, in general pretty good, is occupied in common by the tenants of Craulista, Balnakiel, Tunsgarry, Krista, Balnicol, Penny Donald, Balgriasich Capital, Carrishader, Geshader, Ungshader, Enaclet, Gislay and Kaunreasort and contains 17879. 3. 16
Strome
Landale
Enaclet
P
Scaliscroe
Linshader
Little Loch Roag
A
Kaunhusil a pendicle of Carnish
Gislay
Loch Hamnavay
Collan
Clelachoag
Caultshal Veg
Torray a pendicle of Balnicol
Kaunlochroag
Moorish pasture in general wet, occupied in common by the tenants of the following, viz by the tenants of Little Bernera, by part of the tenants of Meikle Bernera, the tenants of Lienshader, Lyndale, Earshader, Dravernish, Strome Scaliscroe, Clelachoag, Kaunlochroag, Vallos, Berve, Kneep and Viavore and Kaun and contains Acres 28135. 1. 32
LOCH REASORT
Benishall
Loch Loin
ISLAND
Lh. Uledale
Kaunreasort
Avon Hoarsey
Bien Hondwick
Loch Loan
March claimed
Bien voe
Glen Hondwick

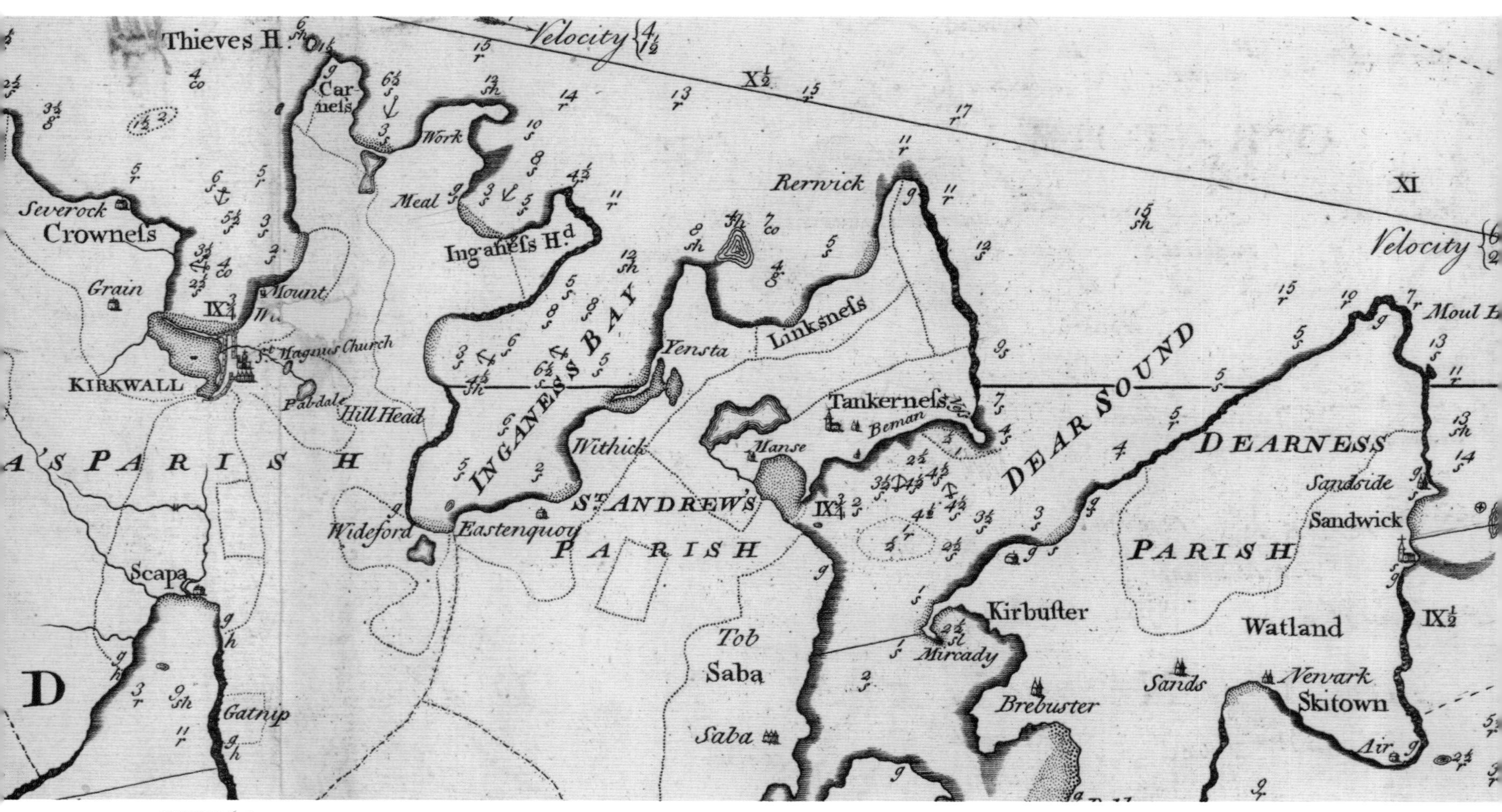

FIGURE 6.5
Mackenzie's charts of Orkney show the prominent turf dykes that surrounded the townships (shown here as dotted lines inland from the bold coastal outline). In addition, Mackenzie annotated his maps with 'g' (grass) and 'h' (heather) to indicate land use, including common grazings and grassland. *Source*: Murdoch Mackenzie, *Pomona or Main-Land* (1750).

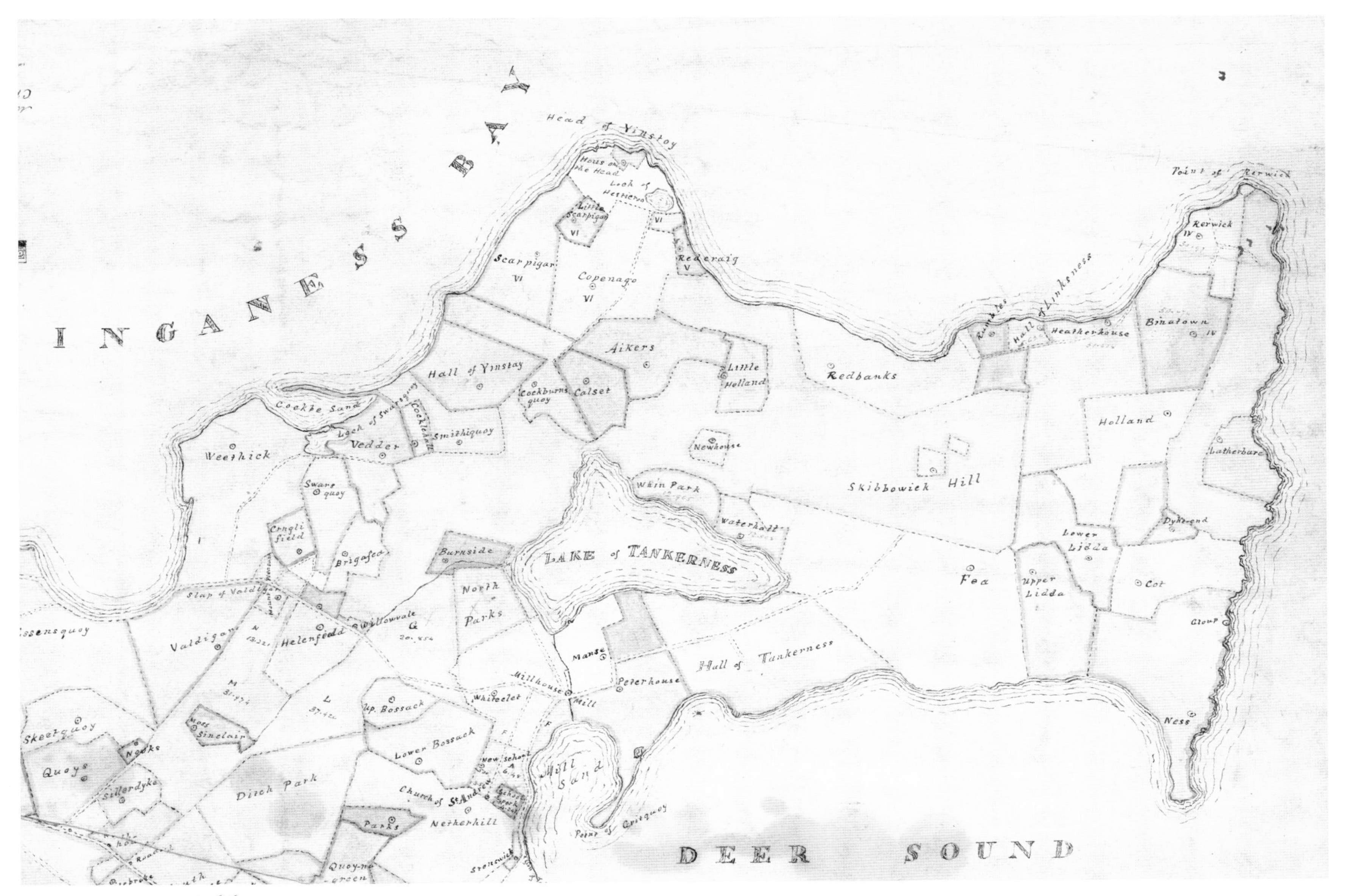

FIGURE 6.6

James Johnston was unusual in being both a landowner (as Laird of Coubister) and a practising land surveyor. Much of his work reflects the need for mapping the changing agricultural scene at this time in Orkney. The Coubister estates were scattered over Orkney, but particularly concentrated around Orphir parish and in Stromness. This map of the fertile Tankerness peninsula, east of Kirkwall, captures the newly enclosed and improved landscape, although it is still in a state of transition. The attractive colouring disguises the extent to which many of the larger fields were still unimproved pasture. The detailed schedule of fields and land confirms that less than 40 per cent was arable, with 58 per cent pasture, the remainder being houses, gardens, etc. Over the next century, there was a steady expansion of arable land with further enclosure and improvement. *Source*: James Johnston of Coubister, *Plan of the Estate of Tankerness* (*c*.1840?).

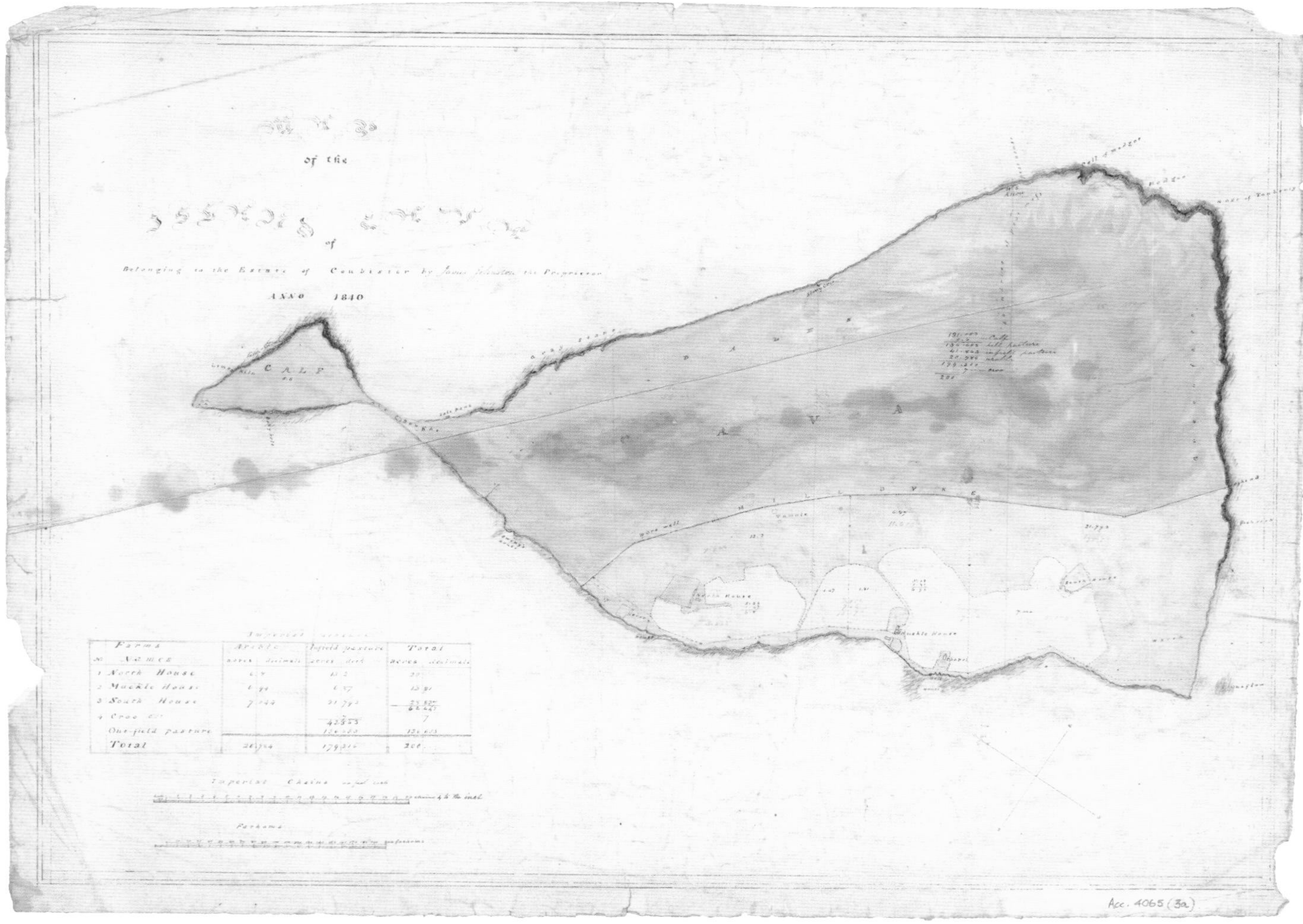

FIGURE 6.7

Measuring only 107 hectares (260 acres) in extent, the island of Cava lies just over a mile to the north-east of Hoy in Scapa Flow, Orkney. It was owned in the 1840s by James Johnston of Coubister, and his plain map of the island is interesting for several reasons. Given the rapid pace of agricultural improvements on Orkney at this time, the map reflects something of the growing economic value of Cava, especially for pasturing livestock. Johnston went to particular trouble to delimit the relatively small proportion (about 26 acres) of arable ground on the west side, from the main pasture land to the east. The prominent hill dyke (marked as a stone wall in places) that separated these two zones, along with the three main farms – North House, Muckle House and South House – within the arable zone are the main features of importance. The scattering of names around the island records more than is found on later Ordnance Survey maps, as well as the lime kiln on the Calf to the north.

The map shows Cava entering a period of prosperity. The 'New' *Statistical Account* reports that there were 21 people living there in the 1841 census, and 'the cultivated soil is a rich black loam, producing excellent crops, both of oats and bear, there being an abundance of sea-weed for manure'. By the late nineteenth century, the arable had extended, with new fields on the west side. As for much of Orkney, however, the late nineteenth century was a period of agricultural decline given the influx of lower-priced commodities. In the twentieth century, Cava's population steadily declined – North House and South House are now roofless – although there were still residents on the island until the 1980s.
Source: James Johnston of Coubister, *Map of the Island of Cava belonging to the Estate of Coubister* (1840).

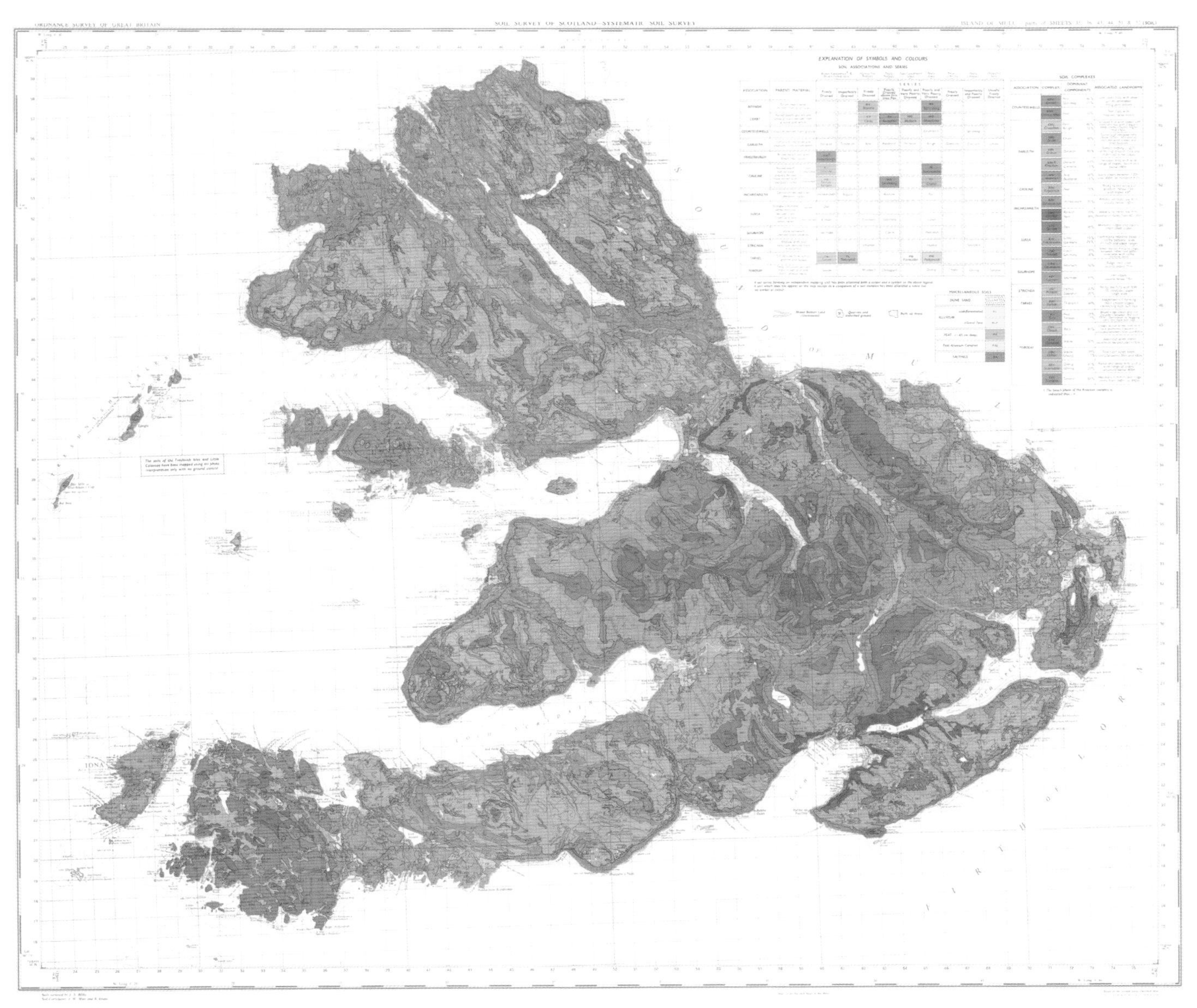

FIGURE 6.8

Soils reflect the complex interplay of geology, geomorphology, climate, land use and time, and this is particularly evident in this strikingly colourful presentation of Mull's soils. Most of Mull consists of volcanic basalt with lavas, but with a scattering of Cretaceous and other sediments, as well as a granite outcrop in the Ross of Mull. Wind speeds are strong. Rainfall is relatively higher around the mountains in the south, but lower elsewhere, and the proximity of the sea gives warmer winters than inland. This soil map of Mull shows the predominant mineral gleys (blue), and peaty gleys (green) on higher ground, giving way to podzols (pink, orange or red) and brown earths (brown) on lower ground. The basalt in general provides more fertile, freely drained soil than other underlying rocks, and although the Mull soils are relatively shallow, there are patches of arable and improvable brown earth in scattered locations around the coast.
Source: Macaulay Institute for Soil Research, *Soil Survey of Scotland, 1:63,360. Isle of Mull. Parts of Sheets 35, 36, 43, 44, 51 and 52* (surveyed 1972, published 1974). By permission of the James Hutton Institute.

seventeenth centuries, but little came of them. Intermittent attempts to establish fishing stations in the islands were made by armed force, along the lines of the Ulster plantations. In 1598, the Duke of Lennox, the king's cousin, with royal permission and the support of twelve lairds of Fife, attempted to attack and occupy the island of Lewis/Leòdhas with a force of about 500 men, in order to displace the local MacLeods and initiate a new burgh and fishing station at Stornoway/Steòrnabhagh. In the eyes of Scotland's Privy Council, the MacLeods were regarded as 'avowed enemies to all lawfull traffique and handling in these bounds', with the result that 'the maist profitable and commodious trade of fishing' was, by 'their barbaritie altogether neglectit and overruin'. The so-called 'Gentleman Adventurers of Fife' were several times defeated by the MacLeods and withdrew after ten years.

Further progress was made through the formation, in 1786, of the British Fisheries Society for 'Extending the Fisheries, and Improving the Sea-Coasts of this Kingdom'. Significant credit here goes to the Scot John Knox, a successful bookseller and writer. From 1764, Knox made a series of tours to the west and east coasts of Scotland – a total of sixteen visits over twenty-three years – each recounted in detail in his *A Tour through the Highlands of Scotland and the Hebride Isles in 1786* (1787). His work and his links with the Highland Society of London, established in 1778, allowed Knox to develop ideas for the improvement of northern Scotland, particularly through fisheries (fig. 6.9). Through Knox's efforts, legislation was implemented, including a premium per barrel on herring, as well as a tonnage bounty on fishing boats, while the British Fisheries Society purchased land for lease to fishermen and curers. Many of Knox's recommendations were later implemented, including the selection of Ullapool, Tobermory on Mull and Lochbay on Skye as suitable places for fishing stations. In the nineteenth and twentieth centuries especially, maps also provide evidence of large-scale engineering, the construction of harbour facilities, wharfs, quays and channels into port to promote specific fishing ports. Many of these were produced by the Stevenson family of lighthouse builders and civil engineers (fig. 6.10).

Planned villages and urban expansion

The new agricultural scene in the islands, which followed the expansion of sheep farming in the nineteenth century, required fewer people to work the land. Many islanders moved to towns or emigrated. Landlords were also keen to encourage settlement in towns; the market for agricultural commodities could be encouraged by related agricultural trades such as textiles or manufacturing, and the feuing of ground was a useful addition to the rent roll. It has been estimated that almost 450 new villages were planned and laid out in Scotland between 1720 and 1750, mostly in the later eighteenth century. A few of these were on the islands.

The thinking and improvement intentions behind these planned villages are evident in several maps – for Bowmore

FIGURE 6.9
(*Opposite*) John Knox's *A Commercial Map of Scotland* was included in his *View of the British Empire, more especially Scotland . . .* (1784), with the main outlines of the coast and islands taken from James Dorret's map of 1750. It is essentially a vehicle for notes by Knox on the poverty and suffering of the Highlanders, laced with practical suggestions of what could be done about it: 'The Hebrides, or Western Islands, are 300 in number, and contained, before the late emigration, 48,000 people. Here, as well as on the opposite coast, human misery resides in all its shapes. The people, cut off on one side from the Low Countries, by impassable mountains, and on the other side, by long dangerous seas; frequently perish thro want, fatigue, grief and epidemic disease, without the possibility of relief during a tedious, severe winter'. As he further reported, 'The numerous channels, bays, and lakes, are one continued receptacle of Turbot, Cod, Ling, Tusk, Haak, Herrings, Whitings, Haddocks, Skate, Soles, Phinocs, Mackarel, Salmon, Trout, Char, Pike, Eels, and other valuable Fishes; also Shell Fish of every denomination; Seals and Whales of various sizes; but from the Mull of Cantire to Thurso on the Pentland Firth; a coast of 300 miles, besides that of the Islands, there are no towns, markets, store houses, manufactures, shipping, or commerce of any sort; neither are the people in a condition to purchase provisions, and the necessary materials, whereby they might avail themselves of this heaven directed bounty.'
Source: John Knox, *A Commercial Map of Scotland* (1782).

A COMMERCIAL MAP
OF
SCOTLAND;
with the
ROADS, STAGES, and DISTANCES,
brought down to 1782.
LONDON, Publish'd by J. KNOX, Sept. 13th 1782.
ORKNEY ISLANDS
SHETLAND ISLANDS
PENTLAND FIRTH
SUTHERLAND SHIRE
CAITHNESS SHIRE
ROSS SHIRE
MURRAY FIRTH
MURRAY SHIRE
ABERDEEN SHIRE
INVERNESS SHIRE
MEARNS SHIRE
ANGUS SHIRE
PERTH SHIRE
ARGYLE SHIRE
MULL ISLE
SKYE
LEWIS
HARRIS
THE MINCH
WESTERN ISLES
THE HEBRIDES OR WESTERN ISLES
ATLANTIC OCEAN
BRITISH SEA
FIRTH OF TAY
FIRTH OF FORTH
FIFE SHIRE
EDINBURGH SHIRE
HADDINGTON SHIRE
BERWICK SHIRE
LANARK SHIRE
RENFREW SHIRE
AYR SHIRE
PEEBLES SHIRE
SELKIRK SHIRE
ROXBOROUGH SHIRE
DUMFRIES SHIRE
KIRKUDBRIGHT SHIRE
WIGTOWN SHIRE
ARRAN
CANTIRE
ISLA
JURA
ENGLAND
IRELAND
SOLWAY FIRTH
EXPLANATION
4 Degrees West from Edinburgh
1 Degr. East from Edinburgh

a.

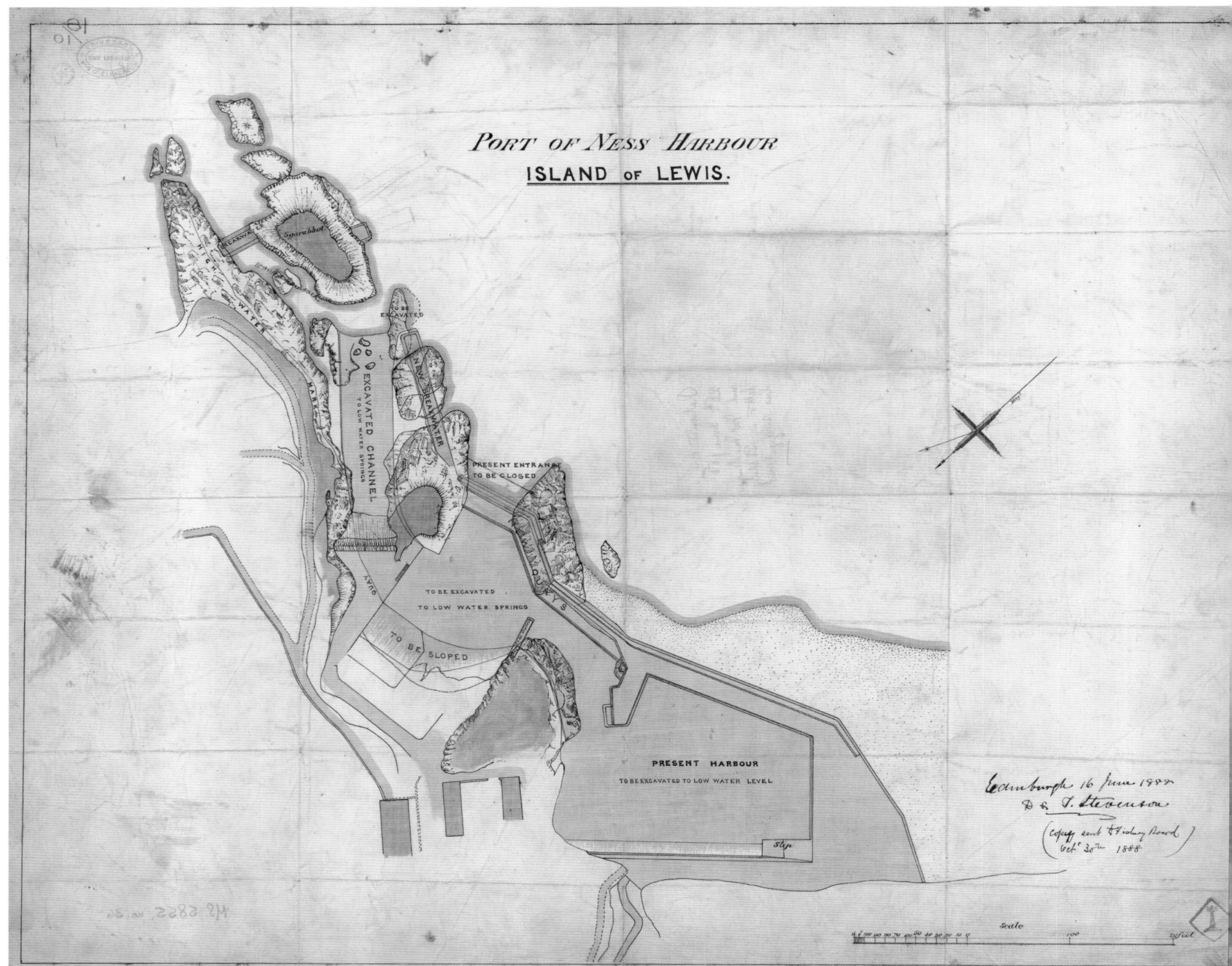
PORT OF NESS HARBOUR
ISLAND OF LEWIS.
Sgarabhot
TO BE EXCAVATED
EXCAVATED CHANNEL
TO LOW WATER SPRINGS
PRESENT ENTRANCE
TO BE CLOSED
TO BE EXCAVATED
TO LOW WATER SPRINGS
TO BE SLOPED
QUAY
PRESENT HARBOUR
TO BE EXCAVATED TO LOW WATER LEVEL
Slip
Edinburgh 16 June 1888
D. & T. Stevenson
(Copy sent to Fishery Board
Oct. 30th 1888)
Scale
100

b.

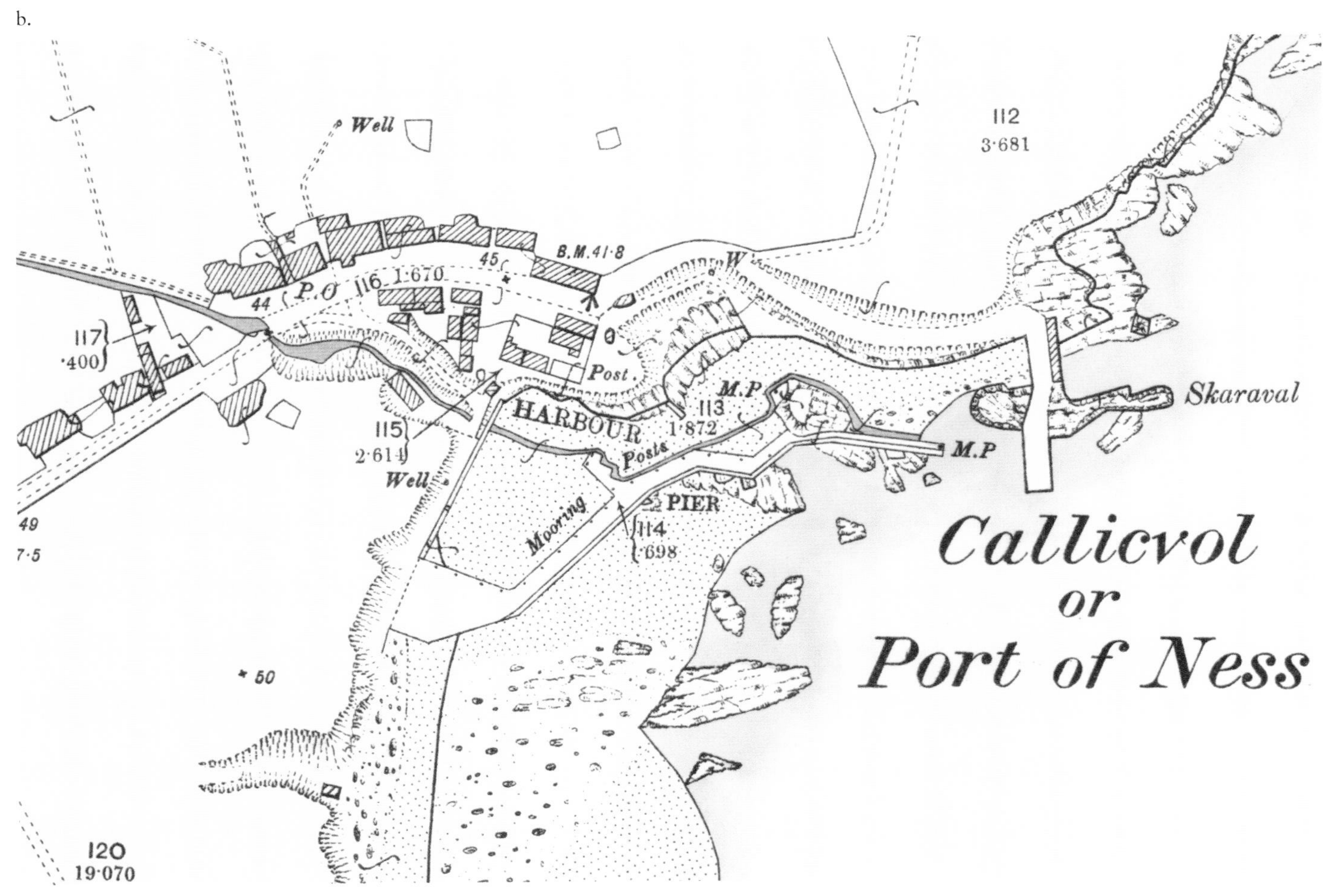

FIGURE 6.10

Port of Ness/Port Nis at Callicvol/Cealagbhal, some two miles south of the Butt of Lewis/Rubha Robhanais, grew particularly in the early nineteenth century as a fishing port, after the collapse of the kelp trade in the 1820s. The harbour, however, was always difficult to enter, with numerous submerged rocks at high tide, and the Niseach fishermen were forced to use small boats to make it through the harbour entrance. A visitor in 1842 recorded about a dozen boats, each with the usual fishing crew of 6–8 men, with two rows of double oars. Although catches at this time were good, competition from larger boats increased substantially by the 1880s.

Visitors also criticised the narrow mouth to the harbour, and the difficulties of launching boats: men usually had to wade up to their waists to launch the craft, and then stay wet at sea for up to two days. A pier had been built in 1835, funded by the Herring Fishery Board and Lord Seaforth, but it afforded little improvement. There were reports of boats lost and drownings in 1836, 1844, 1847 and 1849. In 1862, the 'Bathadh Mor' (Great Drowning) occurred, when five boats and their crews were all lost, with another two boats driven across the Minch and wrecked at Scourie.

The detailed engineering proposal (a), drawn up in 1888 by the renowned Stevenson civil engineering firm, is typical of their numerous infrastructure improvements on Scottish coasts and islands at this time. While their work on lighthouses earned them fame, they also undertook many local schemes of practical improvement. The Port Nis proposal involved completely blocking the former harbour entrance, using part of its breakwaters to create a much larger harbour, and moving the main harbour entrance 80 yards to the east. It also involved using the existing Sgarabhol Rock or Islet as part of the new defences, and substantially dredging the main channels. The work required Parliamentary approval, and, for this, several line drawings were created based on this engineering plan. The work took several years but the completed structure is shown well on the Ordnance Survey revised 25-inch mapping of 1895 (b). *Source*: (a) D. & J. Stevenson, *Port of Ness Harbour, Island of Lewis* (1888). (b) Ordnance Survey, 25 inch to the mile, *Ross and Cromarty (Isle of Lewis), Sheet III.3* (revised 1895, published 1897).

on Islay, for example, which was developed from the late 1760s (fig. 6.11). Maps also show unrealised intentions: in fact, relatively few planned villages developed quite as anticipated. Major expansion of Stornoway/Steòrnabhagh was planned in the 1820s, at a time of considerable optimism, given the income derived from kelp manufacture as well as from fishing. John Wood's *Descriptive Account* gives insight into this:

> Stornoway was, within the last twenty years, only a small fishing Village, but from the spirited and patriotic exertions of Lord Seaforth, the proprietor, and the grant of irredeemable feus for building, it has become a place of considerable importance as a Fishing station. No place in the north of Scotland, and in an insulated situation, also, has made more rapid strides at improvement, both in a domestic and commercial point of view, than Stornaway. The fisheries, especially for white fish, is conducted on a large scale. The number of boats fitted out annually for that fishery, amount on average to 120.

Wood admitted, however, that 'the Herring fishery has of late been on the decline'. In practice, most of these proposed developments shown on the manuscript map did not take place. The collapse of the kelp market, and the continued expansion of sheep farming resulted in serious impoverishment for the majority of people in Lewis/Leòdhas, with the result that Stornoway/Steòrnabhagh's economy was fragile in succeeding decades (figs 6.12 and 6.13).

Gardens

Following the 'pacification' of the Highlands that came in the wake of Culloden in 1746, many Highland and island landowners were keen to develop extensive formal gardens around their houses. This trend, already well underway in the Lowlands, included developing orchards, walled gardens or kitchen gardens for food and for flowers and shrubs, and the laying out of estate policies for amenity with water features and woodland. Some landlords encouraged the planting of woodland, and experimented with new tree species. These were useful as windbreaks, in landscaping, as an amenity and as a source of revenue once felled. The relative violence of island winds and the shorter growing season provided significant challenges to large-scale landscape gardening in many islands, so the results are all the more impressive when we do find them, as at Dunvegan in Skye (fig. 6.14). The most striking results, such as at Mount Stuart in Bute (fig. 6.15), were achieved only through significant investment over generations.

Maps are key tools for understanding improvement because they afford new perspectives in planning change and because they can facilitate attitudes of optimism and enthusiasm in looking at and refashioning the world. Even if plans for the improvement of many Scottish islands often fell short of their original ambitions, the work of improvers and map makers provided the basis for later changes. One of the reasons why the work of William Bald in the Western Isles survives in the form of copies is because the improvements failed, and the estates were sold, necessitating further maps to plan future development. Maps thus provide documentary evidence of intentions and results – the two not always being the same.

FIGURE 6.11

Daniel Campbell the Younger succeeded to his Islay estates in 1753 aged only 16, but with visionary ideas for their improvement, particularly through promoting new agricultural techniques, as well as fishing and mining. His first petition for financial support to the Commissioners for the Annexed Forfeited Estates in 1766 was unsuccessful, but his second, in 1777, resulted in support for the promotion of fisheries and the improvement of quays. The new planned village of Bowmore, constructed from the late 1760s, replaced the long-established core of the estate at Kilarrow, and was designed 'street by street, house by house, and garden by garden, with certain social, economic and architectural considerations always in view'. The new village encouraged non-agricultural employment, with significant investment in housing and with each house given a garden for growing potatoes, oats, flax and grass for a cow. As with other planned villages, the neat geometric plan gives an impression of order, with a wide main street running uphill to the new distinctive round Protestant church with its conical roof. The church was perhaps inspired by a similar one planned (although not constructed) at Inveraray, between 1747 and 1760 by William and John Adam. The associated manse, courthouse and school (for promoting the English language) were also funded by Campbell. Although there were initial difficulties in creating enough alternative employment – fishing and textiles arrived only later, and the famous distillery was only constructed in the late 1770s – the village was well-established by the time of this map, a century later.

Source: Ordnance Survey, 25-inch to the mile, *Argyll and Bute, Sheet CCVIII.10 (Killarrow)* and *Argyll and Bute, Sheet CCVIII.14 (Combined)* (surveyed 1878, published 1882).

FIGURE 6.12

John Wood (*c.*1780–1847) was the most significant surveyor of British towns in the early nineteenth century. Between 1818 and 1846 he drew maps of at least 148 towns in Britain – a monumental achievement given the difficulties of travel and a relatively small market for urban maps. In a few cases, Wood borrowed from the work of earlier map makers, and these borrowings are clearly credited, but he more often undertook original survey. Wood settled in Edinburgh from 1813, and his initial work focused on Scottish towns; forty-eight of these were gathered together into his *Town Atlas of Scotland* (1828), together with a detailed *Descriptive Account of the Principal Towns in Scotland* (1828).

Only Wood's final printed plans usually survive, but for Stornoway/Steòrnabhagh, surveyed in 1821 (shown here), we have an original manuscript map by Wood. The manuscript map is almost twice the size of the printed map and, although surveyed at the same time, contains a number of differences. From its title we are told that it is a 'Plan . . . designed for building'. There are new inland streets proposed, parallel to Kenneth Street and Church Street, along the lines of the future Keith Street and Scotland Street. A note on the map confirms that these 'houses and lines of Streets coloured light Red are intended to be built'. There are outlines of a possible circus or square around the Church (today Stornoway High Church) and a diagonal vista leading from it. Although these streets were never built, the idea was kept alive in later years: Leverhulme's plans for a new canning factory in the town in 1918 used Wood's design as a model. The manuscript map also covers a much larger territorial extent than the printed map, showing the proposed new roads, crofts and fields in lands up to two miles distant.

Source: John Wood, *Plan of Gro[un]d Designated for Building on & Acres &c* (*c.*1820–21).

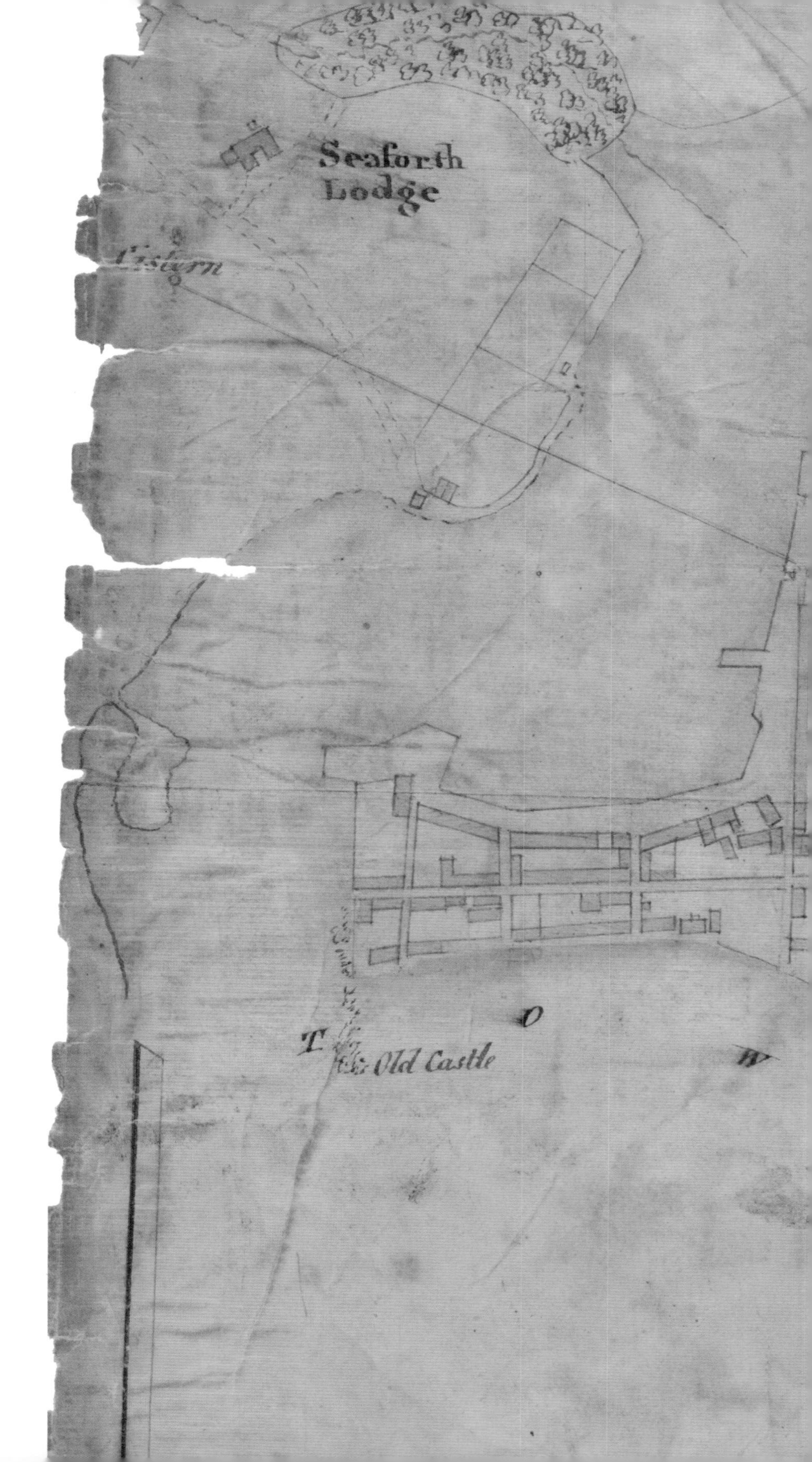

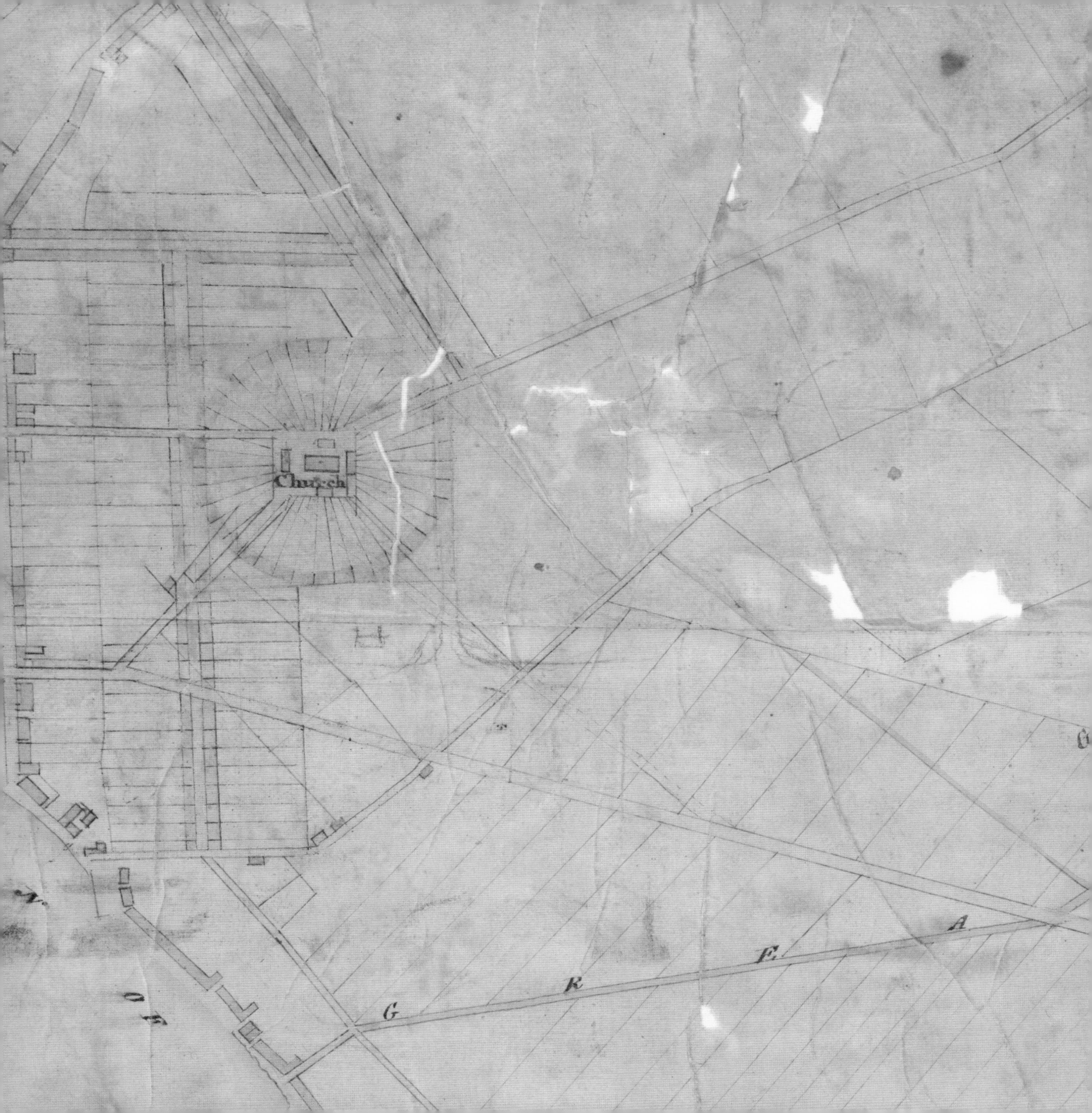
Church
G R E A

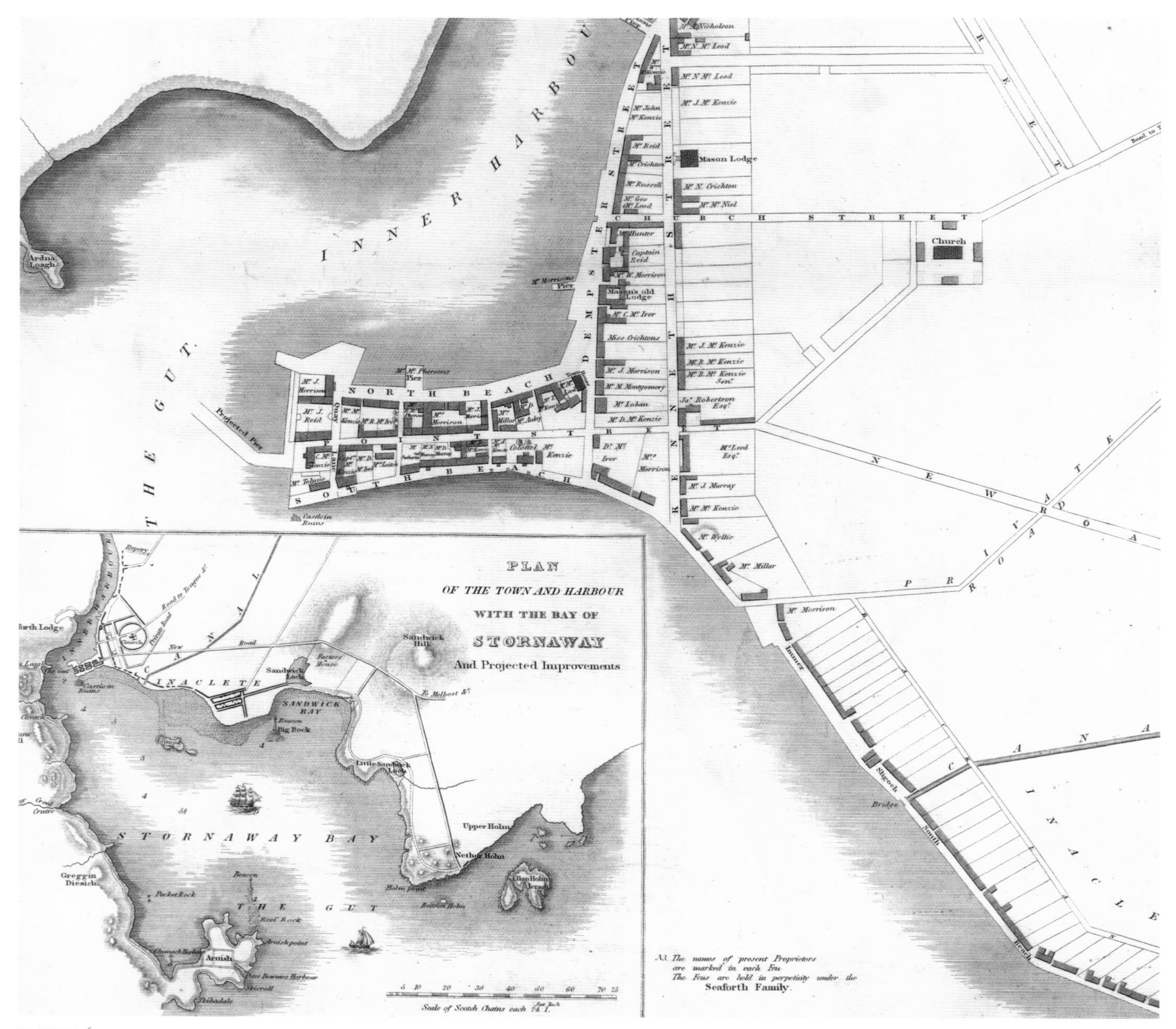

FIGURE 6.13

John Wood's printed plan is particularly valuable for its list of the main property owners in Stornoway/Steòrnabhagh itself, several of whom had had illustrious careers overseas. Captain John Mackenzie (who lived at the corner of South Beach and Quay Lane) had distinguished himself at the defence of Gibraltar against the combined forces of Spain and France (1779–83), and died in 1830 aged 67. Further east along South Beach we can see Carn House, the property of Colonel Colin Mackenzie, who died in May 1821 in Chowringhee near Calcutta, the year this map was published. Mackenzie left Stornoway/Steòrnabhagh for Madras in 1783, and in 1815 was appointed Surveyor General of India, headquartered at Fort William in Calcutta. It is possible that the £30,000 which Mackenzie left his sister in Stornoway/Steòrnabhagh, which she used to support numerous charitable causes, funded the work by Wood. Further east again, at a building marked 'McLeod Esq' (today the site of Martin's Memorial Church), was the birthplace of the Canadian explorer Alexander Mackenzie, remembered for his pioneering traverse across the Rockies to establish a landward route across Canada.
Source: John Wood, *Plan of the Town and Harbour of Stornaway, Island of Lewis, from Actual Survey* (1821).

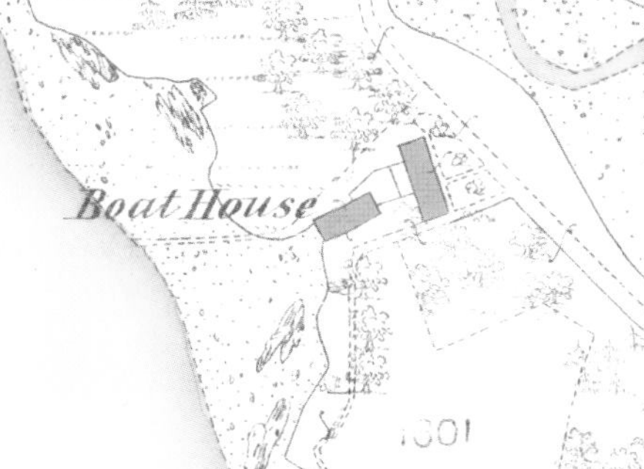

FIGURE 6.14

Dunvegan Castle on the Isle of Skye stands on a mass of columnar basalt which was once an island within Loch Dunvegan – its landward entrance was only constructed in the mid eighteenth century – and most communication and travel around the estate before this time was by sea rather than land. Occupied by the Chiefs of Macleod since the thirteenth century, James V visited the Castle in 1540 as part of his circumnavigation of Scotland and he was reputedly entertained on Macleod's Tables, the flat-topped hills nearby. The Castle has a complex history of rebuilding and extension – described by one commentator as 'an amorphous mass of masonry of every conceivable style of architecture' – but its appearance today largely dates from a Romantic restoration by Robert Brown of Edinburgh in the 1840s, which gave the building its Victorian battlements and ornamental turrets all along the roof line.

As with many eighteenth-century Highland landlords, estate improvement involved investment in the policies and gardens around the main residence or country seat, and so it was with Dunvegan. The Castle Gardens were originally laid out in the later eighteenth century. Although the warming effect of the Gulf Stream helped, immense perseverance and expertise lies behind the mature gardens we see on this map of the 1870s: like much of the rest of Skye, this was formerly a largely treeless and windswept moorland.

The first edition 25 inch to the mile Ordnance Survey maps capture the detail of the Castle and gardens with informative hand-colouring. The red buildings are of stone or brick, while the blue-hatched buildings are glasshouses. The W by the Castle marks its famous freshwater well – a factor in the Castle's former impregnability – and we can see the layout of the formal gardens, including the Round Garden and Walled Garden (further south), and the Croquet Garden to the east. The surrounding trees are a mixture of broad-leaved and various conifers, with a network of paths to explore and enjoy.

Source: Ordnance Survey, 25 inch to the mile, *Inverness-shire (Skye), Sheet XXI.7 (Duirinish Parish)* (surveyed 1877, published *c.*1881).

FIGURE 6.15
A detail of Mount Stuart on Bute, from the Roy Military Survey of Scotland. The islands were largely excluded from the Survey, 'excepting some lesser ones near the coast', as Roy noted. While the primary intention behind the survey was to depict selected topography of military value, the close attention to and superb portrayal of designed grounds around selected country seats may also reflect the patronage and aristocratic support for the work. The map exemplifies the characteristic features of Roy's militarily minded cartography: topography is shown with the use of hachure lines and shading (the darker and closer together the lines the steeper the slope – contour lines had not yet been invented). Managed woodland and estate policies are shown, the larger estate buildings picked out in red. Agricultural land is hinted at by the use of stippled shading.
Source: William Roy, Military Survey of Scotland, 1747–55. © The British Library Board

Island of Askog
Lochley
Loch of Askog
Tayvalley
Mid Askog
Lower Askog
Askog
Woodend
Loch Fad
Bardaroch
Blackburn
Gronach
Manse
Barmore
Barnald
Ambrisbeg
Birgadale Knock
Quien Loch
Scoulack
30
Upper Scoulack
Kirk of Kingarth
Scoulack Burnfoot
Birgadale-na-Chruc
McBrides Farm
Ambrismor
Muirbutt
Mount Steuart
Gallochan
Whinbutts
Kirrylamond
Killeatton butt
Killeneanach Steuart
Butt Lieny
Coulevan Butt
Killeatton-mor
Langlebunoch
Coulevan
Kerrymeanach
Broughag
Killeatton beg
Quochag
Middle Kirk
Killeatton mill

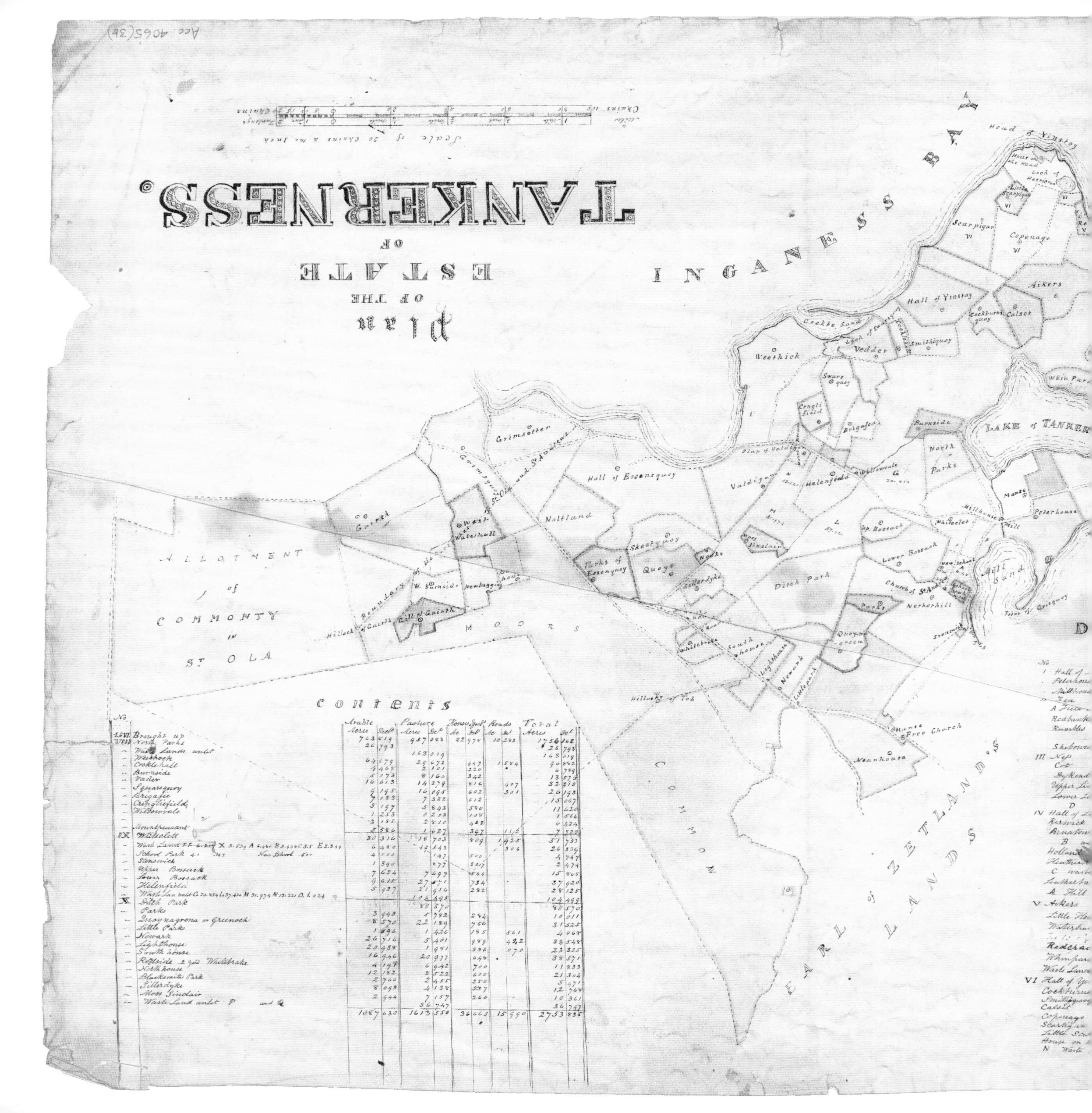

Contents

No.		Arable Acres	Dec.	Pasture Acres	Dec.	Houses, Yards Ac.	Dec.	Roads Ac.	Dec.	Total Acres	Dec.
I. & VI.	Brought up	763	819	957	083	22	974	10	282	1754	162
VIII	North Parks	26	793							26	793
–	Waste Lands unlet			163	019					163	019
–	Weethick	64	179	29	672		447	1	584	96	882
–	Cocklehall	4	468	2	101		220			6	789
–	Burnside	5	073	8	160		342			13	575
–	Vader	16	613	14	379		816		407	32	215
–	Squarsquoy	9	195	16	095		602		301	26	193
–	Brigafie	7	133	7	322		612			15	067
–	Cringliefield	5	197	5	843		580			11	620
–	Willowvale	1	253	0	203		108			1	564
–		3	182	2	810		432			6	424
–	Mountpleasant	5	586	1	627		397		112	7	722
IX	Whiteclett	30	316	18	703		809	1	925	51	753
–	Waste Land FF 6.217 X 5.539 A 4.440 B 3.934 C 3.5 E 2.349	6	480	19	543				306	26	329
–	School Park 4.1 .147 New School .500	4	100		147		500			4	747
–	Stenswick	1	390		877		2017			2	474
–	Upper Bossack	7	624	7	697		544			15	865
–	Lower Bossack	9	615	27	571		734			37	920
–	Helenfield	5	927	21	916		282			28	125
–	Waste Lan unlet C 20.854 L 37.424 M 31.974 N 13.221 O. 1.024			104	495					104	499
X	Ditch Park			80	570					80	570
–	Parks	3	943	5	782		284			10	011
–	Quoynagreena or Greenock	8	570	22	189		760			31	525
–	Little Parks	1	896	1	426		185		561	4	068
–	Newark	26	716	5	401		989		442	33	548
–	Lighthouse	20	938	1	981		336		070	23	325
–	South house	16	946	20	977		648			38	571
–	Rottside .2 and Whitebrake	4	197	6	942		700			11	833
–	Northhouse	12	182	8	522		600			21	304
–	Blacksmiths Park	2	766	2	455		250			5	471
–	Sillerdyke	8	093	4	138		537			12	768
–	Moss Sinclair	2	944	7	157		260			10	361
–	Waste Land unlet P and Q			36	747					36	747
		1087	630	1613	550	36	665	15	990	2753	835

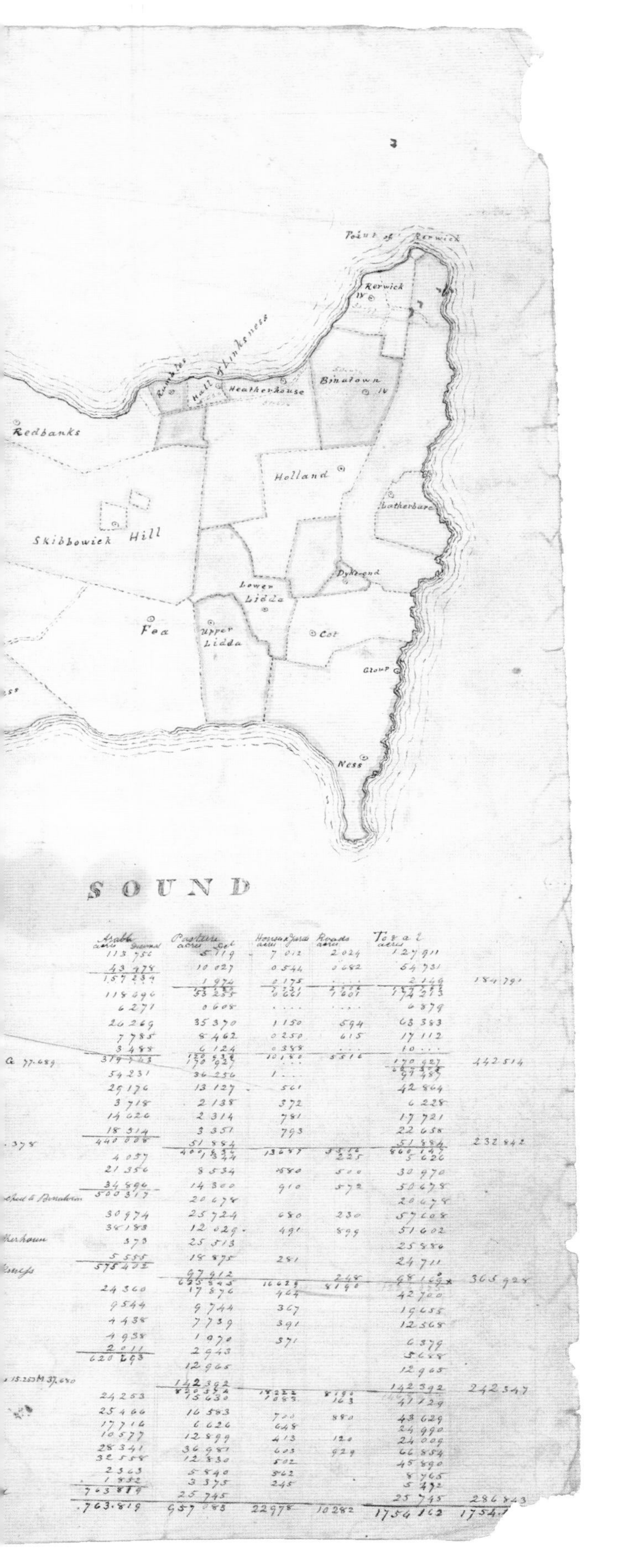

James Johnston of Coubister, *Plan of the estate of Tankerness* (*c.*1840?).

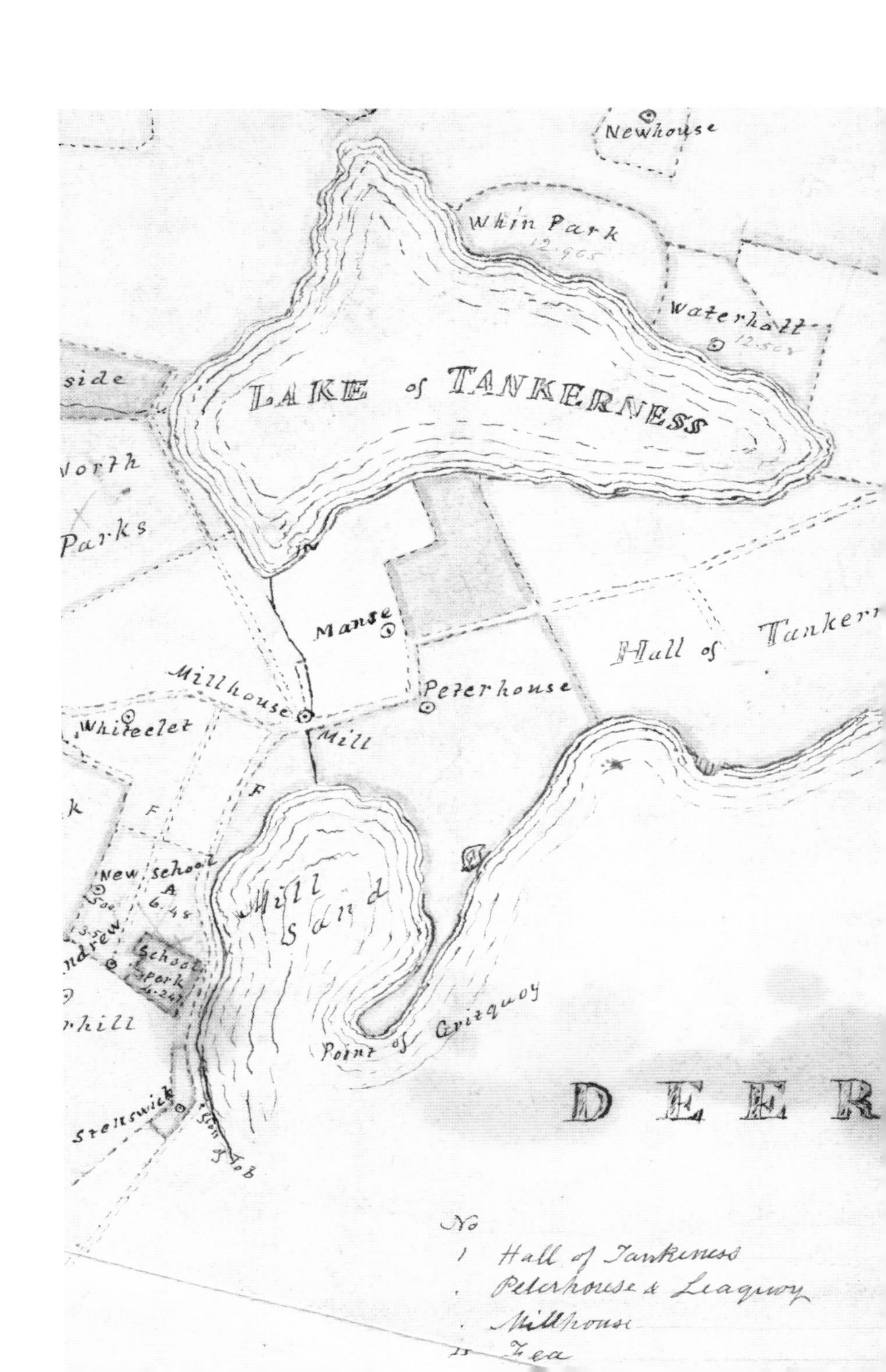

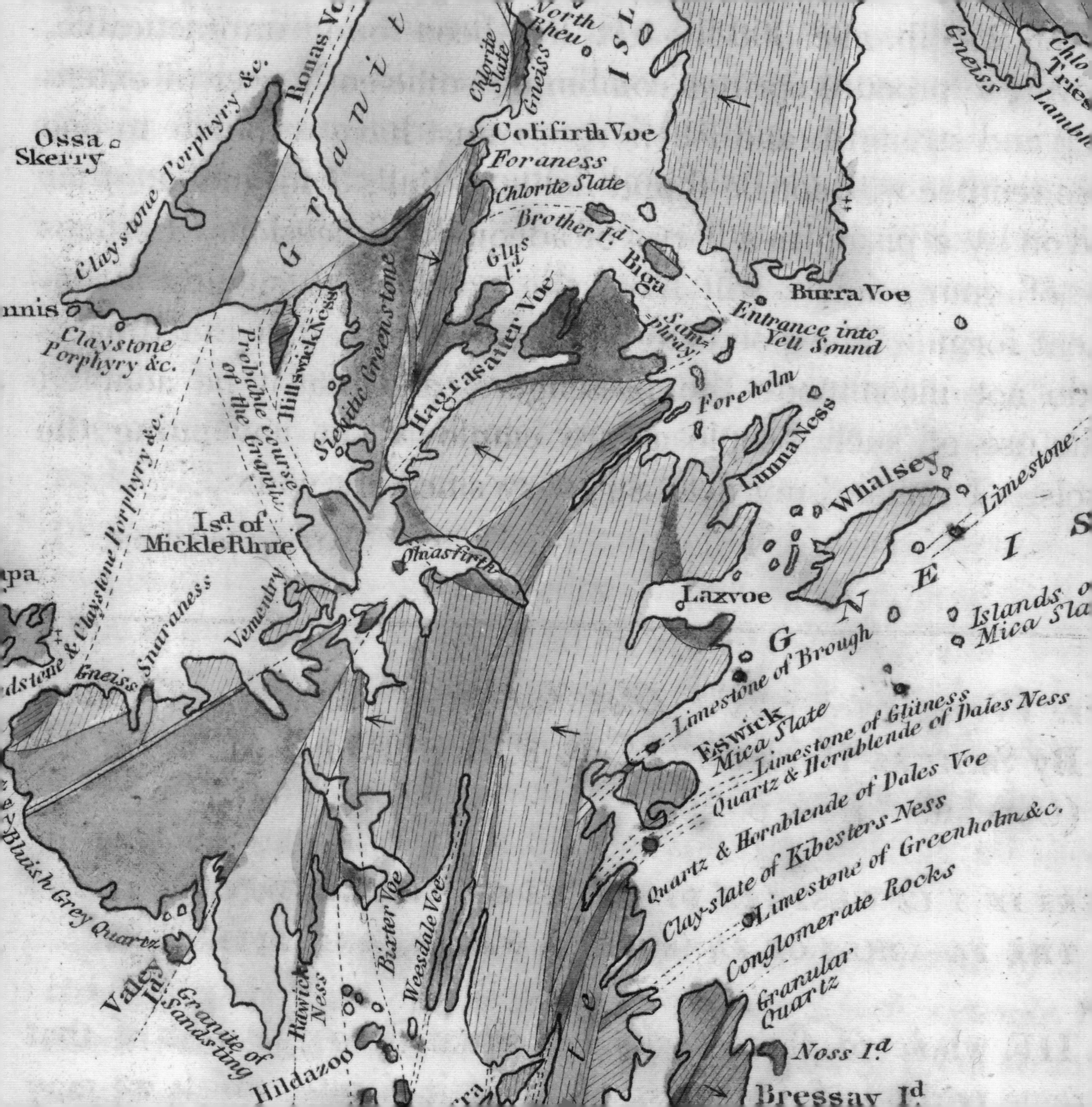

Ossa Skerry
Claystone Porphyry &c.
Ronas Voe
Granite
North Rheu
Chlorite Slate
Gneiss
Colifirth Voe
Foraness
Chlorite Slate
Brother Id.
Glus Id.
Biga
Burra Voe
Entrance into Yell Sound
Samsphray
Hagrasatter Voe
Sienitic Greenstone
Hillswick Ness
Probable course of the Granite
Claystone Porphyry &c.
Foreholm
Lunna Ness
Whalsey
Limestone
Isd. of Mickle Rhue
Olnasfirth
Laxvoe
Gneiss
Islands of Mica Slate
Vemenbry
Snaraness
Limestone of Brough
Eswick
Mica Slate
Limestone of Glitness
Quartz & Hornblende of Dales Ness
Quartz & Hornblende of Dales Voe
Quartz
Clay-slate of Kibesters Ness
Limestone of Greenholm &c.
Conglomerate Rocks
Granular Quartz
Noss Id.
Bressay Id.
Bluish Grey Quartz
Valley Id.
Granite of Sandsting
Rawick Ness
Bixter Voe
Weesdale Voe
Hildazoo

CHAPTER SEVEN

EXPLOITING

The exploitation of natural resources on Scotland's islands, and in its seas and waters, has been of immeasurable importance for the long-term survival of island life, and the independence and initiative of islanders have often rested on this understanding. The earliest inhabitants practised fishing, and later basic agriculture, but over time new types of exploitation, especially of underground mineral resources, have also become important. In recent centuries, the growing influence of commercial markets for natural resources has increased the breadth, quantity and range of natural resource exploitation in the islands, redefining island products as commodities for extraction and sale. Globalised economic activities, often organised and managed from far away, have also redefined notions of place and home and had profound local consequences for island life.

Maps are closely associated with the processes of exploitation, as a common motive behind mapping has been the desire to manage natural resources: think of marine charting in this light, or geological mapping under land and sea. As well as locating natural resources, and describing their qualities, maps are also essential tools in allocating rights and ownership. The rise of cadastral mapping – that is, of maps specifically showing boundaries and ownership of parcels of land – often reflected and drove the need to distinguish the possession of geographical resources. At larger and more detailed scales, maps and plans have often been used to manage resource exploitation: think of mines, oilfields, or timber extraction from woods and forests. At smaller scales, maps have been used to distribute resources to customers. Because resource exploitation is often contested, maps have been produced either by proponents of or objectors to particular proposals, harnessing the persuasive power of mapping. We can distinguish resource exploitation maps – maps created with a specific purpose to exploit a resource – from more general topographic maps, such as those produced by Ordnance Survey, that depict resource exploitation as part of a broader remit. This chapter looks in turn at these maps and the subjects they show, grouping them thematically: first, at exploiting the seas, then at exploiting the rocks, and, finally, at exploiting the land.

Opposite. Detail from Samuel Hibbert, 'Geological Map of the Shetland Isles', *Edinburgh Philosophical Journal* 2 (1820).

Exploiting the seas

It is possible to trace the rise and fall of the great maritime nations of Europe through their sea charts. In the sixteenth and seventeenth centuries, the Dutch had a virtual monopoly on catching herring in the North Sea, and their huge fishing fleets, often numbering thousands of boats, regularly swept by the coasts of Shetland. The Dutch invention of gibbing in the late fifteenth century – preserving partially gutted herring in barrels with salt – allowed catches to be preserved at sea, and enabled longer voyages to follow the fish. The main catches of North Sea herring usually came from their wintering grounds near Norway, but, during the summer, the fish could be found feeding and spawning over a wide area near to Shetland and Orkney and down Scotland's east coast. The Dutch successfully adapted a variant of the original Viking longship, popularly known as the *bŭza*, as an open-topped cargo boat, and the Dutch *haringbuis* or herring buss was a 'factory vessel' ideally suited to storing large drift nets, barrels and salt. Together with their acknowledged expertise in catching and curing fish, the Dutch developed an onshore processing industry, allowing export of the salted herring all over Europe, and developed related trades of making nets, shipbuilding, navigation and drawing charts.

Lucas Jansz. Waghenaer published the first European sea atlas in 1584 – on the basis of information from his own experience as a seaman as well as from other Dutch mariners. From his work, the anglicised term 'waggoner' came to be used for a standard volume of sailing charts. Waghenaer borrowed some aesthetic elements from earlier portolan charts, but introduced several innovative features of his own – occasional depths in fathoms, profiles of the land from the sea, a three-dimensional representation of the coast, and an indication of safe harbours. These were later adopted and used by other chart makers, and are still standard features on charts and accompanying pilot guides today.

Given the absence of credible rival surveys, and intense competition among Dutch map and chart publishers, Waghenaer's coastal outlines reappear again and again on later Dutch charts for the following century, often with fresh discoveries and information (fig. 7.1). Sea atlases by Janssonius and Blaeu in the early seventeenth century closely followed Waghenaer's outlines, and even as late as the 1670s John Seller, Hydrographer to King Charles II, issued practically the same charts as Waghenaer (see chapter 4). As Seller's *The English Pilot* and *Atlas Maritimus* were reprinted through the eighteenth century, Waghenaer's coastal outlines can thus be traced over at least 200 years.

Attempts to encourage British fisheries were only intermittent and faltering until the late eighteenth century, and it is from this period that we can note a significant growth in the number of British private chart publishers. Scotland's islands' coastlines and seas were only comprehensively mapped through the state-funded work of Ordnance Survey and the Hydrographic Office of the Admiralty. The island of Sula Sgeir, for example, some 40 miles north of the Butt of Lewis/Rubha Robhanais, appears on several early Dutch charts of the north-west of Scotland, but it is only with the detailed work of Ordnance Survey in the 1870s that we have the first accurate depiction of its real shape and size (fig. 7.2a–c). Captain William Henry Otter of the Hydrographic Office, whose surveying work focused on the Hebrides and north coast of Scotland, only included Sula Sgeir at a smaller scale, with the result that the earliest detailed survey of Sula Sgeir by the Hydrographic Office was by Captain Moore in the Surveying Ship *Research* in 1897. For centuries, the Lewis/Leòdhas men from the Port of Ness/Port Nis district have sailed to Sula Sgeir in the late summer or autumn as part of the annual cull of 'guga', or young gannet. In a task involving considerable dexterity and skill, the men work in pairs over two weeks, catching the young gannets from their nests with poles or loose ropes, then killing the birds with a blow to the head. Admiralty charts and Ordnance Survey maps are also of immense value today for their successive editions, allowing change over time to be visualised. Even as Scotland's expertise with fishing increased in the eighteenth and nineteenth centuries, the fish were notoriously unpredictable in their number and location, resulting in regular boom and bust cycles (fig. 7.3).

a.

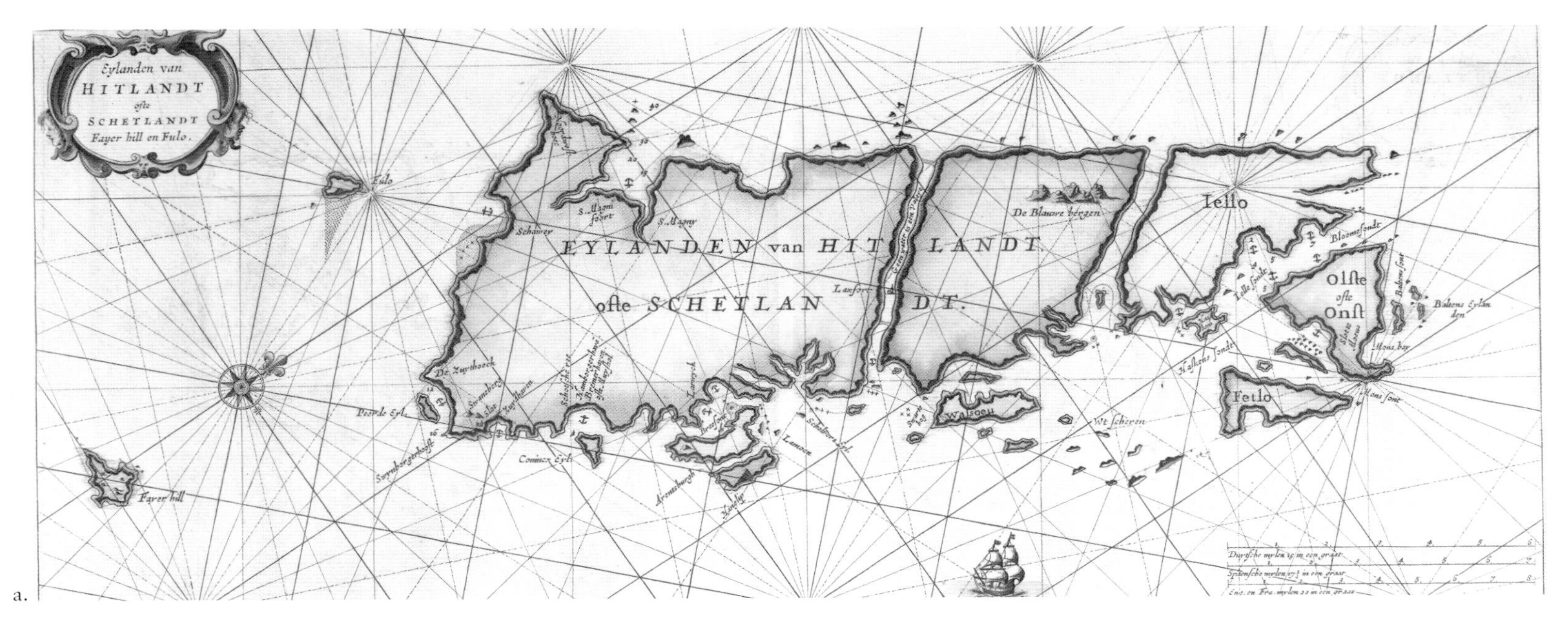

b.

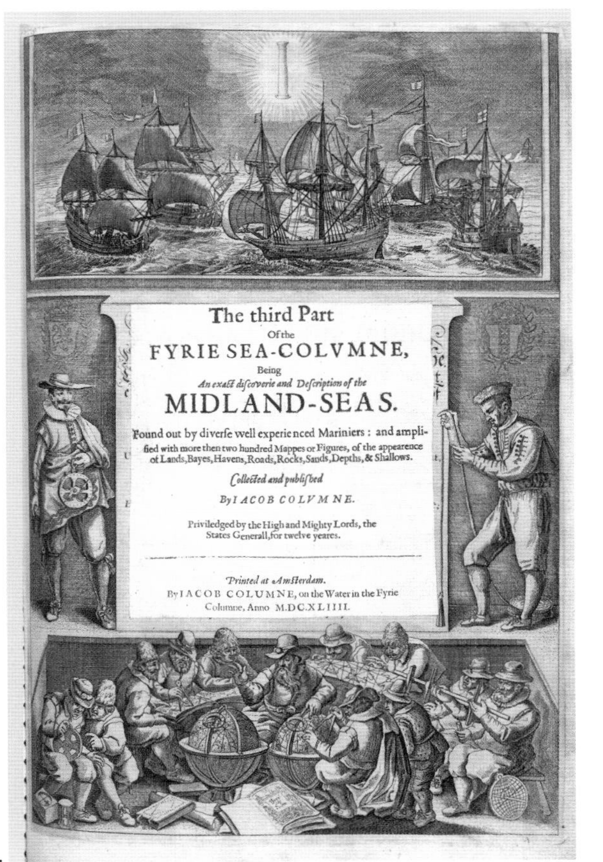

FIGURE 7.1

(a) This chart, by Jacob Aertsz Colom (1600–1673), a printer and bookseller who settled in Amsterdam from 1622, shares the outlines and place names of the earlier Waghenaer family of Dutch charts. The chart shows how some original Norse names were rendered into Dutch (such as the 'Ve Skerries' as 'Wt Scheren' or 'Balta Sound' as 'Baltens Sont'), as well new Dutch names for places (such as 'De Blaeu bergen' for St Ronas Hill, 'De Zuythoeck' for Sumburgh Head). Names such as Hamburgerhaven/Bremerhaven indicate the importance of the Dutch fisheries and trade in Shetland at this time.

(b) The chart shown in (a) was included in a pilot volume *De Vyerighe Colom* (1632), which was translated into English as *The Fyrie Sea Columne* (1633) – a pun on the author's name, as well as on the name of his house on Mandemakersteeg. Colom's title (which he retained in his later editions of the pilot: *The Upright Fyrie Colomne* (1648), *The True and Perfect Firie Colom* (1668)) was also an allusion to the biblical 'pillar of fire' which guided the Israelites out of Egypt. The splendid title page shows this brilliantly illuminated pillar guiding several ships in full sail. Below, the cartouche is flanked by two men holding an astrolabe on the left and, on the right, a plumb line. Further nautical surveying instruments, including dividers, a cross staff and astrolabes can be seen at the foot of the page, their expertise distilled into a copy of the *Vyerighe Colom*, open at the title page in the centre. Colom also chose in his preface to criticise the pilot book of his main rival, Joan Blaeu, resulting in a furious response from Blaeu in a hastily prepared *Havenwyser van de Oostershce, Noordshce en Westershce Zeen* (1634): 'Most experienced and trialed seamen, no doubt you will have heard by hearsay about a certain chart-book . . . recently published under the name of the Fiery Colomne. A certain person, who calls himself the author of that book, has undertaken to copy our Chartbook the Sea Mirror. In order that those who buy it may think that they are obtaining something new and particular, he has brought it out in a different size and given it a new title, in which he unabashedly and falsely says that the errors and mistakes of the Light and Mirror have been plainly demonstrated and corrected.' Blaeu went on to detail the new errors by Colom, but his criticisms were overstated. In fact, Colom's chart of Shetland gathered significant extra information over and above the charts of Janssonius and Blaeu, with the islands of Unst, Fetlar and Walsay named for the first time, as well as Lerwick. *The Fiery Colomne* continued to be published for the next thirty years.
Source: (a) Source: Jacob Colom, *Eylanden van Hitlandt oste Schetlandt Fayer hill en Fulo* (1632). (b) Jacob Colom, title page from *The Lighting Colomne or Sea-Mirrour, containing The Sea-Coasts of the Northern, Eastern and Western Navigation* (1644).

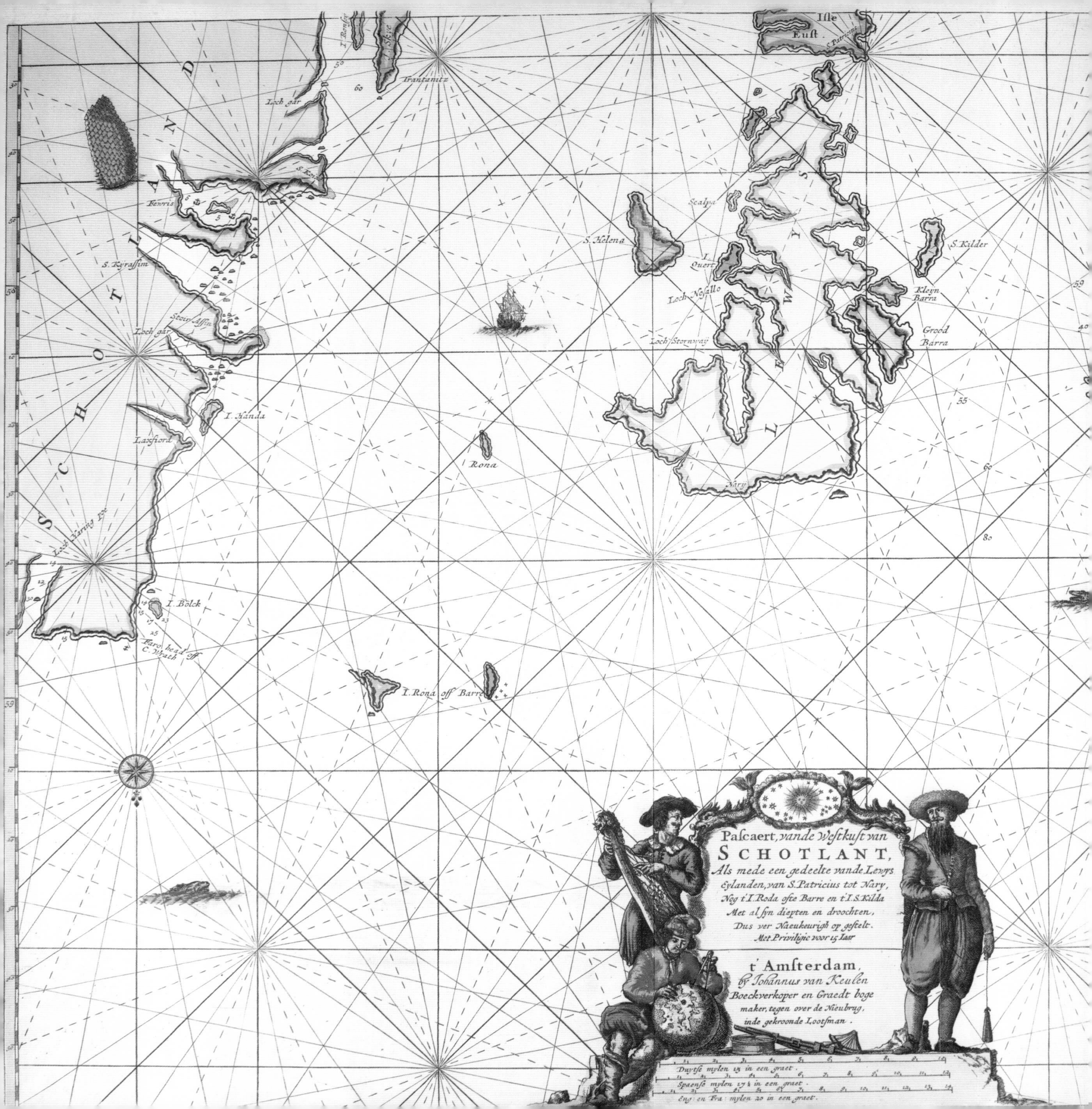
Paſcaert, vande Weſtkuſt van
SCHOTLANT,
Als mede een gedeelte vande Lewys
Eylanden, van S. Patricius tot Nary,
Nog t'I. Roda ofte Barre en t'I. S. Kilda
Met al ſyn diepten en droochten,
Dus ver Naeukeurigh op geſtelt.
Met Privilegie voor 15 Iaar
t' Amſterdam,
by Johannus van Keulen
Boeckverkoper en Graedt boge
maker, tegen over de Nieubrug,
inde gekroonde Lootſman.
Duytſe mylen 15 in een graet.
Spaenſe mylen 17½ in een graet.
Eng: en Fra: mylen 20 in een graet.
Isle Eust
S. Patricius
I. Skye
I. Ronsey
Trantanitz
Loch gar
S. Kona
Fenris
S. Kyrassim
Stoir Assin
Loch gar
I. Hinda
Laxfiord
Loch Haring Eye
I. Bolck
Faro head off C. Wrath
I. Rona off Barre
Rona
S. Helena
Scalpa
I. Quert
Loch Nosallo
Loch Stornway
Nary
S. Kilder
Kleyn Barra
Grood Barra
SCHOTLAND
LEWIS

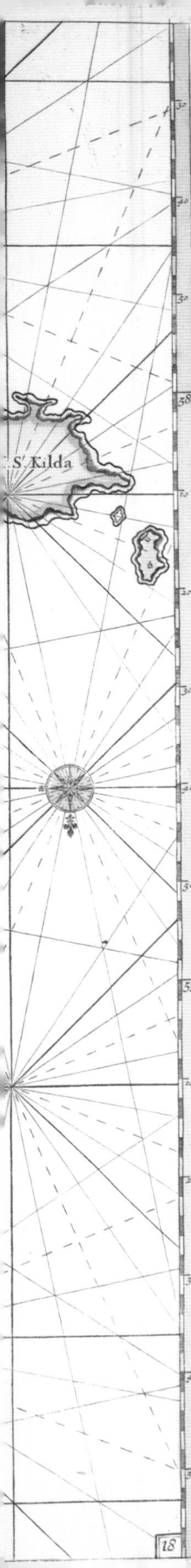

a.

FIGURE 7.2

The rocky island of Sula Sgeir is one of the remotest of the British Isles. Unlike its neighbouring island of Rona/Rònaidh, eleven miles to the east, Sula Sgeir has never been permanently inhabited, although there are traces of temporary shelters or bothies, such as Taigh Beannaichte on the east headland. Its name, deriving from the Old Norse words *súla* (gannet) and *sker* (skerry) signify the main human interest in the island, then as now. The island was a significant feature for mariners, both as a hazard and a navigational aid, and appears on many early sea charts, such as (a). It was only during the nineteenth century through the detailed work of Ordnance Survey (b) and the Hydrographic Office (c) that its proper shape and detail emerges. At this one-inch to the mile scale, Ordnance Survey's expert engraving with 'waterlining' – parallel concentric circles around the islands – creates a mesmeric effect around the islands, although the depths in fathoms and annotations to show the sea bed on the Hydrographic chart (crl coral, r rock, s sand, sh shells) were of more value for mariners.

Joan Blaeu, in his *Atlas Novus* (1654), describes the island thus, in a text probably deriving from George Buchanan: 'Sixteen miles from this to the west lies the island of Suila Sgeir, one mile long; it produces no grass, nor even heather. Only black cliffs rise up, of which some are covered with black moss. Sea-birds lay and hatch their eggs everywhere there. The nearest people from the island of Lewis sail here for those not yet mature enough to fly, and devote about a week to collecting them, until they fill their boats with the wind-dried flesh and feathers.' Everything necessary for such expeditions had to be taken with them, including drinking water, and the entire catch had to be collected and stored for safe return to Lewis/Leòdhas. Although the guga is considered by some to be a great delicacy, it is an acquired taste!

Source: (a) Johannes van Keulen, *Pascaert vande westkust van Schotlant . . .* (*c*.1712). (b) Ordnance Survey, One-Inch to the mile, *Scotland, Sheet 113* (surveyed 1873, published 1882). (c) Hydrographic Office, *Thurso Bay to the North Minch*. Admiralty Chart 1954 (revised 1897, published 1900).

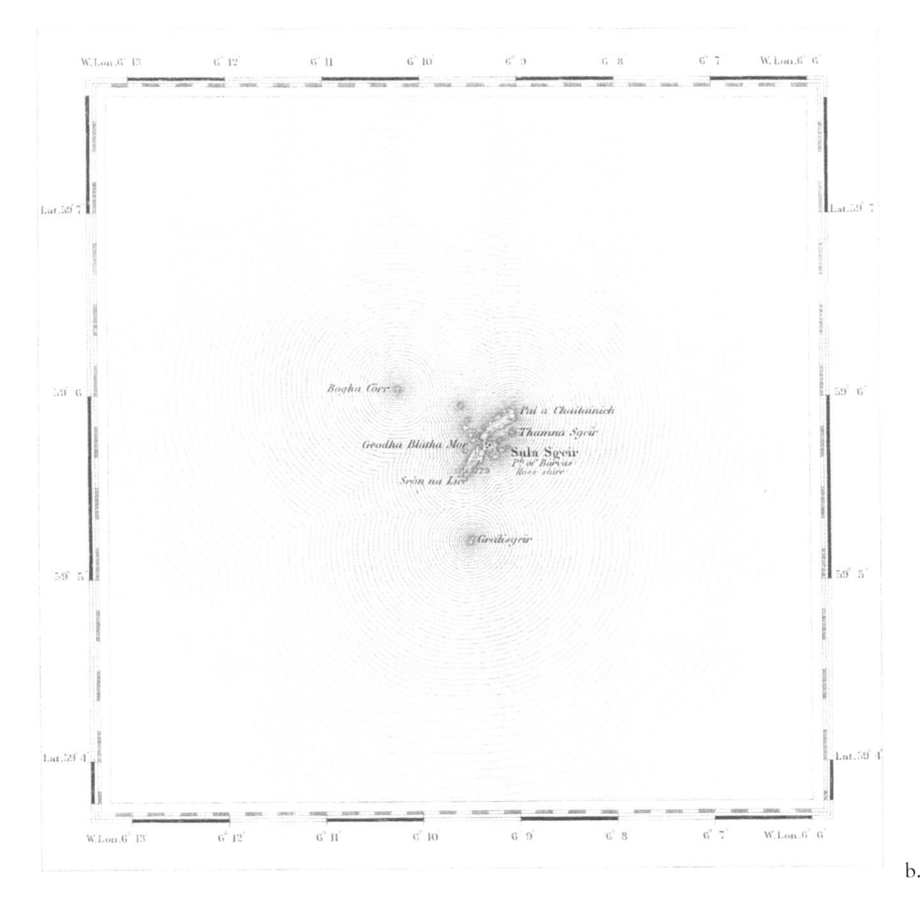

b.

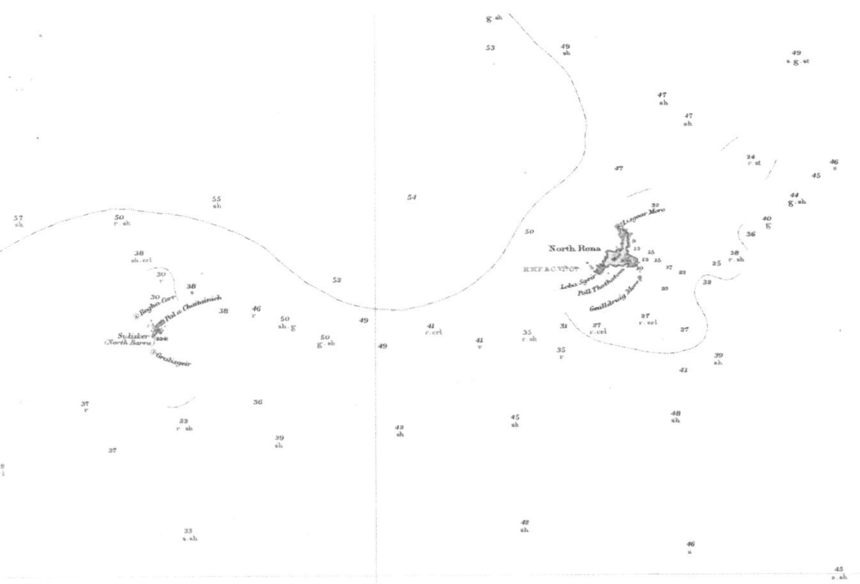

c.

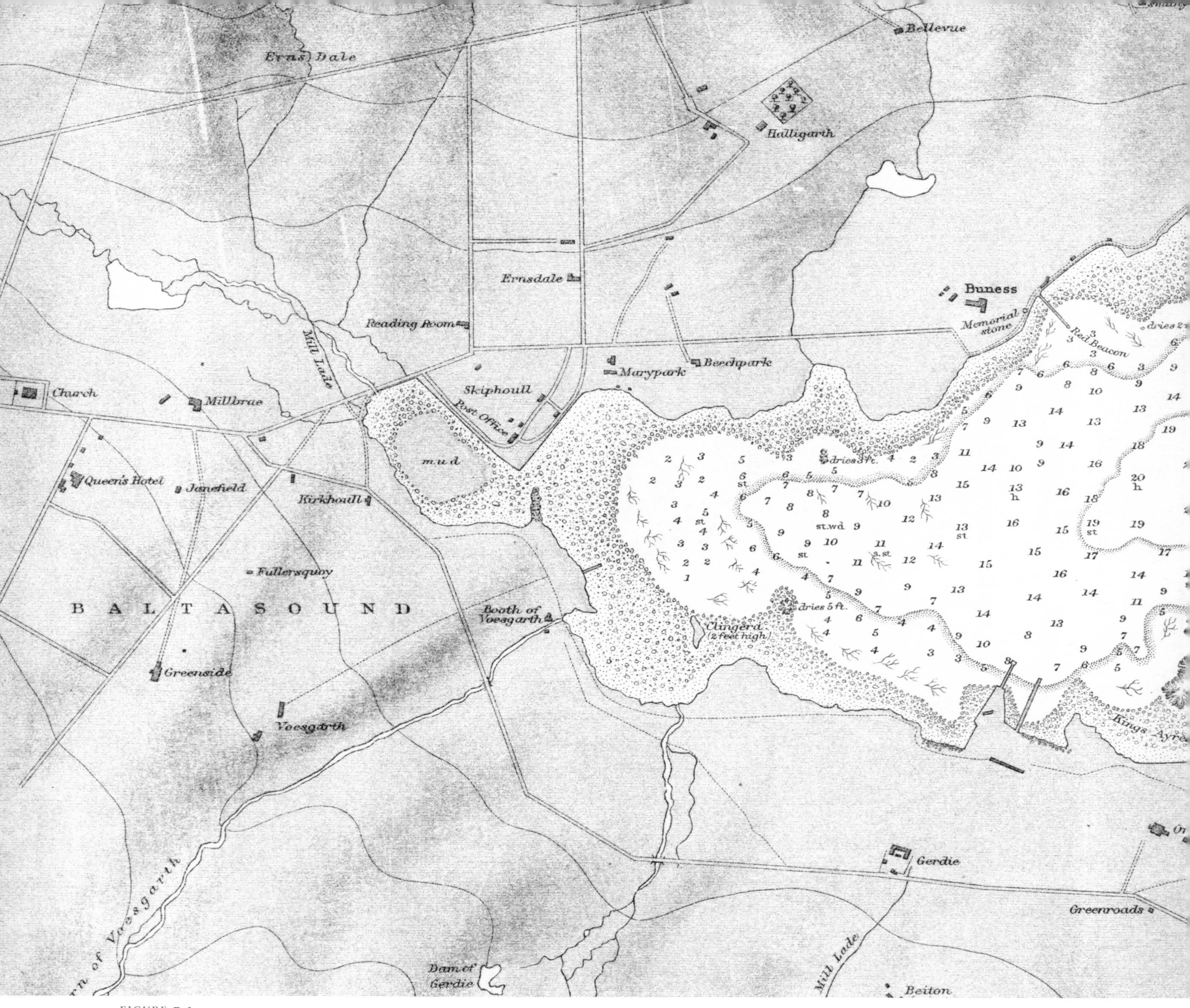

FIGURE 7.3

This Admiralty chart shows Balta Sound, near the northern end of the northernmost Shetland Island of Unst, at the height of its prosperity during the herring boom of the early twentieth century. Balta Sound had been used by the whaling industry from the early 1800s, and the curing of seal skins also took place, as did boat building, but, from 1880 to 1925, it was witness to an incredible herring boom. During the 'season', the resident population of 500 swelled to over 10,000, many living in temporary wooden shacks around the coast. Over 600 fishing boats operated from the forty-six or more herring stations spread around Balta Sound and Balta Island, which can each be clearly seen here, along with piers and processing sheds. Note the 'Fish Manure Factory' with its conspicuous chimney (marked for mariners) to the south of the Sound, and a little further to the east, a Mission Church for the many new souls to be saved. In a typical summer, about 250,000 barrels of herring would be cured, and in the peak

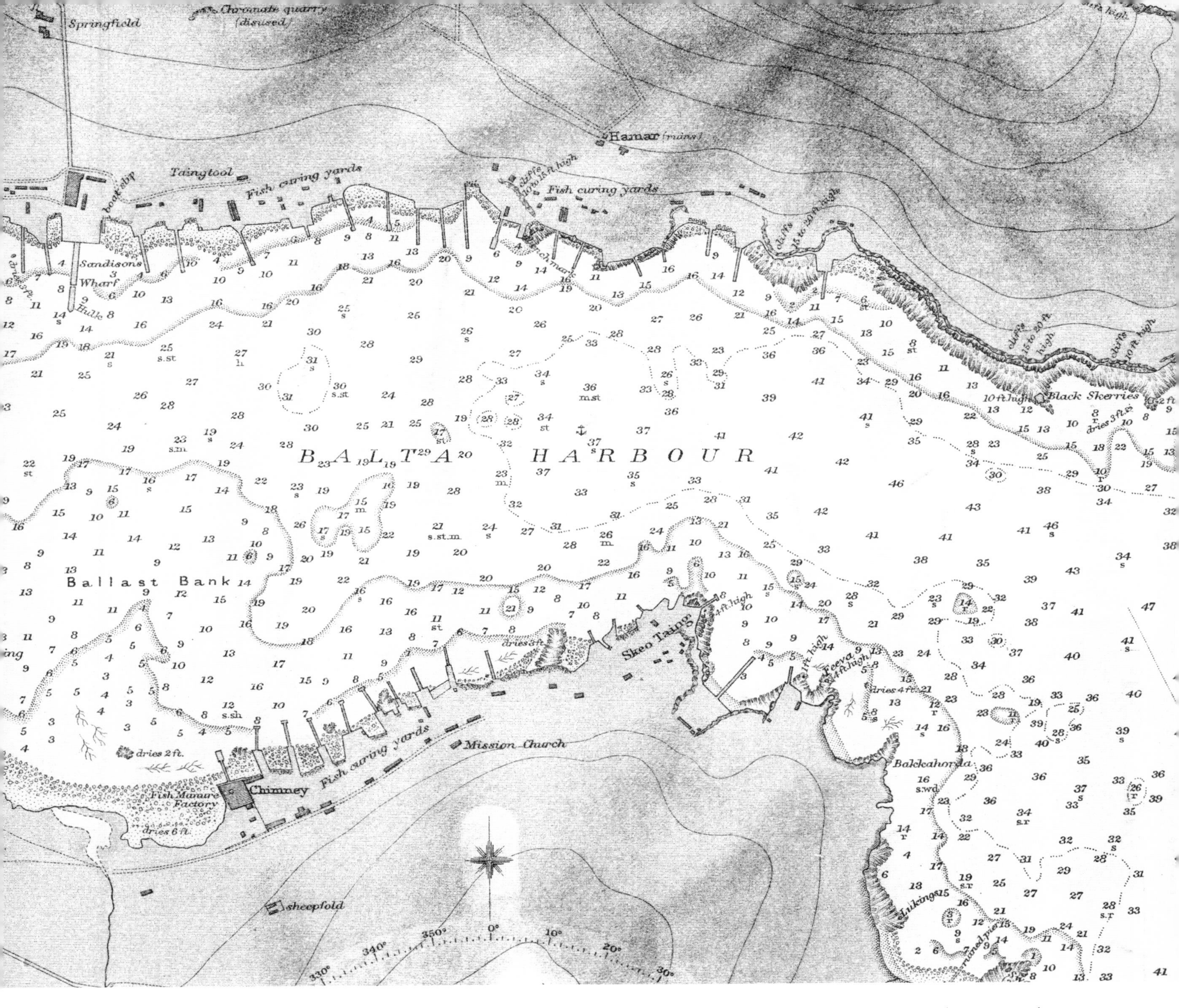

year of 1905, some 700 boats brought back a total catch of 28,000 crans or 30 million herring. A steamer service operated from 1877, carrying passengers to and from Lerwick via Mossbank on Shetland's mainland.

Just as unexpectedly as they had arrived, the herring left. By 1939, most of the herring stations had closed. Virtually all the piers and curing houses were gone by then and related industries such as boat building ceased by the 1980s. The large St John's Church (on the far left-hand edge of the map), built in the nineteenth century to accommodate 2,000 people was partly demolished in 1959 to create the more modest church of today. *Source*: Hydrographic Office, *Balta Sound*. Admiralty Chart 3643 (revised 1912, published 1914).

Although nations and individual state enterprises have appropriated and exploited the seas for centuries, only in recent decades have significant tranches of the sea bed been appropriated for economic use. In the twentieth century, with the growing technical capacity to exploit deep-sea resources, Exclusive Economic Zones were developed, extending from the edge of the territorial waters to 200 nautical miles offshore. These were formally ratified by the United Nations Convention on the Law of the Sea in 1982. Following successful exploratory wells in the North Sea in the 1940s and 1950s, the five countries bordering on the North Sea divided the sea bed into spheres of national interest as far as they could offshore by the 1960s.

The exploitation of North Sea oil and gas has had a profound transformative effect on the economy and society of the Northern Isles from the 1970s (fig. 7.4). The Sullom Voe oil terminal in Shetland, constructed between 1975 and 1981, became one of the largest oil terminals in Europe, and in 2015 nearly one quarter of all UK crude oil was landed there. In 2015, the smaller Flotta terminal in Orkney landed about 7 per cent of UK crude oil from the Piper, Claymore and Tartan fields. In the last thirty years, the extent of these oil and gas fields has continued to expand geographically, most recently to the north-west of Shetland, with new oil and gas pipelines from the Clair, Laggan and Tormore fields. In spite of continuing concerns over the expense of operations, the decommissioning of former oil wells and a declining annual output from the late 1990s, the North Sea was still, in 2015, the most active offshore drilling region in the world with 173 rigs in production.

Exploiting the rocks

One of the few facts we have relating to the late sixteenth-century chorographer and map maker Timothy Pont is that he was commissioned in 1592 by Lord Menmuir 'Master of the Metals' to undertake a mineral reconnaissance of Orkney and Shetland. It is thought that his mapping of the Northern Isles, published later by Hondius (1636) and Blaeu (1654), dates from this time. The more systematic surveying of islands' geology is largely a nineteenth-century affair. Although proprietors of Scottish islands may have looked with envy at other landed estates, particularly those with rich coal seams in parts of Fife, the Central Belt, Midlothian and Ayrshire, the islands boasted other valuable minerals. Limestone was quarried on Lismore, Shuna and Tiree (fig. 6.1), lead was mined on Islay and Coll, granite was extracted from Ailsa Craig and Berneray/Bernaraigh, and iron ore from Raasay.

Proprietors' interests in exploiting the natural wealth of their islands was not confined to minerals. Several estate maps confirm a broader interest in fishing, farming and quarrying. William Bald's original survey of Harris/Na Hearadh in 1804 and 1805 provides a detailed record of attempts to exploit the island's agricultural, fishing and mineral potential. In common with many other islands in the Outer Hebrides in the late eighteenth century, the MacLeod clan chiefs, with family ties to their tacksmen and tenants, sold their estates to new landlords, many of whom had interests and investments elsewhere. In 1779, Harris/Na Hearadh was sold by MacLeod of Dunvegan to Captain Alexander MacLeod, who had made his money in the East India trade, and, after retiring to Harris/Na Hearadh in 1782, he invested heavily in improvements. From John Knox's visit in 1786, there are details of MacLeod's restoration of St Clement's Church in Rodel/Roghadal, the construction of a parochial school and inn, and (as shown on the 1829 copy of Bald's map, fig. 7.5a) the laying-out of good cart roads from the harbour to the village and beyond to An t-Ob (later Leverburgh). Although it seems the trees did not make a lasting impression on the landscape, we learn that 'he has done something in the planting way, and he finds that the hazel and sycamore thrive best'. MacLeod's main focus was encouraging fisheries, deepening the channel into Rodel/Roghadal Harbour (marked as Basin above the island of Vallay by Bald), constructing two new quays and a graving bank, a new store house and a textile mill. His success with fishing may perhaps have been overstated by Knox, but in March and April 1786, he was reported to have caught '4,400

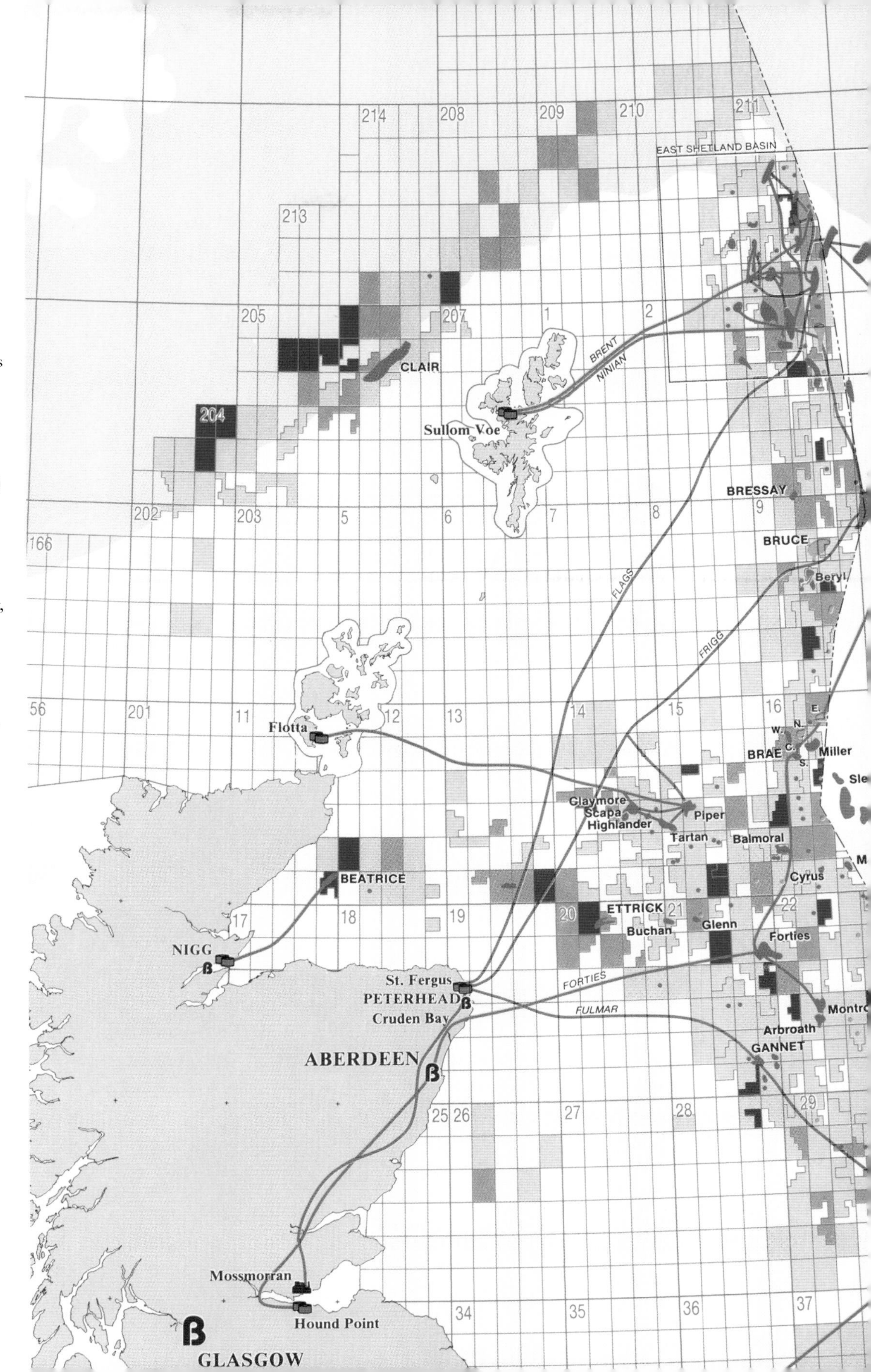

FIGURE 7.4
This map showing UK continental shelf licence interests up to March 1986 shows not only the limits of UK sea-bed rights, bordering those of Norway, Denmark, Germany and the Netherlands, but also how these countries licensed these rights on to other companies to exploit. Essentially, the UK continental shelf was divided into quadrants of 1 degree latitude by 1 degree longitude, with each quadrant then subdivided into 30 blocks measuring 10 minutes of latitude by 12 minutes of longitude. Companies were invited to bid for exploiting these quadrants on a periodic (usually annual) basis. Note how the Lambert Conformal Conic Projection (named after the eighteenth-century Swiss mathematician Johann Heinrich Lambert) stretches these square quadrants and blocks in a north–south direction.

As well as showing oil pipelines in green and gas pipelines in red, the map shows different private and public ownership of licence rights. Britoil was a short-lived company, formed through the privatisation in the 1980s of the former British National Oil Corporation (BNOC), founded in 1975. The BNOC attempted to stabilise prices and investment, allowing a mixed economy of state control and private interests along similar lines to other national oil industries. By 1986, however, its blue-licensed areas were already overshadowed by areas under multinational control (shown in purple), hinting at future policy directions of the prevailing British Conservative Party. Britoil was acquired by BP in 1988, and the North Sea oil industry has been dominated by international companies and the market ever since. That said, successful negotiations by the Councils of Shetland and Orkney over the landing of oil have allowed a very significant boost to public sector incomes, investment and employment in the islands.
Source: Britoil, *Britoil U.K. Continental Shelf Licence Interests* (20 March 1986).

a.

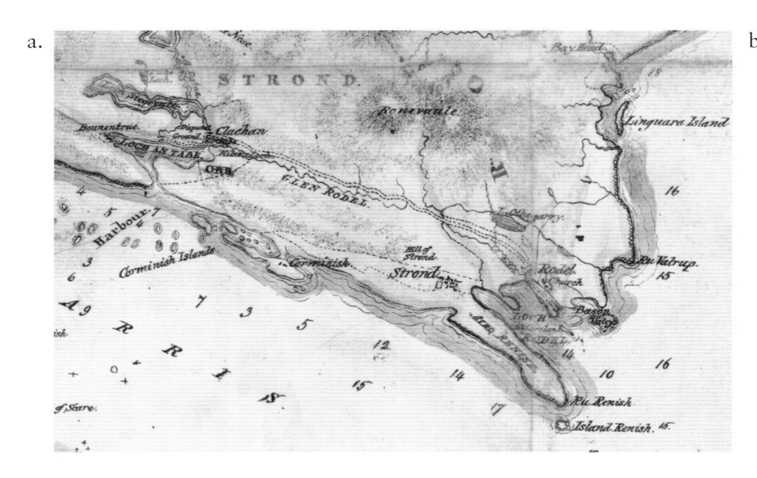

b.

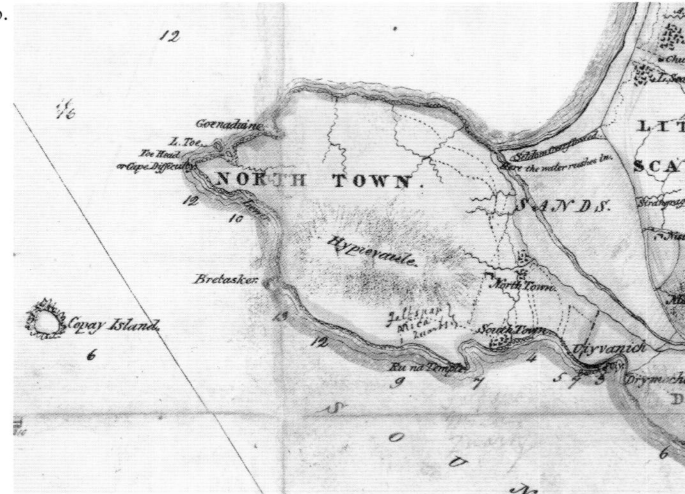

c.

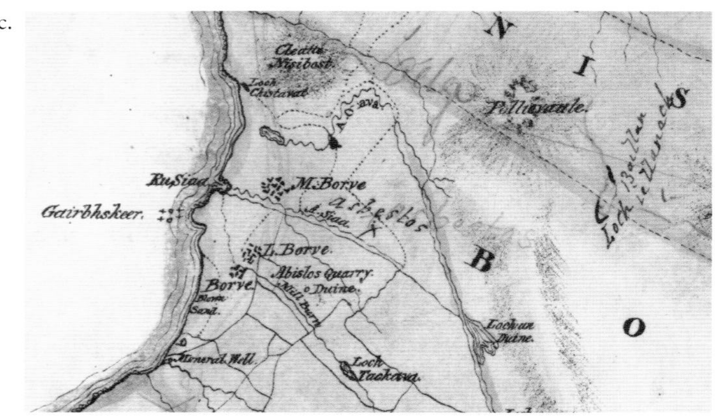

FIGURE 7.5
These details from William Bald's *Map of Harris* illustrate different attempts to exploit its potential. Detail (a) shows the laying out of roads from Rodel Harbour to An t-Ob carried out under Captain Alexander MacLeod in the 1780s. There are also later annotations on the map, made after 1829, referring to possible mineral resources: 'Feltspar, Mica, Quartz' near Northton (North Town) (b) and 'asbestos' near Finsbay (c).
Source: William Bald, *Map of Harris* (surveyed 1804–05, published 1829).

large cod and ling; 4 or 500 skate; innumerable quantities of dog fish, large eels, and many boat loads of cuddies'. MacLeod also took soundings for 30 miles around St Kilda (part of his estate), and, as a result, correctly calculated that the fisheries extended beyond this range.

Alexander MacLeod's son, Alexander Hume MacLeod – to whom Bald's map is dedicated – inherited Harris/Na Hearadh in 1790. He had different priorities from his father. Despite the fact that he settled there in 1802, he left much of the running of the estate to his factor and tacksmen. Bald's map is primarily a record of agricultural potential, with carefully delimited farms and townships, and a detailed table of land use for each; 90 per cent of Harris/Na Hearadh was described as 'Moor and Pasture', with only 5 per cent 'Arable with the Spade or Pasture' (the cas-chrom or crooked spade), and less than 2 per cent 'Arable with the Plough'. Later, as the kelp industry collapsed, many townships were forcibly cleared, including Northton on Harris/Na Hearadh, which is shown here as a large cluster of buildings, but which was reduced to a single farm by the time of Ordnance Survey mapping in the 1870s. Bald's map is uniquely important in confirming the original pattern of farming settlement along the western Atlantic coast, and his map also has annotations relating to minerals. For example, near Northton (fig. 7.5b) is a note 'Feltspar, Mica, Quartz'; further north at Borve (fig. 7.5c) is the note 'asbestos' near Abistos Quarry; near Finsbay are the notes relating to asbestos, serpentine and talc. These notes are not by Bald, and although there is no evidence of any later commercial exploitation at these sites, the presence of these minerals has been confirmed by later geologists. Bald's original manuscript plan has never been found, so we are fortunate that the proposed sale of the island (which took place in 1834) probably prompted copies of this lithographed map to be made.

The emergence of geology as a practical, theoretical and professional pursuit in the early nineteenth century attracted many thinkers, not least in Edinburgh with the controversies between the Neptunist ideas of Abraham Gottlieb Werner, endorsed there by Robert Jameson, Professor of Natural History, and the rival plutonist or vulcanist theories as advanced by James Hutton and James Hall. Practical testing of these ideas in the field was the key to progress, and the geology of several Scottish islands was surveyed in relation to these competing ideas regarding the Earth's history. Arran was mapped by the clergyman James Headrick in 1807, and Samuel Hibbert published geological maps of Shetland in 1820 (fig. 7.6) and of Foula (1822). The surgeon and geologist John MacCulloch was employed by the Ordnance Trigonometrical Survey as a geologist at this time (1814–21), to examine whether underlying rocks around the survey stations might

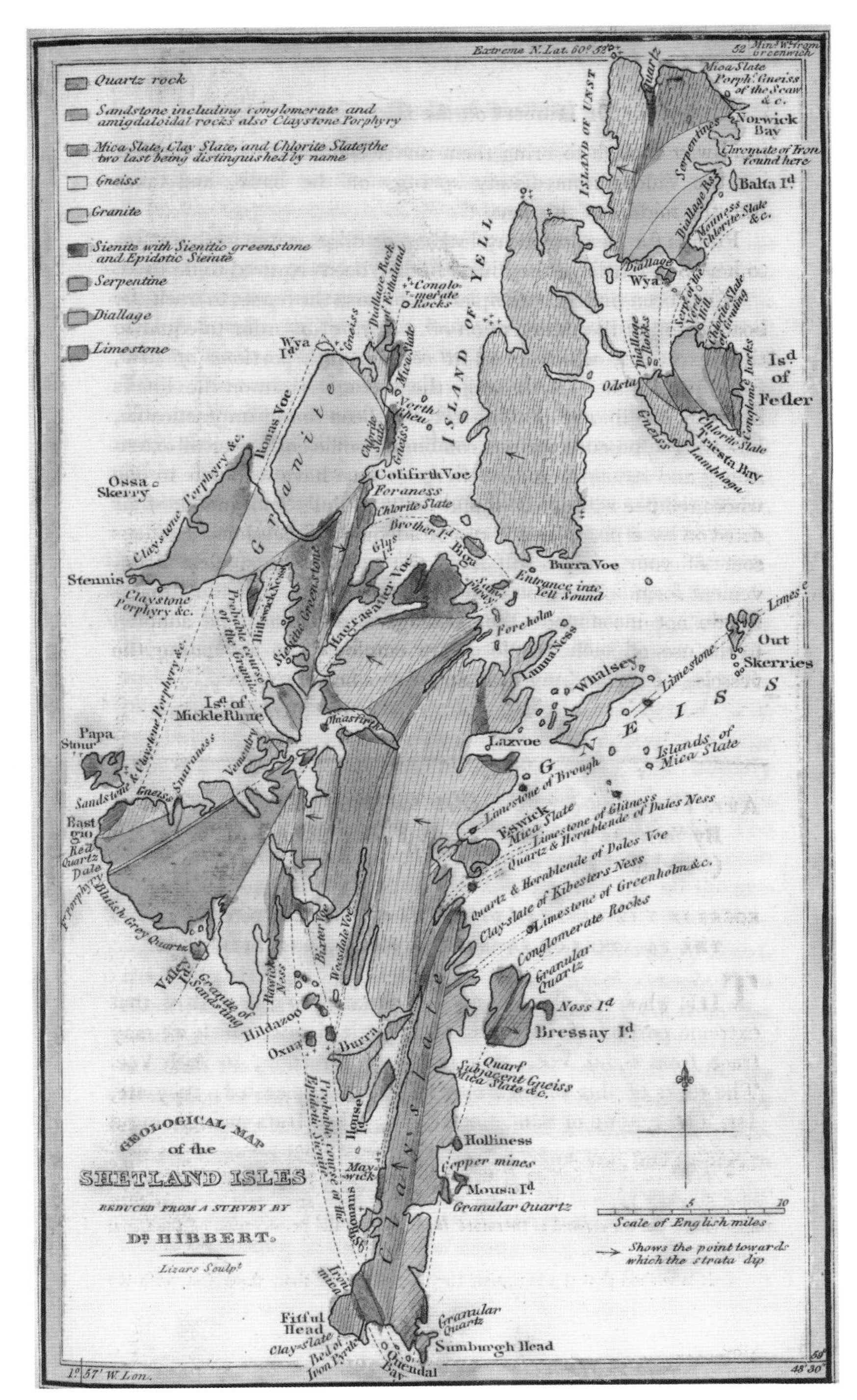

FIGURE 7.6

This is the first detailed map of Shetland's geology. Samuel Hibbert (1782–1848) was the son of a Lancashire linen yarn merchant. Following medical studies at Edinburgh, he became fascinated by geology, and made two voyages to Shetland in 1817–18 to research the islands' geology. One particular early excitement was his discovery of chromate of iron, north of Balta Sound, and which was subsequently quarried commercially: at the time, chromate of iron was imported at great expense from America. He published this coloured geological map of Shetland to accompany his paper in the *Edinburgh Philosophical Journal* of 1820. The map was engraved by Lizars, the famous Edinburgh engravers, and hand-coloured, with the rocks grouped into nine categories with annotations. Hibbert made much of the difficulties of the work, owing to the problems of travelling in the islands, most of which had no roads, and the poor topographic mapping of Shetland at the time. Hibbert later published a larger-scale geological map of Shetland in 1822 to accompany his *A Description of the Shetland Islands*, although, unlike this map, it was uncoloured.
Source: Samuel Hibbert, 'Geological Map of the Shetland Isles', from 'Sketch of the Distribution of Rocks in Shetland', Plate V, *Edinburgh Philosophical Journal* 2 (1820), facing page 224.

distort a vertical plumb line and so introduce error. Much of this work necessarily extended over the islands of Scotland: MacCulloch published detailed written accounts of the Highlands and Western Isles; his geological work on the islands was included in these works and other journals, and brought together in his 1836 geological map of Scotland.

Some islands show quarrying on an industrial scale. The quarrying of slate from Easdale and Seil in the Firth of Lorn dates to the early seventeenth century, and mining there expanded significantly in the following century. With backing from the Earls of Breadalbane, about 500,000 slates were quarried in 1745, 5 million slates in 1795, and, in 1869, a peak of production with 9 million slates. As most of the good-quality slate lay at or below ground level, workings from the early nineteenth century depended upon the development of more efficient pumps. By the 1820s, over 200 men were directly employed in these slate quarries, with many more working on transporting the slates. In 1825 alone, 526 ships, including 245 steamers, used the harbour. As the first edition Ordnance Survey map surveyed in 1871 clearly shows (fig. 7.7a), the quarries had an elaborate system of railway tracks, with power for the rail wagons and pumps provided from the Engine House. The map also records the original name of Eilean nam Beathach (Island of the Birches) as it was formerly an island separated from Seil, but the narrow channel had been filled in with the accumulation of quarry waste. The main quarry to the south of the two rows of workers' cottages was protected from the sea only by a narrow wall of unquarried rock, strengthened by a masonry wall. By the 1870s, the quarry was 240 feet or 73 metres below sea level.

Disaster struck in November 1881 when a major storm breached this wall, flooding the quarry and destroying the workings. Production continued intermittently until the Second World War on other quarries, but Easdale's heyday was over, and the second edition map (from a revision of 1898) shows the dramatic change (fig. 7.7b). At first glance, the abandoned 'Old Quarries' look alike, but the lack of a parcel number and acreage in the main quarry confirms it as being flooded by the sea. It remains so today. Although the maps clearly show the industrial exploitation of the island and its dramatic end, Easdale has since the 1970s managed to exploit tourist interest in its industrial archaeology. The population had fallen to only four people in the 1960s, but it has since risen to over seventy houses, thirty of which are occupied by permanent residents.

The theme of quarrying as exploiting may be brought up to the present day with the proposed Harris/Na Hearadh 'superquarry', a site only a mile north of Rodel/Roghadal village, surveyed by Bald two centuries earlier. The main proposal, submitted in 1991, was the largest mineral extraction planning application in the UK – to remove 600 million tonnes of anorthosite (used chiefly for road building and in concrete), over a period of sixty years, from the mountain of Roineabhal, in the south of Harris/Na Hearadh. The quarry was to cover an area of 1 by 2 kilometres, extending 370 metres above sea level and 180 metres below. After the site had been exhausted, the plan was to blast out a sea loch that would have had the highest sea cliffs in the British Isles, six times the height of the white cliffs of Dover. These, it was suggested, would become a tourist attraction.

With the promise of employment and economic regeneration, locals were initially in favour of the quarry (by 62 per cent to 48 per cent in a 1993 poll). Later polls evidenced a growing opposition, especially in the local Obbe ward. The Western Isles Council (Comhairle nan Eilean Siar from 1997) originally supported the application, but had changed its mind by 1995. Numerous objections were raised by environmental groups, criticising the quarry's location in a designated National Scenic Area, raising concerns over marine pollution from the increased shipping traffic, the negative impact on the area's biodiversity, and questioning the environmental wisdom of

FIGURE 7.7
(*Opposite*) These details from the first and second editions of the Ordnance Survey 25 inch to the mile maps show the quarries at Easdale, first at their height in the 1870s (a), but then in decline in the 1890s (b), following the disastrous flooding of 1881.
Source: (a) Ordnance Survey, 25 inch to the mile, *Argyll and Bute, Sheet CXXI.7* (surveyed 1871, published 1875). (b) Ordnance Survey, 25 inch to the mile, *Argyll and Bute, Sheet CXXI.7* (revised 1898, published 1899).

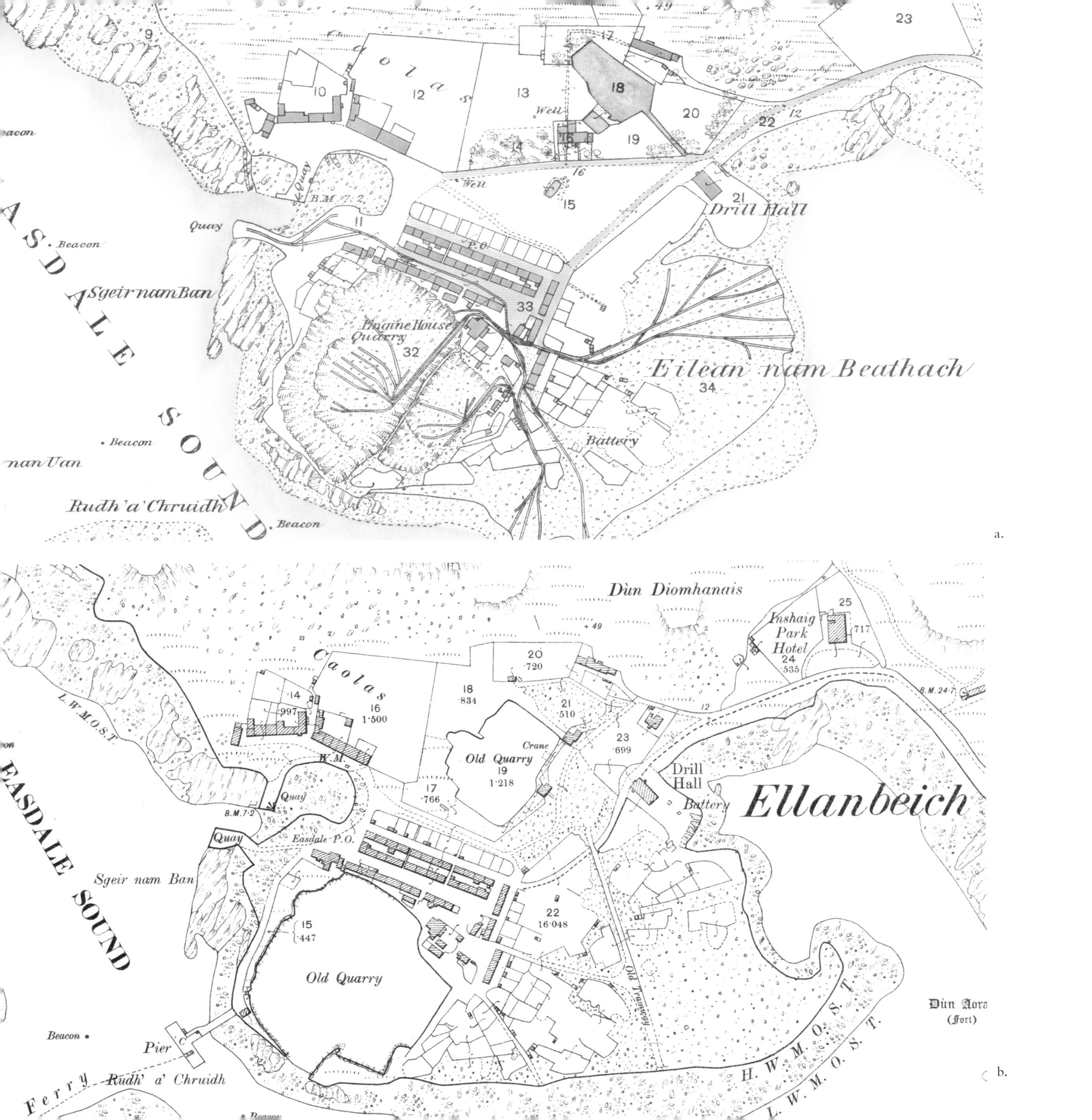
Caolas
Well
Quay
B.M. 7.2
Beacon
Sgeir nam Ban
EASDALE SOUND
Engine House
Quarry
Drill Hall
Eilean nam Beathach
Battery
nan Uan
Rudh' a' Chruidh
a.
Dùn Diomhanais
Inshaig Park Hotel
Old Quarry
Crane
Easdale P.O.
Ellanbeich
Old Tramway
Pier
Ferry
Dùn Aora (Fort)
H. W. M. O. S. T.
L. W. M. O. S. T.
b.

extracting vast quantities of rock for motorway construction.

A montage of what the quarry might have looked like (fig. 7.8) was prepared by Envision 3D for Scottish Natural Heritage in the 1994–95 public inquiry and used to great effect in raising opposition to the quarry proposal. The inquiry lasted five years, before, in 2000, the Scottish Executive turned the application down. Even then, the battle was not over. By this time, Redland had been taken over by Lafarge, the world's biggest producer of building materials, and they prompted a second inquiry in 2001 over the legal validity of consent granted in 1965 for a smaller quarry in the area. Following an appeal and further deliberations, in April 2004 Lafarge withdrew the original application and dropped their proposals for the site.

Exploiting the land

The later nineteenth century witnessed a dramatic expansion in the number and size of Highland sporting estates. This followed a long-term decline in agricultural commodity prices, and diminishing returns from sheep farming, combined with technological developments such as the expansion of the railway network and improvements to the design of the rifle. Before 1811, there were fewer than ten estates in Scotland actively managed for hunting. By 1873, this had grown to seventy-nine, and, by 1911, it had risen to 170, covering nearly 3 million acres. Eighteen of these estates were on various islands, from Arran, Jura and Scarba in the south, to Mull, Rum and Skye further north, and large parts of Lewis/Leòdhas and Harris/Na Hearadh.

The conversion of good grazing land into hunting grounds caused considerable resentment among island crofting populations. At Park/Pairc in Lewis/Leòdhas, owned by Major Matheson, there were famous land raids in November 1887 when crofters, driven by famine and poverty, took possession of part of the Park estate and killed several hundred deer. This event, part of a longer-run history of land protest and riot in the west Highlands and islands, followed Lady Matheson's refusal to listen to crofters' requests to restore the hunting grounds to crofting. She continued to lease Park/Pairc to Mr and Mrs Joseph Platt, who still rented the estate in 1911 (fig. 7.9). Shooting estates dominated several islands even in the twentieth century. From May 1918, however, the island of Lewis/Leòdhais saw quite different and ambitious plans for its economic exploitation when it was purchased by the successful industrialist and soap magnate William Lever, 1st Viscount Leverhulme.

Leverhulme's main priority was the revival of the fishing industry through capital investment, better transport and marketing following his acquisition of the Mac Fisheries chain of fishmongers. Imaginative plans for rural transformation involved the industrial-scale conversion of peatlands into arable lands to make Lewis/Leòdhas a food-producing island, and a forecast increase in population from around 31,000 to nearly 200,000. In August 1918, as part of these plans, Leverhulme invited the renowned landscape architect Thomas Hayton Mawson to the island to discuss schemes for agricultural improvement: farming willow trees for the basket trade; creating experimental gardens growing plants such as camomile, mint and other medicinal herbs for distilling; producing soft fruit, such as raspberries, strawberries, blackcurrants and gooseberries, in sufficient quantities to open a Hebridean jam factory. Mawson had designed gardens around several of Leverhulme's properties in London and Lancashire, but his optimistic schemes, however well intentioned, seriously underestimated the severity of the Hebridean climate.

Leverhulme's contacts with the sociologist and planner Patrick Geddes led to an invitation to his son, Arthur Geddes, in the autumn of 1919, and to the scientist and farmer Dr Marcel Hardy, to undertake a detailed agricultural reconnaissance and vegetation survey of Lewis/Leòdhas and Harris/Na Hearadh. Fisheries, it was assumed, would be successful and the main source of income. Hardy and Geddes were asked to explore ancillary activities such as commercial peat-cutting, the planting of commercial forestry and the development of market gardening. The initial survey, dividing the land into eight vegetation categories (cultivation, sandy pastures, sedge

FIGURE 7.8
The superquarry on the Island of Harris/Na Hearadh, proposed from 1991, was the largest mineral extraction planning application in the UK, and the subject of an acrimonious battle between its supporters and opponents for the next fourteen years. Using an original air photograph of the site taken by Aerographica (a), Envision 3D then prepared a photomontage (b) to illustrate the massive visual impact of the quarry, a graphic that was successfully used in campaigning against the quarry proposal.
Source: Aerographica/ Envision 3D, *Proposed Lingerbay Superquarry Montage*, 1995. Original photograph by permission of Patricia and Angus Macdonald/Aerographica; photomontage by permission of Envision 3D Ltd.

a.

b.

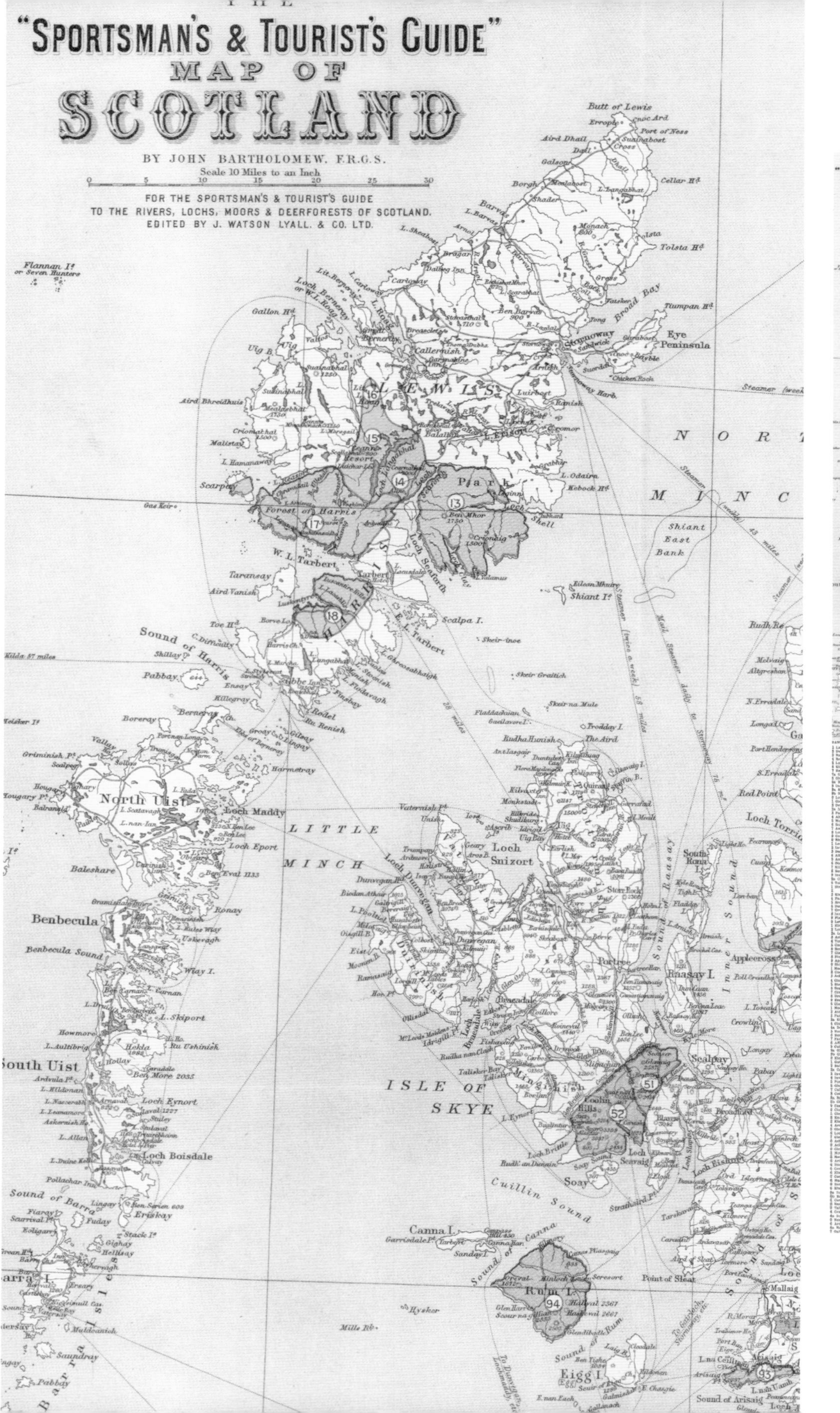
THE
"SPORTSMAN'S & TOURIST'S GUIDE"
MAP OF
SCOTLAND
BY JOHN BARTHOLOMEW, F.R.G.S.
Scale 10 Miles to an Inch
0 5 10 15 20 25 30
FOR THE SPORTSMAN'S & TOURIST'S GUIDE
TO THE RIVERS, LOCHS, MOORS & DEERFORESTS OF SCOTLAND.
EDITED BY J. WATSON LYALL. & CO. LTD.
Butt of Lewis
Port of Ness
Cellar Hd.
Tolsta Hd.
Tiumpan Hd.
Broad Bay
Eye Peninsula
Stornoway
Flannan Is. or Seven Hunters
Gallon Hd.
Uig B.
LEWIS
Callernish
Park
Forest of Harris
Scarpay
Taransay
Tarbert
Loch Seaforth
Loch Shell
NORTH
MINCH
Shiant East Bank
Shiant Is.
Scalpa I.
Sound of Harris
Pabbay
Berneray
North Uist
Loch Maddy
Loch Eport
LITTLE
MINCH
Benbecula
Benbecula Sound
Wiay I.
L. Skiport
Hekla
Ben More 2035
South Uist
Loch Eynort
Loch Boisdale
Sound of Barra
Eriskay
Barra I.
Vaternish Pt.
Loch Snizort
Dunvegan Hd.
Loch Dunvegan
Duirinish
Bracadale
Portree
Raasay I.
South Rona I.
Sound of Raasay
Inner Sound
Applecross
Loch Torridon
Red Point
Rudh Re
ISLE OF
SKYE
Coolin Hills
Soay
Loch Scavaig
Cuillin Sound
Canna I.
Sound of Canna
Rum I.
Eigg I.
Point of Sleat
Scalpay
Pabay
Loch Eishort
Sound of Arisaig
Mallaig
Arisaig

THE
"SPORTSMAN'S & TOURIST'S GUIDE"
MAP OF
SCOTLAND
BY JOHN BARTHOLOMEW, F.R.G.S.
FOR THE SPORTSMAN'S & TOURIST'S GUIDE
TO THE RIVERS, LOCHS, MOORS & DEERFORESTS OF SCOTLAND.
EDITED BY J. WATSON LYALL. & CO. LTD.
NORTH
MINCH
LITTLE
MINCH
ISLE OF
SKYE
ARGYLL
INDEX TO DEER FORESTS

FIGURE 7.9
The *Sportsman's and Tourist's Guide* had been intermittently published under an earlier title before 1890. From 1895 to 1915 it was regularly published, twice a year, in the summer and autumn, with this Bartholomew map as a standard foldout inside the front cover. Its publisher, J. Watson Lyall & Co. Ltd, based in Pall Mall in London, was Shooting and Fishing Agent for the 'Letting and Selling of Scotch Deer Forests, Grouse Moors, Mansion Houses, Low-Ground Shootings, Salmon Fishings, &c., &c.' The *Guide* was packed with information and liberal advertisements, with train timetables, steamer routes, hotels, the Ground Game Act (1880), and, importantly, well-indexed lists to the main deer forests and salmon fisheries. The Park/Pairc Deer Forest is, for example, described as 'a very extensive and excellent deer forest in the Lews. It extends to 75,000 or 80,000 acres, and yields first-rate deer-stalking, in addition to a good bag of grouse, woodcock, snipe and wildfowl.' There are further notes on the quality of the fishing, the nearest place for supplies, and access by rail, steamer and car.

The *Guide* also describes the shooting rights on each estate, and whether they were held by the proprietor or the lessee. For example, Major Duncan Matheson, who owned Lewis/Leòdhas from 1899, was also the main tenant with shooting rights to the Morsgaill estate (no. 15), north of Loch Langabhat: he leased the shooting rights on his other Lewis/Leòdhas shooting estates. As one might expect – and prefiguring arguments made today by landowners over the economic benefits of Highland estates – the optimistic tone left no doubt regarding the benefits of the sporting estate: 'The importance to Scotland of its shootings may be estimated by the fact that the rental paid for them by sportsmen is more than £400,000 per annum . . . If to [this total] be added the money spent by sportsmen in other ways in connection with them, such as cost of transit, cost of living, &c., it will be seen that the shootings of Scotland are the means of pouring into that country year by year a golden stream of no small magnitude, and one which benefits all classes.' The Pairc Community Trust was finally allowed to buy the Pairc estate in 2014, under the crofting 'Right to Buy' provisions of the Land Reform (Scotland) Act 2003. The Trust formally took over the land in 2015.
Source: John Bartholomew & Co., *Sportsman's and Tourist Guide Map* (1911).

and cotton grass, marshy grass moors, hill and fair pasture, woods, rocky pastures and heaths) was hand-coloured onto Ordnance Survey one-inch to the mile base mapping as part of a detailed written report to Leverhulme (fig. 7.10a). Ten years later, Arthur Geddes was encouraged by the Scots-Canadian philanthropist T. B. Macaulay to return to Lewis/Leòdhas to update the survey. Geddes – who became a distinguished sociologist and historian of rural Scotland in addition to having interests in India and medical geography – would spend many subsequent seasons on the island.

Years later, Geddes reviewed his early work and Leverhulme's failure in Lewis/Leòdhas. While the decline in fish consumption and collapse of the fish market after the First World War proved catastrophic for Leverhulme's plans, Geddes also attributed their failure to a lack of understanding by Leverhulme of the culture and society of Lewis/Leòdhas. Leverhulme's virulent opposition to granting land for crofting, which he saw as being backward, and his determination to create factories with wage-earners as vehicles of economic growth, were fiercely opposed by the islanders. His schemes were largely abandoned by 1923.

These several examples illustrate the value and power of mapping in revealing the exploitation of islands' natural resources. They show the changing geographies of such exploitation over time, and how 'new' resources – anorthosite in Harris/Na Hearadh, oil off Shetland – have potentially dramatic consequences for island life and livelihoods. These maps also facilitate an outsiders' view, encouraging a 'detached' understanding of islands as resources. Such a perspective underscores that dominant economic perception of islands in recent decades, views which have come to reflect the decisions of institutions or shareholders made from far away. Local influence is possible, of course, but it is always hard won. As Scotland's islands have become the destination for increased levels of tourism, many island communities have exploited this 'new' scenic resource. Maps are at the heart of this commodification of islands too – presenting heritage and scenery to be enjoyed, walked through, and looked at by visitors from afar, even if not physically removed and sold.

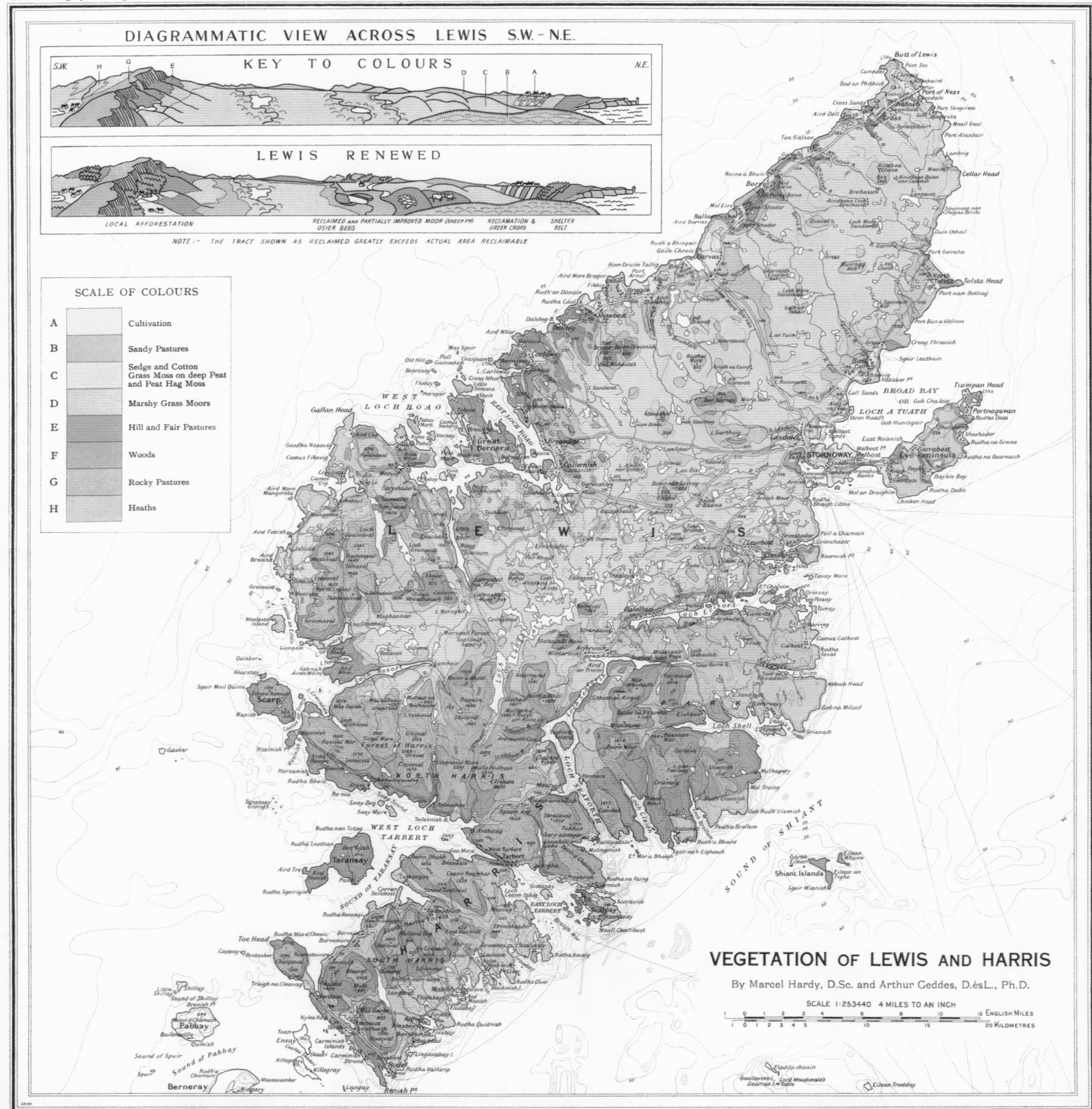

DIAGRAMMATIC VIEW ACROSS LEWIS S.W.-N.E.
KEY TO COLOURS
S.W.
N.E.
H
G
E
D
C
B
A
LEWIS RENEWED
LOCAL AFFORESTATION
RECLAIMED AND PARTIALLY IMPROVED MOOR (SHEEP Fm)
OSIER BEDS
RECLAMATION & GREEN CROPS
SHELTER BELT
NOTE:- THE TRACT SHOWN AS RECLAIMED GREATLY EXCEEDS ACTUAL AREA RECLAIMABLE
SCALE OF COLOURS
A Cultivation
B Sandy Pastures
C Sedge and Cotton Grass Moss on deep Peat and Peat Hag Moss
D Marshy Grass Moors
E Hill and Fair Pastures
F Woods
G Rocky Pastures
H Heaths
VEGETATION OF LEWIS AND HARRIS
By Marcel Hardy, D.Sc. and Arthur Geddes, D.èsL., Ph.D.
SCALE 1:253440 4 MILES TO AN INCH
ENGLISH MILES
KILOMETRES
L E W I S
P A R K
NORTH HARRIS
H A R R I S
SOUTH HARRIS
Forest of Harris
Butt of Lewis
Port of Ness
Cellar Head
Tolsta Head
Tiumpan Head
BROAD BAY
LOCH A TUATH
Eye Peninsula
STORNOWAY
WEST LOCH ROAG
EAST LOCH ROAG
Great Bernera
Gallan Head
Scarp
LOCH SEAFORTH
Loch Shell
WEST LOCH TARBERT
Taransay
Tarbert
EAST LOCH TARBERT
SOUND OF TARANSAY
SOUND OF SHIANT
Shiant Islands
Toe Head
Pabbay
Sound of Pabbay
Berneray

b.

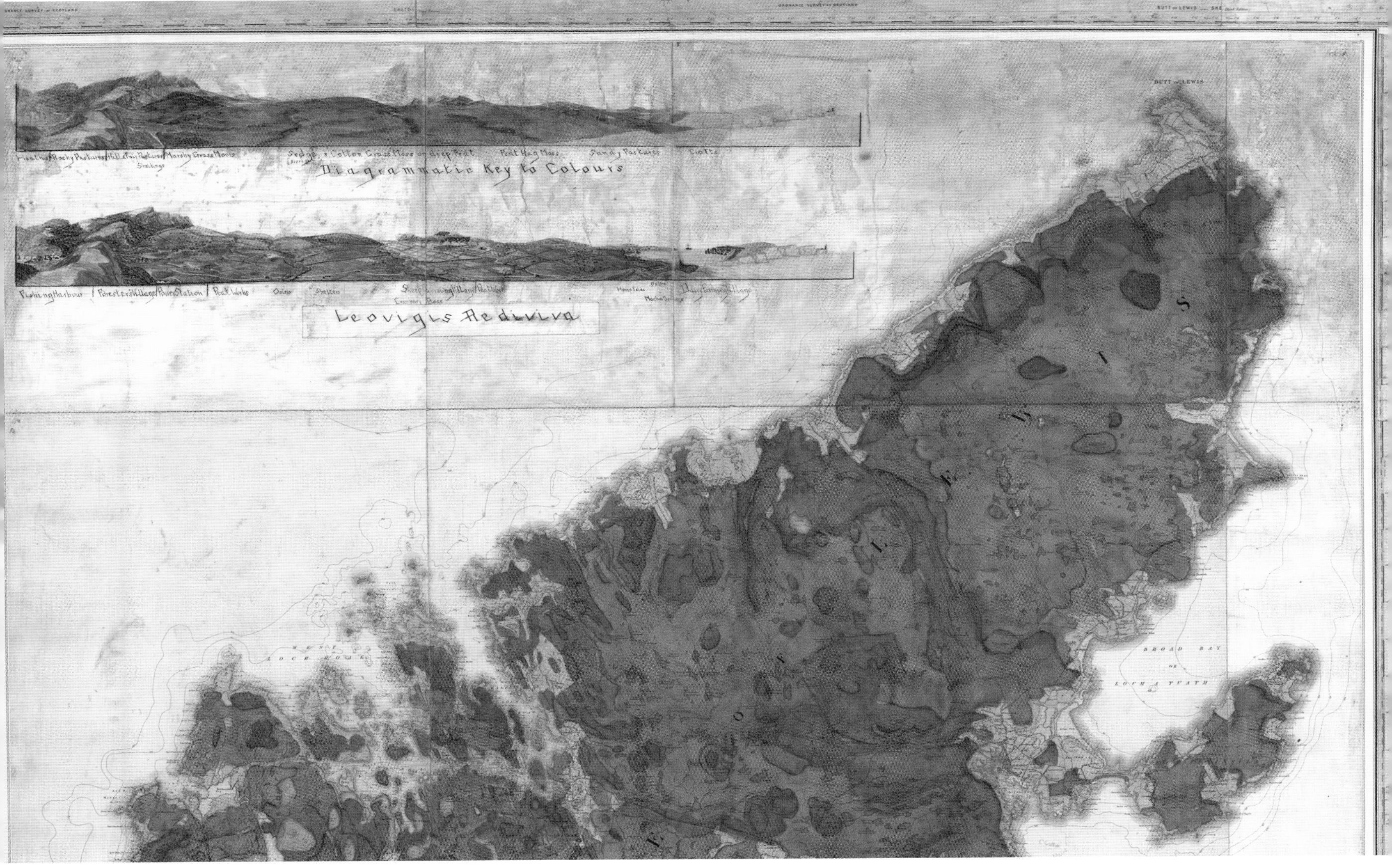

a.

FIGURE 7.10

Detail (a) (*above*) shows part of the original survey of land use on Lewis/Leòdhas, carried out by Marcel Hardy and Arthur Geddes at the request of Lord Leverhulme in 1919, with the different land-use categories hand-coloured onto Ordnance Survey one-inch to the mile mapping. Geddes went on to publish a printed map at a smaller scale of 4 miles to one inch (b) (*opposite*), drawn by John Bartholomew & Son. The map itself is a rare early example of vegetation and land-use mapping in Scotland, and the agricultural changes that it promoted (see inset, 'Lewis Renewed') were more realistic, based on a sound understanding of ecology and climate. See also pp. 186–7.

Source: (a) Arthur Geddes and Marcel Hardy, *Lewis and Harris: Main Aspects of the Vegetation* (1919). (b) Marcel Hardy and Arthur Geddes, *Vegetation of Lewis and Harris* (1936).

DIAGRAMMATIC VIEW ACROSS LE

NOTE :- THE TRACT SHOWN AS RECLAIMED GREATLY EXCEEDS ACTUAL

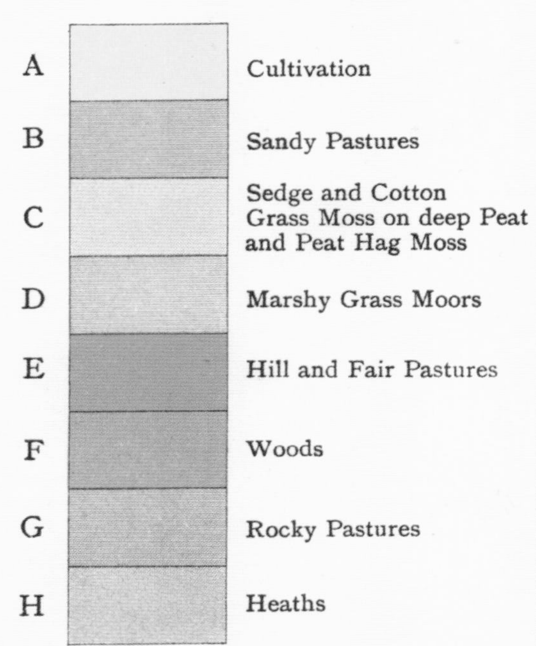

Old Hill
Bearasay
Floday
Harsgeir
WEST LOCH ROAG
Gallan Head
Pabay More
Kyles Pabay
Aird Uig
Geodha Nasavig
Camus Fihavig
Forsnaval
Valtos
Croulista
Camus Uig
Uig Ch.
Uig Lo.
Meavig
Aird More Mangersta
Carnshader
L. Staesavat
Ardroil
Suainaval
Mangersta
Loch Suainaval
Aird Fenish
Islivick
Aird Brenish
Mealisval
Raonasgail
Tahaval
Brenish
Cracaval
North Lingal
Greineim
Mealista
Tamanaisval
Loch Grunavat
Enaclete
Skeun
Beinn Mheadhonach

Marcel Hardy/Arthur Geddes, *Vegetation of Lewis and Harris*, 1936.

c.

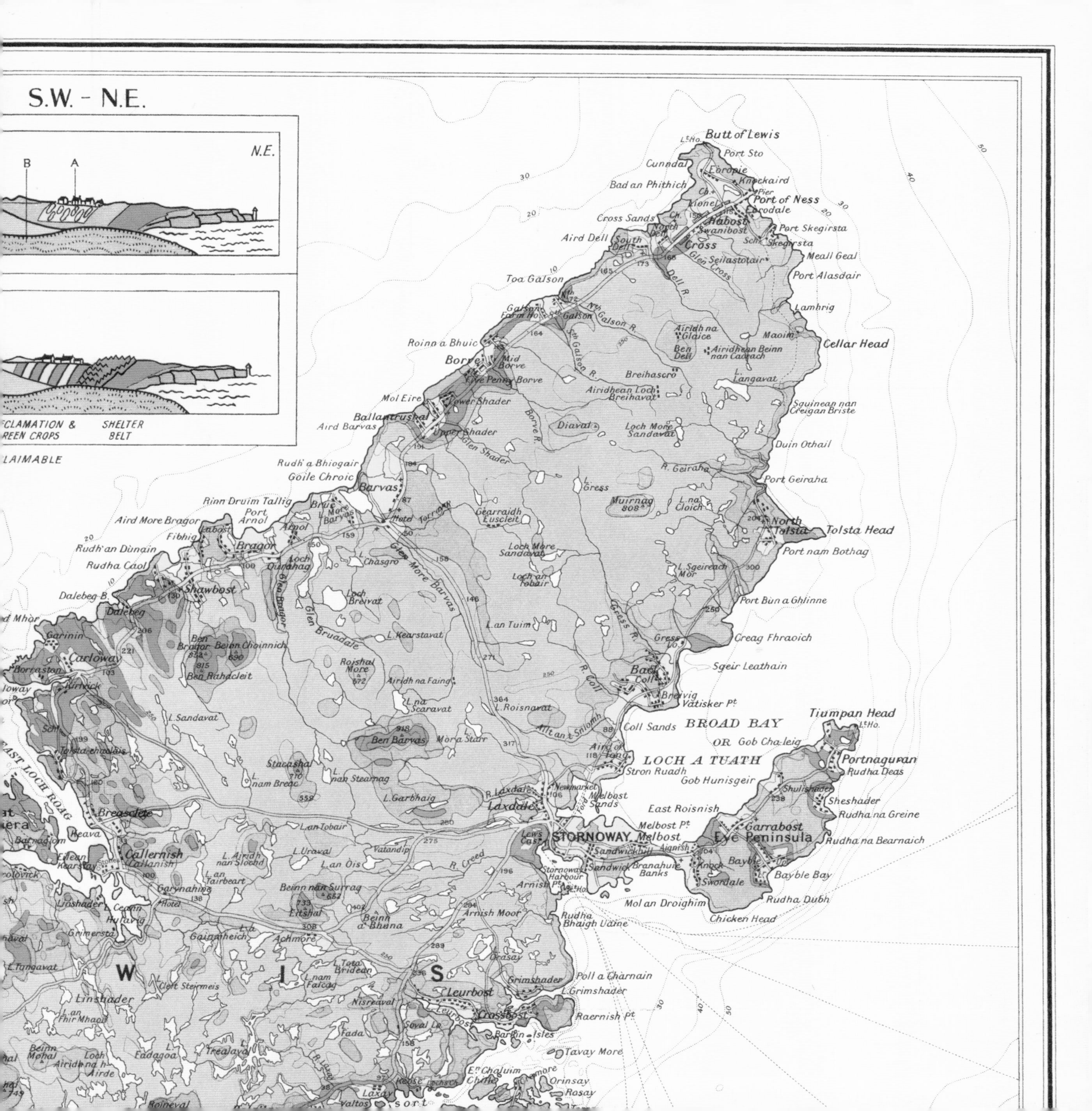

S.W. - N.E.
N.E.
B
A
ECLAMATION &
REEN CROPS
SHELTER
BELT
LAIMABLE
Butt of Lewis
Port of Ness
Cellar Head
Tolsta Head
Tiumpan Head
BROAD BAY
OR
LOCH A TUATH
Eye Peninsula
STORNOWAY
Barvas
Carloway
Callernish
Shawbost
Bragor
Laxdale
Leurbost
Crossbost
Ben Barvas
Muirnag
Glen More Barvas
EAST LOCH ROAG
W
I
S

CHAPTER EIGHT

PICTURING

All maps are pictures in part. That is, all maps, in one way or another, are partial depictions or representations of a particular part of the world. All maps are also works of art in the general sense that mapping is a skill – in selection, depiction, technical production, in the use of colour, line and symbol, and so on – to show the world and its regions and places in map form so that users can understand what is being shown. By convention, four dimensions are reduced to two: lines (contours and, earlier, hachures) stand for height; at sea, soundings (numbers) stand for depth. Even modern electronic mapping which in several ways and in soft-tone colours 'visually flattens' the world it purports to depict (see fig. 1.12 for example) is a work of art.

These brief observations are to make an important point concerning how we understand map history – of Scotland's islands, of Scotland, or more widely and generally. Once, map historians held the view that map history was a narrative of progress, a story told as a chronology of improvement in accuracy and in completeness of coverage. A key element of this interpretation was that maps got 'better' over time, became more readily useful documents, because, over time, they shed the colours, lines and symbols of artistic embellishment and replaced them by a language of plainness. Proper map making reflected the triumph of science over art, of mathematical precision over painterly ornamentation. This view is no longer subscribed to. Map history is now understood not in terms of linear development towards a final end but as a more complex and diverse subject in which we always need to understand the content of maps in relation to the context of maps: to ask not 'what is a map?', but 'what job of work is that map doing?'; not 'is that an accurate map?' but 'What was the purpose of that map – is it even appropriate to think of accuracy?' Any map is always to be understood as a compromise between the technologies used to produce it, the science (and the politics) that lie behind the map, and the art and artistry of its maker. This chapter looks at how maps are works of art, at connections between map making and the world of art, and at the idea of the map as a form of world picturing.

Opposite. Detail from D. McKenzie/William Home Lizars, 'A Comparative View of the Heights of the Principal Mountains of Scotland', from John Thomson's *Atlas of Scotland* (1832).

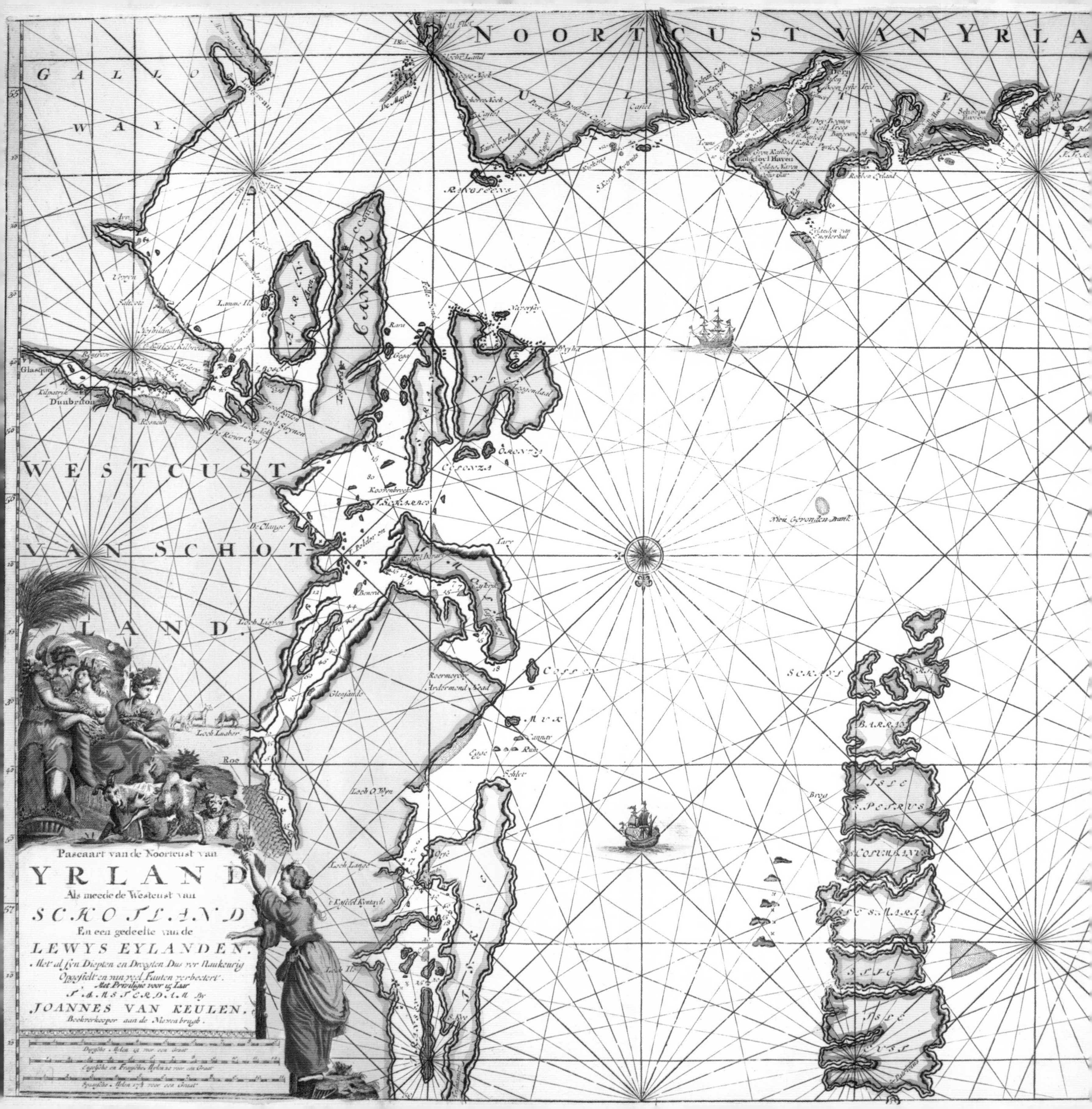

NOORT CUST VAN YRLA
GALLO WAY.
WESTCUST VAN SCHOT LAND.
Pascaart van de Noortcust van
YRLAND
Als meede de Westcust van
SCHOTLAND
En een gedeelte van de
LEWYS EYLANDEN.
Met al syn Diepten en Drogten Dus ver Naukeurig
Opgestelt en van veel Fauten verbetert.
Met Privilegie voor 15 Jaar
t'AMSTERDAM By
JOANNES VAN KEULEN.
Boekverkooper aan de Nieuwenbrugh.
Duytsche Mylen 15 voor een Graat
Engelsche en Fransche Mylen 20 voor een Graat
Spaansche Mylen 17½ voor een Graat
Glasque
Dunbriton
Ayr
Rangeleens
Coll
Mulk
Tarr
Cosser
Barray
Isle S. Petrus
Isle S. Marta
Rum
Egge
Cannay
Brog
Nieu Gevonden Bank
Loch Lange
Glenfande

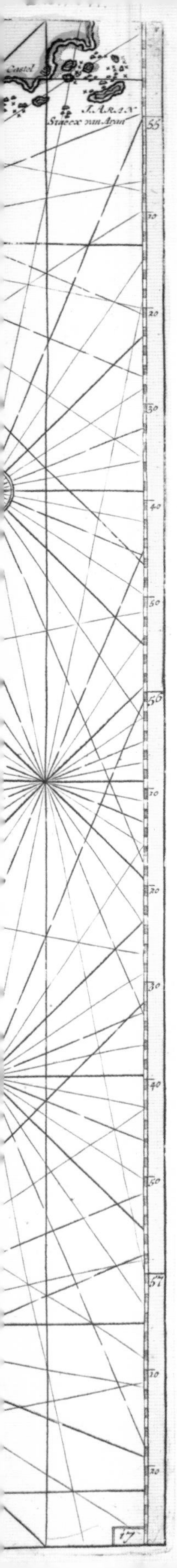

FIGURE 8.1
Johannes van Keulen's map is interesting for several reasons. It is clearly not accurate in any formal geometric or topographic sense. Rather than adopt what we think of as the standard convention of orienting the map with north to the head of the map, van Keulen has here 'reversed' the west coast of Scotland and the Outer Isles/Innse Gall, showing them from north to south. In this sense, it is as if the Western Islands/Na h-Eileanan an Iar are being shown 'in front and below', as might be the case for a Baltic or Low Countries' trader entering these waters having crossed the North Sea and sailed west along the Pentland Firth. Of the several islands pictured, Skye is especially aberrant, being made to run 'long and thin'; note also the Shiant Islands and the small crosses nearby, perhaps to indicate shipwrecks or navigational hazards.
Source: Johannes van Keulen, *Pascaart van de noortcust van Yrland als meede de westcust van Schotland* (*c*.1712).

Maps and the geography of art

As one illustration of the point made above, consider the work of the Dutch map maker Johannes van Keulen in his maps and charts of Scotland in the early eighteenth century (figs 8.1, 8.2 and 8.3). Here, the importance of the maps rests hardly at all in their accuracy of depiction of Scotland's islands and coastlines and more in van Keulen's demonstration of technical proficiency: the maps' cartouches, symbolically depicting the power of the map maker, in one case (fig. 8.3) use a female figure to represent 'Geographia', who is shown inscribing and picturing the world in map form. Van Keulen's map would have been of little use at sea, none on land: this is a map whose purpose is ornamentation, whose role would have been in domestic decoration rather than in practical navigation. That van Keulen's maps are different in style but close in date to other very different and 'plainer' pictures of the same islands (cf. fig. 8.1 with fig. 8.4 for example) exemplifies this point about map history as an account of complexity and difference rather than a narrative of progress understood as a move away from artistic subjectivity to scientific objectivity.

The creative process is common to all artists. Map makers have to merge several requirements in making their art. No more and no less than a landscape painter, say, who may emphasise one feature to lend a certain perspective to the painting, or even invent a particular view to make an aesthetic point, map makers have to effect balance in what and how things are included. There are close parallels between the demands upon the artist, the map maker and the emergence of the art market in Western Europe from the sixteenth century onwards. With the invention of perspective in European art from the early fifteenth century, first in Italian painting and then in the many *landschip* (landscape) paintings of Dutch and Flemish painters, artists were able to give depth to their portrayal of space, drawing the viewer's eye into the landscape. The art of map making and the making of art were closely connected, first in medieval Italy, then in the Low Countries and, later, in the eighteenth century, in France and Britain. This was because of the patronage needed by both, the

FIGURE 8.2
In this image showing the cartouche from the above map in more detail, artistic skill has been put to use to make a political point. Ireland, symbolised by the figure stretching upwards and with a harp at her feet, is presenting a red rose to the woman above her, perhaps as a figurative gesture of political loyalty.
Source: Johannes van Keulen, *Pascaart van de noortcust van Yrland als meede de westcust van Schotland* (*c*.1712).

markets provided by the moneyed mercantile classes and the close and trade-based associations between the allied crafts of engraving, colouring, paper making and printing. In an age when maps and pictures were both commonly seen as objects of cultural status and were commonly produced by people whose artistic and technical skills were brought to bear on canvas and paper alike, we should expect there to have been such close connections.

Yet there is no reason to expect that these artistic and contextual connections would have been especially close, for example, when Scotland and its islands were not the centre of detailed cartographic attention and were being pictured by Italian map makers (cf. fig. 1.2). As Scotland and its islands start to appear in closer focus from the mid and later sixteenth century, and to be produced in Antwerp and in Amsterdam, then both key centres in the European geography of art making and art buying, there is every reason, given the associations between art, commercial activity and cultural status in these places, to expect maps to have an emphasis on colour, ornamentation and design. At the turn of the sixteenth century, those cartouches used on maps of Scotland produced in the Low Countries were heavy and formal in form (fig. 8.5a and pp. 204–5). By the mid seventeenth century, the influence of the Baroque on paintings, sculpture and architecture had found its way into maps and charts. Strong forms, often well-endowed female figures, small children, or winged cherubs or 'putti', all found their way into these exuberant title pieces. Animals and agricultural implements were often included to make the point about national wealth and productivity, actual or potential (see fig. 8.5d in particular in this respect).

Other types of map are no less artistic despite their different purposes and styles. This is especially true of the eighteenth-century maps and plans produced by military engineers and draughtsmen. These maps and plans have already been discussed in chapter 5 in terms of their application to defending the realm. But there was what we might think of as a regulated and standard colour code for this sort of work: a form of artistic regimentation necessary for these maps and plans made to a common purpose to be understood

FIGURE 8.3
(*Above*) In this detail of the title frame and surrounding ornamentation in his map of Scotland's east coast between Berwick and the Orkney islands, Johannes van Keulen shows the arts and instruments of geography and map making necessary to picture the world in map form (set square, dividers, perhaps even a small way-wiser or perambulator used to measure linear distance in the hands of the figure to the far left). Geographia is shown sketching a map. In placing her foot upon the globe, van Keulen wants us, the map's reader, to be under no illusions that maps are authoritative documents and of his authority as a leading map maker: as is noted in the title frame, he has held the privilege (production and trading rights in this respect) for fifteen years.
Source: Johannes van Keulen, *Nieuwe Pascaert van de oost cust van Schotlandt beginnende van Barwyck tot aen de Orcades Ylanden* (*c.*1712).

FIGURE 8.4
(*Right*) Tiddeman's map from around 1730 depicting Scotland's western islands is altogether different in style from van Keulen's works of 1712 (and from many others shown throughout this book). Plainer and not coloured, it is no less a work of art, however.
Source: Mark Tiddeman, *To the Honorable Sr Charles Wager, this Draught of Part of the Highlands of Scotland is humblely presented by Mark Tiddeman . . .* (1730).

a.

b.

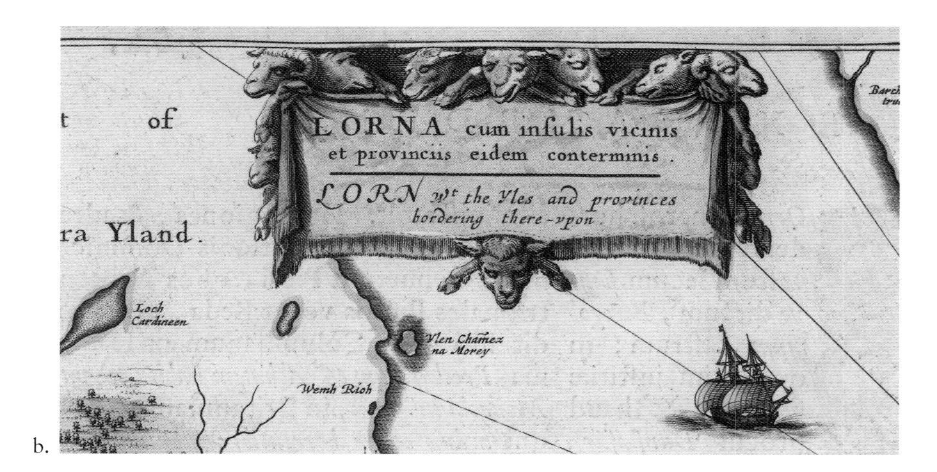

c.

d.

FIGURE 8.5

In these four cartouches, taken from maps published between 1595 and 1710, a variety of figures and devices have been used to lend the map artistic adornment. In (a), from Mercator in 1595, the title frame serves little purpose: it is as if the map maker wanted only to fill the space that might otherwise be blank or empty sea. In the title cartouche to his map of Lorn (b), set against a portion of Jura and the islands of Luing, Seil and Kerrera among others, Joan Blaeu would appear by his portrayal of the many sheep to be signalling the wealth of the district. The Blaeu map of the Northern Isles of Orkney and Shetland (c) makes the same point about the islands' wealth but this time with fish instead of sheep – but leaves enough of a gap in doing so to depict the Fair Isle. The title frame and cartouche to Homann's map of 1710 (d) is at once suggestive of political loyalty and agricultural prosperity and industriousness. In the years soon after the union of the parliaments of Scotland and England in 1707, national unity had to be worked at: here, a child passes an apple from the Tree of Knowledge towards the lion and the unicorn and other heraldic symbols – the crown and arms – of Britishness. To the left of the image, the beehive speaks both to industriousness and collective endeavour while, to the right, a spade in the hands of the child-like figure stands for hard work in the fields, the garland on his head a gesture towards peace. Note, too, that as the clouds of an earlier gloomier age roll away in the face of this new political climate, the map maker has left space for the Outer Isles/Innse Gall to the left and, to the right, the north-west mainland.
Source: (a) Gerard Mercator, *Scotia regnum* (1595). (b) Timothy Pont/Joan Blaeu, *Lorna cum insulis vicinis et provinciis eidem conterminiis* (1662). (c) Timothy Pont/Joan Blaeu, *Orcadum et Shetlandiae insularum accuratissima descriptio* (1654). (d) Johann Baptist Homann, *Magnae Britanniae pars septentrionalis qua Regnum Scotiae* (c.1710).

in different places. Red was used for masonry and buildings, blue for water, green in different shades to depict different ground conditions (marshy ground, enclosed formal grounds, arable fields, for example), and gamboge or shades of yellow for other notable features. Trainee draughtsmen were often given the task of copying the work of a more experienced draughtsman as part of their training, as was the case for artists' apprentices. A cartographic language and style that required a degree of fidelity to the objects it represented – the heights of ramparts, the position of revetments, and so on – could be accomplished with artistic licence. Paul Sandby, one of the Board of Ordnance's official draughtsmen and associated as a topographical artist with William Roy's Military Survey of Scotland from 1747 to 1755 was later to be known as the 'Father of English Watercolour'. Much of what his paintings brought to the drawing rooms of Britain's Georgian aristocracy he learnt in the fields and woods of Scotland.

Maps and the art of geography

With the exception of hand-drawn and hand-coloured maps whose creation is, of necessity, a work of individual artistry, map making in general shares something with the work of the landscape artist in that both are subject to the demands of picturing the object in question according to certain stylistic conventions. The similarity between art making and map making is further strengthened when the individual art or map work reaches the stages of being engraved for further production. Here, map makers and engravers – if they were not the same person – needed an element of shared technical skill, not least in terms of the design of the final image. Those persons involved in the creative process of engraving map information onto a copper plate, or involved in comparable later processes such as lithography, were required to have an artistic eye if the finished product were to please and be effective in its communication.

This sense of artistic and technical skill is widely evident in the world of map making and in the production of topographic and other scenes in the eighteenth century, and was especially apparent in the nineteenth century. A good example of the talents of the painter, engraver, topographical artist, map maker and printer combined in one individual is the Edinburgh-based William Home Lizars, who was active in the first part of the nineteenth century. William Home Lizars was initially apprenticed to his father Daniel as an engraver within the family firm but changed direction upon the death of his father in 1812. Working as an engraver meant working on maps and on topographic art. Lizars was involved in the production of John Thomson's *Atlas of Scotland*, which was published in 1832. At that time, there was an interest in using stylised representations of topography as visual aids in the teaching of geography and the natural sciences: pictures of mountains and upland scenery, for example, were commonly employed to illustrate the connections between climate, altitude and vegetation. In his *Atlas*, Thomson used Lizars to engrave an image of the relative heights of Scotland's mountains (fig. 8.6). This is not a 'real' geography: it is a sort of topographic collage designed for visual effect and to instruct the reader. Left of centre – seemingly some distance inland – Thomson and Lizars have placed the distinctive Paps of Jura on the island of that name in the Inner Hebrides (fig. 8.7).

Where some artist-engravers used art in mapping to provide a comparative context for the features of Scotland's islands, other forms of mapping with a specific primary purpose sometimes used artistic elements simply to adorn the work. For example, the earliest British Admiralty Hydrographic Chart of the Sound of Iona, first published in 1860 (fig. 8.8), is a sea chart to facilitate navigation of the waters, reefs, rocks and other hazards along the island's coastline and the western margins of Mull. The inclusion on this otherwise 'scientific' navigational chart of a miniature picture of the ruins of Iona Abbey, backed by Ben More, is a welcome if somewhat atypical artistic feature (fig. 8.9).

The case of the Paps of Jura, as shown in Thomson's *Atlas,* presented island topography in what is properly called scenographic style, that is, as a backdrop to a theatrical

A COMPARATIVE VIEW OF THE HEIGHTS OF THE PRINCIPAL MOUNTAINS OF SCOTLAND.

performance. In such work, the gaze of the viewer is drawn into the landscape. Topography is pictured as if from the ground level, certainly from the standpoint of the artist or map maker 'looking in' to the view. But a common convention in map work is the planimetric view, the 'God-like' view from above, where variations in topography have to be represented because they cannot be seen from above. Contours and spot heights might serve for the approximate symbolic measurement of height. The skilful application of colour variation – a common feature of nineteenth-century topographic mapping – can add artistic detail to the map image and serve as a surrogate for measurement, training the eye to 'read the ground' by virtue of the shades used in its depiction. This combination of colour, line and symbol in the art of geography was one of the distinguishing features of the work of the leading map maker and map publisher John George Bartholomew of the firm of that name. If we examine a portion of his half-inch to the mile maps encompassing Mull and Iona (fig. 8.10a–d), we can see how Bartholomew over time developed his colour-based method of depicting Ben More by layers of carefully graded colour between contour intervals, and, for the sea, differentially shaded blues for depth, expressed in fathoms. The four images show Bartholomew's artistic experimentalism in how best to portray the topography, or, perhaps better, how best to show variations in topography within one chosen artistic colour-based method.

The Bartholomew firm was not the only body at the leading edge of map production and publication at the end of the nineteenth century and in the early years of the twentieth century. In the introductory chapter to the 1912 edition of *The Survey Atlas of Scotland* by John George Bartholomew, which was drawn, engraved, printed and published at the Bartholomew's Edinburgh Geographical Institute in Duncan Street in Edinburgh, the leading Edinburgh-based geologist Professor James Geikie who, like John George Bartholomew, was a co-founder of the Royal Scottish Geographical Society, made the following observations about the artistry of Ordnance Survey employees:

FIGURE 8.6
(*Opposite*) This representation of Scotland's mountains formed a stunning preliminary plate to the major county atlas of Scotland in the early nineteenth century. This style of diagram was increasingly used in the natural sciences from the early nineteenth century as part of the emergence of scientific diagrams and other forms of what were called 'statistical graphics'.
Source: D. McKenzie/William Home Lizars, 'A Comparative View of the Heights of the Principal Mountains of Scotland', from John Thomson's *Atlas of Scotland* (1832).

FIGURE 8.7
(*Above*) This painterly depiction of the Paps of Jura is a geographical and atlas-based example of an 'ectopic print', that is, in the wrong place. It is used here to make a point about the relative height of individual mountains within Scotland's overall topography. Perhaps because he had no ready access to them, McKenzie has chosen not to represent the Cuillin on Skye within this comparative geography.
Source: D. McKenzie/William Home Lizars, 'A Comparative View of the Heights of the Principal Mountains of Scotland', from John Thomson's *Atlas of Scotland* (1832).

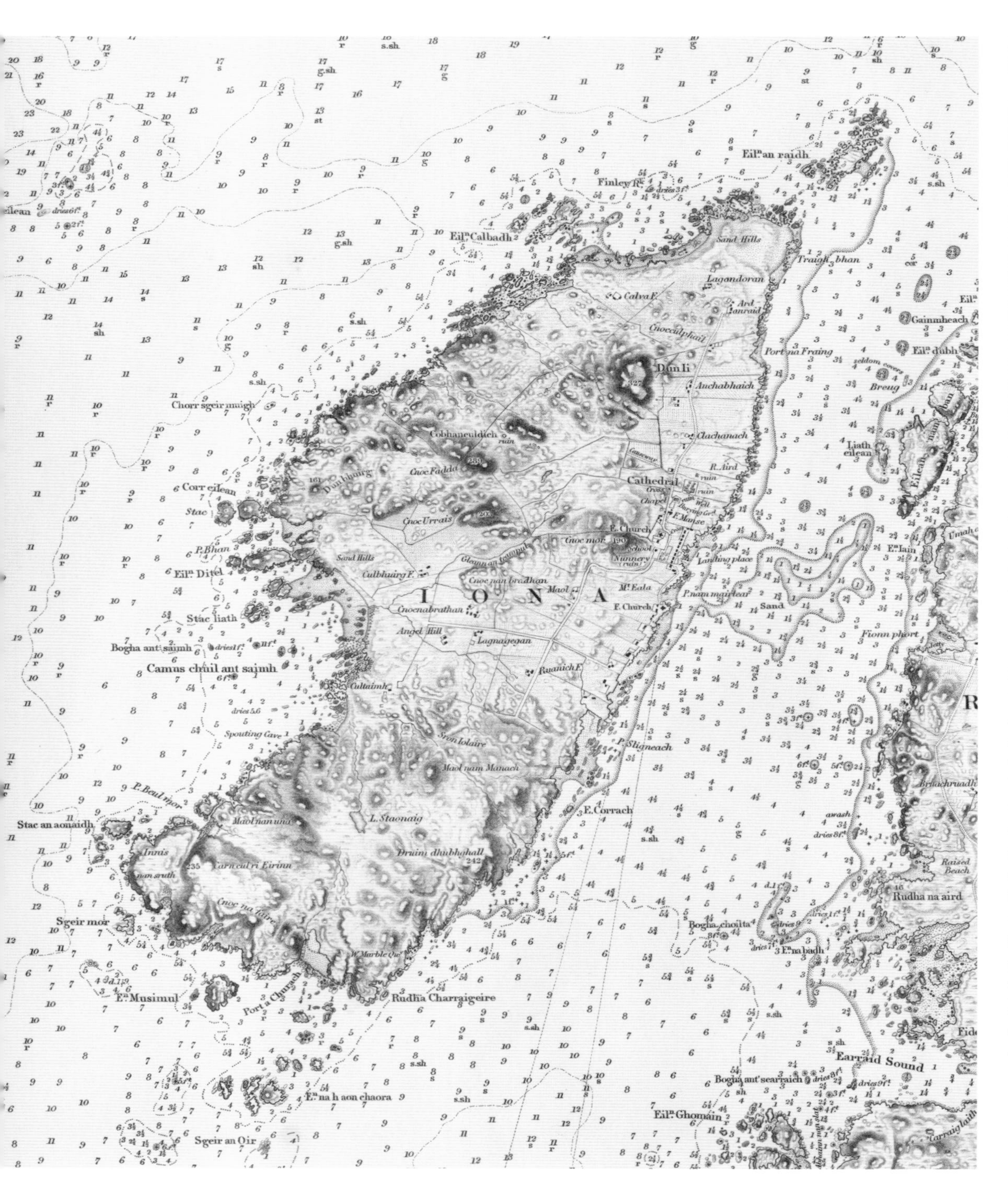

FIGURE 8.8
(*Left*) The island of Iona and the western tip of Mull as depicted in the Admiralty chart of 1860.
Source: Hydrographic Office, *Iona*. Admiralty Chart 2617 (1860).

FIGURE 8.9
(*Above*) The detail of the 'Ruins of Iona Cathedral from the N.W.' from the Admiralty chart of 1860. This choice of view by the naval artist has allowed him to incorporate Ben More on Mull as part of the image, giving both topographic depth and placement to the architectural ruins in the foreground. Since there is no obvious navigational significance to this depiction of the ruined Cathedral, we may presume it was a conscious choice on the part of the naval hydrographer and, perhaps finally, of the Hydrographic Office as a whole, to include this artistic element in deference to its historical and ecclesiastical significance.
Source: Hydrographic Office, *Iona*. Admiralty Chart 2617 (1860).

> In the case of the Ordnance Survey map, on the scale of 1 inch to a mile, the varying forms of the surface are so faithfully delineated as frequently to indicate to a trained observer the nature of the rocks and the geological structure of the ground. The artists who sketched the hills must indeed have had good eyes for form. So carefully has their work been done, that it is often not difficult to distinguish upon their maps hills formed of such rocks as sandstone from those that are composed of more durable kinds.

Even the effects of glaciation had, to his discerning eye, been made clear by the map makers of Ordnance Survey. They had, he observed, 'been reproduced with marvellous skill on the shaded sheets issued by Ordnance Survey. And yet the artists were not geologists. The present writer is glad of this opportunity of recording his obligations to those gentlemen.' So should we be, to the many unsung map engravers of W. & A. K. Johnston, Bartholomew and Ordnance Survey and other map-making bodies who have rendered Scotland's island geographies in ways at once artistic and scientific.

Maps and word pictures: prose and poetry

According to the English landscape historian W. G. Hoskins, 'Poets make the best topographers.' Poetry is indeed a powerful means to place depiction, allowing the reader-cum-listener to imagine that sense of place connoted by the poet's imaginative economy with words. In fiction, too, words carefully chosen may provide a form of 'picturing' and maps can help with this. To illustrate this, let us return to the 1860 Admiralty chart of the Sound of Iona but to a different detail. Robert Louis Stevenson, world-famous author and Edinburgh contemporary of Bartholomew and Geikie, had a background and heritage in the design and building of lighthouses through his father Thomas, and his grandfather Robert, as well as other members of the Stevenson dynasty (see fig. 4.11 for example). Robert Louis Stevenson, we may presume, was familiar with sea charts and with the literary 'power' of maps. The dramatic scene he evokes in his famous tale *Kidnapped*, first published in 1886, when his principal character, David Balfour, is shipwrecked on the Torran rocks to the south-west of the island of Erraid, is based on close personal observation of the terrain in question and, probably, of the relevant sea chart (fig. 8.11). Stevenson spent three weeks on Erraid in August 1870 as an apprentice civil engineer at the time when his father Thomas and uncle David were engineers for the Northern Lighthouse Board and were overseeing construction of a lighthouse on the isolated rock of Dubh Artach, some fifteen miles south-west of Erraid. As Stevenson at one point noted, 'The shores of Earraid were close in; I could see in the moonlight the dots of heather and the sparkling of the mica in the rocks.' In *Kidnapped*, David Balfour, after spending four days apparently marooned on this island in appalling weather conditions, suddenly realises it is only a tidal island and that it is linked by sand to the Ross of Mull twice a day at low tide: 'In about half an hour I came out upon the shores of the creek; and sure enough, it was shrunk into a little trickle of water, through which I dashed, not above my knees, and landed with a shout on the main island. A sea-bred boy would not have stayed a day on Earraid; which is only what they call a tidal islet, and, except in the bottom of the neaps, can be entered and left twice in every twenty-four hours.' Perhaps, if David Balfour had had a sea chart with him, he would have known that he could have readily escaped from his temporary prison. Maps and charts do indeed prompt action – but you have to know of them first.

We know that Robert Louis Stevenson, after abandoning civil engineering and emerging as a writer, was a map lover, for he records his fascination with maps: 'I am told that there are people who do not care for maps, and find it hard to believe . . . he who is faithful to his map, and consults it, and draws from it his inspiration, daily and hourly, gains positive support, and not merely negative immunity from accident.' For Stevenson, the map was much more than a reference tool, a paper geography. In his *Treasure Island*, first published in

MORVEN
SOUND OF MULL
LOCH TUA
Treshnish Isles
Gometra
ULVA
LOCH NA KEAL
Little Colonsay
Inch Kenneth
MULL
ARDMEANACH
LOCH SCRIDAIN
IONA
Brolass
ROSS OF MULL
LOCH BUY
FIRTH OF LORN

a.

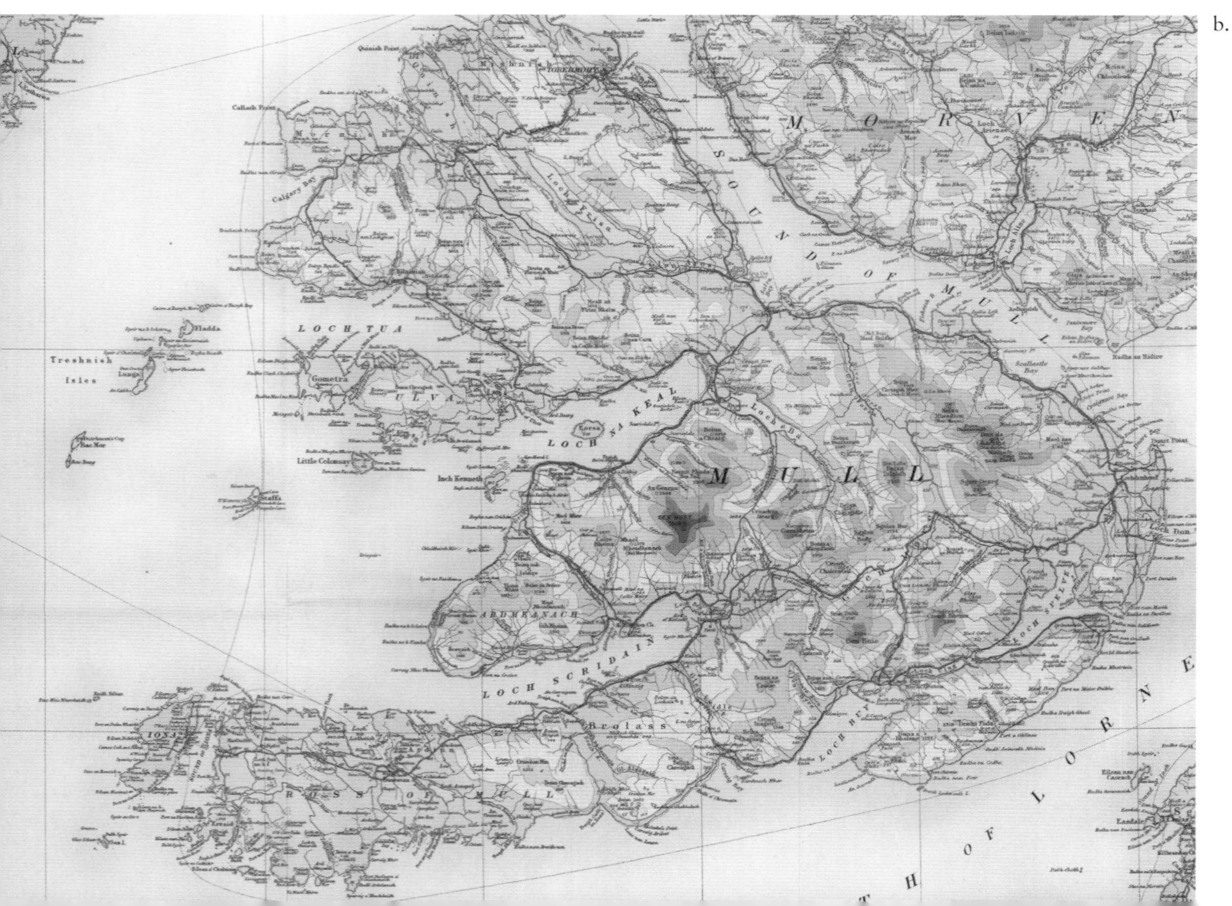
MORVEN
SOUND OF MULL
LOCH TUA
Treshnish Isles
Gometra
ULVA
LOCH NA KEAL
Little Colonsay
Inch Kenneth
MULL
ARDMEANACH
LOCH SCRIDAIN
IONA
Brolass
ROSS OF MULL
LOCH BUY
FIRTH OF LORNE

b.

FIGURE 8.10
The technique of layer colouring – employing different shades of green for low ground closest to sea-level, and shades of brown for higher ground – had been developed on the continent during the nineteenth century, but the Bartholomew family and firm developed it in a way that was both innovative and influential. John Bartholomew Junior (1831–1893) first used layer colouring in maps for a guide book for John Baddeley in 1880, and his son, John George Bartholomew (1860–1920), went on to refine the printing process and aesthetics continually over the next four decades. This can be illustrated very well by these four layer-coloured half-inch to the mile maps of Mull, dating between 1886 and 1911. The earliest sheets were simpler, often with less precise registration of multiple colours, and requiring fewer printing 'pulls' of the sheet to build up colour depth. However, by the early twentieth century, Bartholomew had overcome these problems, creating a complex and subtle map that was both easy to read and strikingly attractive. During these years, we also see Bartholomew's consolidation of their original half-inch 'District Sheets' just covering specific areas, into a regular numbered series covering all of Scotland. With full coverage of Great Britain at this scale complete by 1903, Bartholomew's half-inch series was enduringly popular through most of the twentieth century, particularly for leisure pursuits such as cycling, motoring and walking.
Source: (a) John Bartholomew, *Bartholomew's Reduced Ordnance Survey of Scotland, Mull Sheet* (1886). (b) John Bartholomew & Co., *Bartholomew's Reduced Ordnance Survey of Scotland, New Series, Sheet 10* (1892). (c) John Bartholomew & Co., *Half Inch to the Mile Maps of Scotland, New Series, Sheet 10* (1905). (d) John Bartholomew & Co., *Half Inch to the Mile Maps of Scotland, New Series, Sheet 10* (1911).

c.

d.

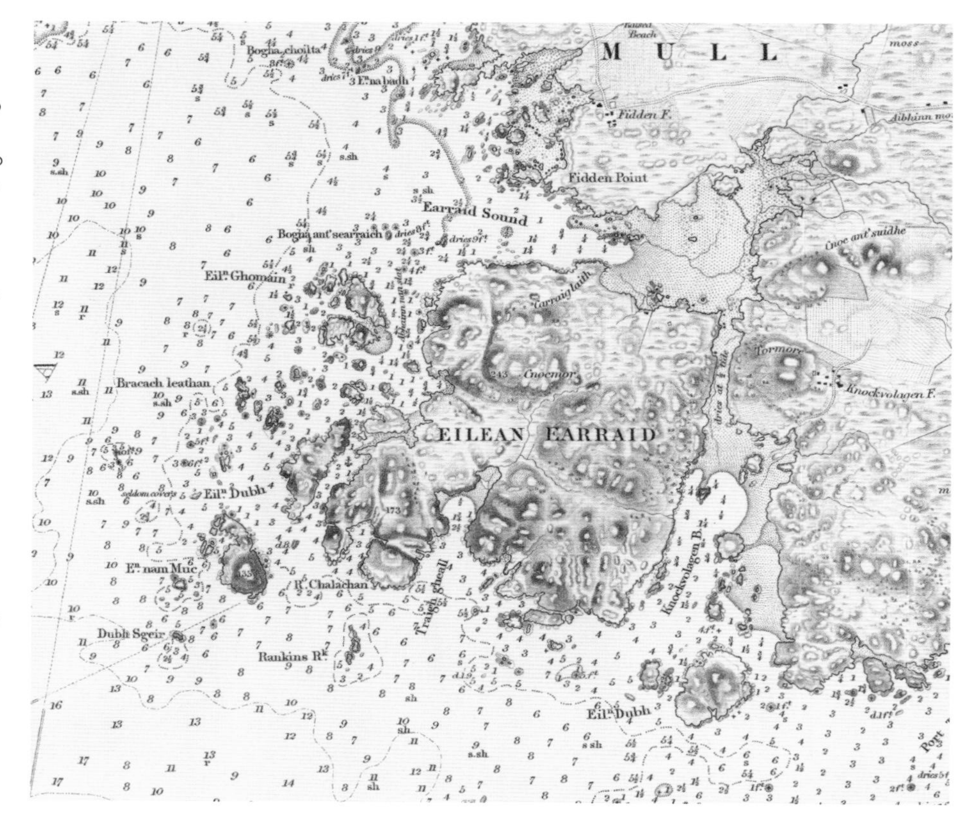

FIGURE 8.11
This image shows a detail of the small tidal island of Erraid, located at the extreme south-western tip of the Ross of Mull and less than two miles due south of Fionnphort, the well-used ferry point for the short crossing to Iona. In 1860 when this Admiralty chart was published there is no evidence of a permanent population living on the island, though there are scattered settlements shown on the adjacent areas of Mull. Some ten years later Erraid was bustling with activity in its role as shore station during the building of the far offshore lighthouse of Dubh Artach. Red granite from Erraid was quarried to provide stone for the lighthouse, tracks were laid to transport this stone and a temporary population of workmen quartered on the island in cottages and bothies. Interestingly, the next issue of this chart in 1886 (to which large corrections had been made in January 1872) shows no evidence of the remains of this temporary workplace – the Hydrographic Office as official chart-maker was more appropriately concerned with recording the changing water and shoreline features along this very tricky coastline.
Source: Hydrographic Office, *Sound of Iona*. Admiralty Chart 2617 (1860).

1883, he treats it as an aid to artistic creation, even as a central 'character' in the unfolding of the narrative itself.

Several Scottish poets make reference to maps and to geography in poetry relating to Scotland's islands: Sorley MacLean – whose work on Hallaig we have noted (see chapter 3 and fig. 3.12) – is one, Norman MacCaig and George Mackay Brown two distinguished others. Norman MacCaig was especially entranced by the theme of islands, by the geographies conjured up in the mind by the power of names to 'fix' things on the map:

> Round my head
> She releases a flicker of names
> That come out of geography but emerge, too,
> From myth – Muckle Flugga, Taransay, Sule Skerry.
> The Old Man of Hoy.

But, as MacCaig also astutely observed, mapping was about the 'in-filling' of space, about the art of authoritative inscription in place of 'uncivilised' blankness, a trope he skilfully reverses:

> there are spaces to be filled
> before the map is completed –
> though these days it's only
> in the explored territories
> that men write sadly,
> *Here live monsters.*

In his *An Orkney Tapestry*, the native Orcadian George Mackay Brown conjures a vivid word picture of the end of the ferry trip across the Pentland Firth from Scrabster in Caithness to Stromness in Orkney:

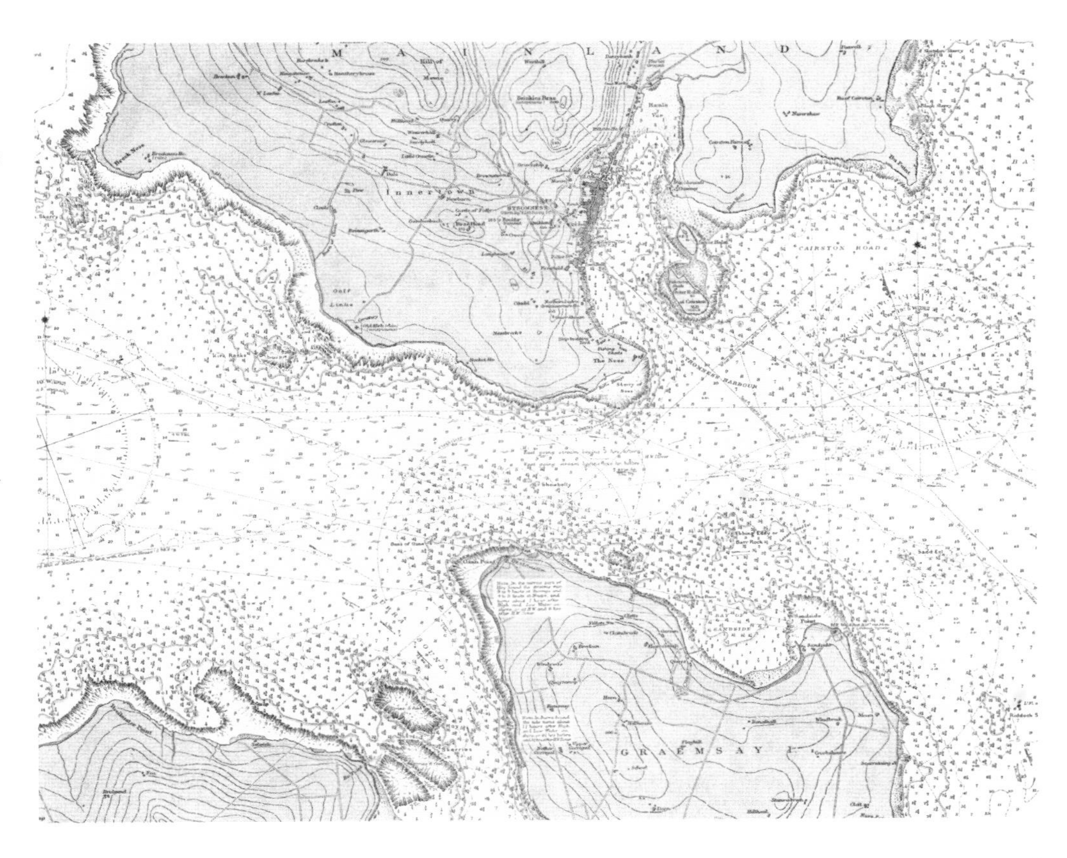

FIGURE 8.12
Travellers from Scrabster in Caithness to Stromness in Orkney may feel apprehensive when crossing the Pentland Firth given its reputation as an occasionally hazardous stretch of water where conflicting tidal races from east and west meet. But unless they are sailors, travellers may not be aware of the other dangers of this coast – the rocks, holms and skerries as well as the many shallows and conflicting tidal streams. This detailed chart of the approaches to Stromness contains all salient navigational information to ensure a safe passage. But safety and accuracy require maps to be kept up-to-date. This newly revised version of a chart first surveyed in 1905–06, was produced six years before the birth in Stromness of George Mackay Brown, during the early part of the First World War.
Source: Hydrographic Office, *Hoy Sound*. Admiralty Chart 2568 (1915).

> The cliffs of western Hoy rise up, pillars of flame . . . The Kame of Hoy, and Black Craig in the south-west of the main island, are the wide pillars of the doorway. The ferry-boat turns between them into Hoy Sound, past the green island of Graemsay with its two lighthouses, the dumpy octagon and slim dazzling cylinder. Small brown-and-green humps appear, Cava and Fara, the Scapa Flow islands; and the hills of Orphir. A tall iron beacon wanly winking, stands in the tide-race. The *St Ola* turns into Stromness harbour. The engines shut off. The boat glides towards the pier, the gulls, the waiting faces.

Getting to the island is a geography of movement, here described in shades of colour and light. The Orkney archipelago is indeed a living tapestry of sorts. Its changing moods and landscapes cannot ever be reduced to the dimensions of the map. One sea chart of Hoy Sound, prepared just before George Mackay Brown's birth in 1915 (fig. 8.12), presents a less bright and certainly a static picture. But we should not, in thinking of this as a 'true' geography, dismiss the work that has gone into its naming, or diminish the artistry of the map maker in translating this and other islands to paper. Nor should we overlook the information that this and other maps contain which allows map makers and map users to know what is an island and what not.

LORNA cum insulis vicinis
et provinciis eidem conterminis.
LORN wt the Yles and provinces
bordering there-vpon.
Part of
Iura Yland.
Loch Cardineon
Wemh Rich
SKARBA.
Miliaria Scotica communia
1 2 3 4 5 6
Meridies
LVING
SEIL
KREIGENES
MEL
FORT.
LAERN-IKRACH.
Part
of
Knap-
dail.
Invidos virtute torqueбо.
Nob. Dño.
D. IACOBO BALFOVRIO
Militi Baroneti di Kynaird, Leoni Armorum
Regis et Regni Scotiæ, nec non Insularum
adjacentium, etc. Tab. hanc D.D.
J. Blaeu.

Joan Blaeu, *Lorna cum insulis vicinis…* (1662).

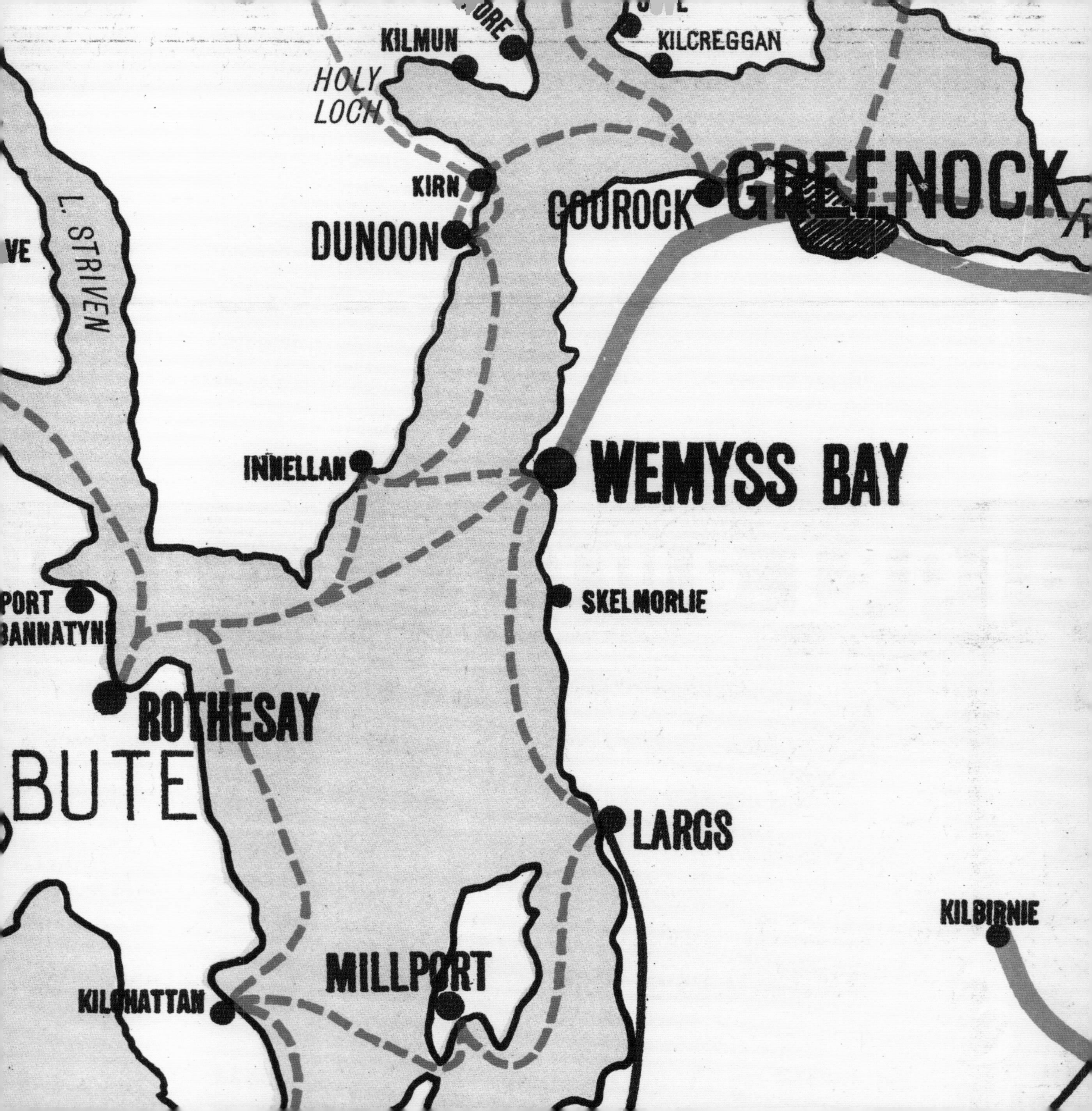
KILMUN
KILCREGGAN
HOLY LOCH
KIRN
GOUROCK
GREENOCK
DUNOON
L. STRIVEN
INNELLAN
WEMYSS BAY
PORT
BANNATYNE
SKELMORLIE
ROTHESAY
BUTE
LARGS
KILBIRNIE
MILLPORT
KILCHATTAN

CHAPTER NINE

ESCAPING

Ev'ry island is a prison
Strongly guarded by the sea;
Kings and princes, for that reason,
Pris'ners are, as well as we. Anon.

James Boswell, in his published journal describing the tour of the Hebrides undertaken in early autumn 1773 with his learned if somewhat irascible companion, Dr Samuel Johnson, recounted the above song when they were staying at Talisker on the Isle of Skye. Johnson – seemingly somewhat out of sorts – suddenly asked Boswell if he remembered the above song. From his drift, Boswell supposed Johnson was thinking of their confined situation on the island and that he wanted to escape – in spite of being pleased by the company and the hospitality – although that would have proved impracticable because of the weather.

Samuel Johnson's recollection of the song points to an interesting paradox: by their very nature islands have been used as places on which to contain or restrain people against their will as a result of the deliberations of others. Think, for example, of Mary, Queen of Scots, first held for her own safety, as a four-year-old child, on Inchmahome in the Lake of Menteith (fig. 9.1), and later in her life, as a prisoner on Castle Isle in Loch Leven in 1567 and 1568. Think, too, of Lady Rachel Grange, imprisoned on St Kilda in 1734 for seven years, or of the island of Rona (North Rona) which was offered, although rejected by the government, as a penal colony by Sir James Matheson, or of Italian prisoners of war held on Lamb Holm, Orkney, during the Second World War (fig. 5.11). Whether queen, titled lady or prisoner of war, it is likely that all will have thought about and may actually have sought to escape from their island prisons – although of these three only Queen Mary was to succeed in escaping, in 1568, albeit only briefly. In contrast those like Johnson, visiting an island with no intent to hold him against his will, may have felt constrained nevertheless, perhaps even agoraphobic, by the simple fact of being separated and isolated from what they saw as the more comfortable familiarity of that known, larger and more connected world beyond the island in question. The paradox lies in the fact that while being constrained on an

Opposite. Detail from John Bartholomew & Co., *Caledonian Railway: Clyde Watering Places and the Western Highlands* (1888).

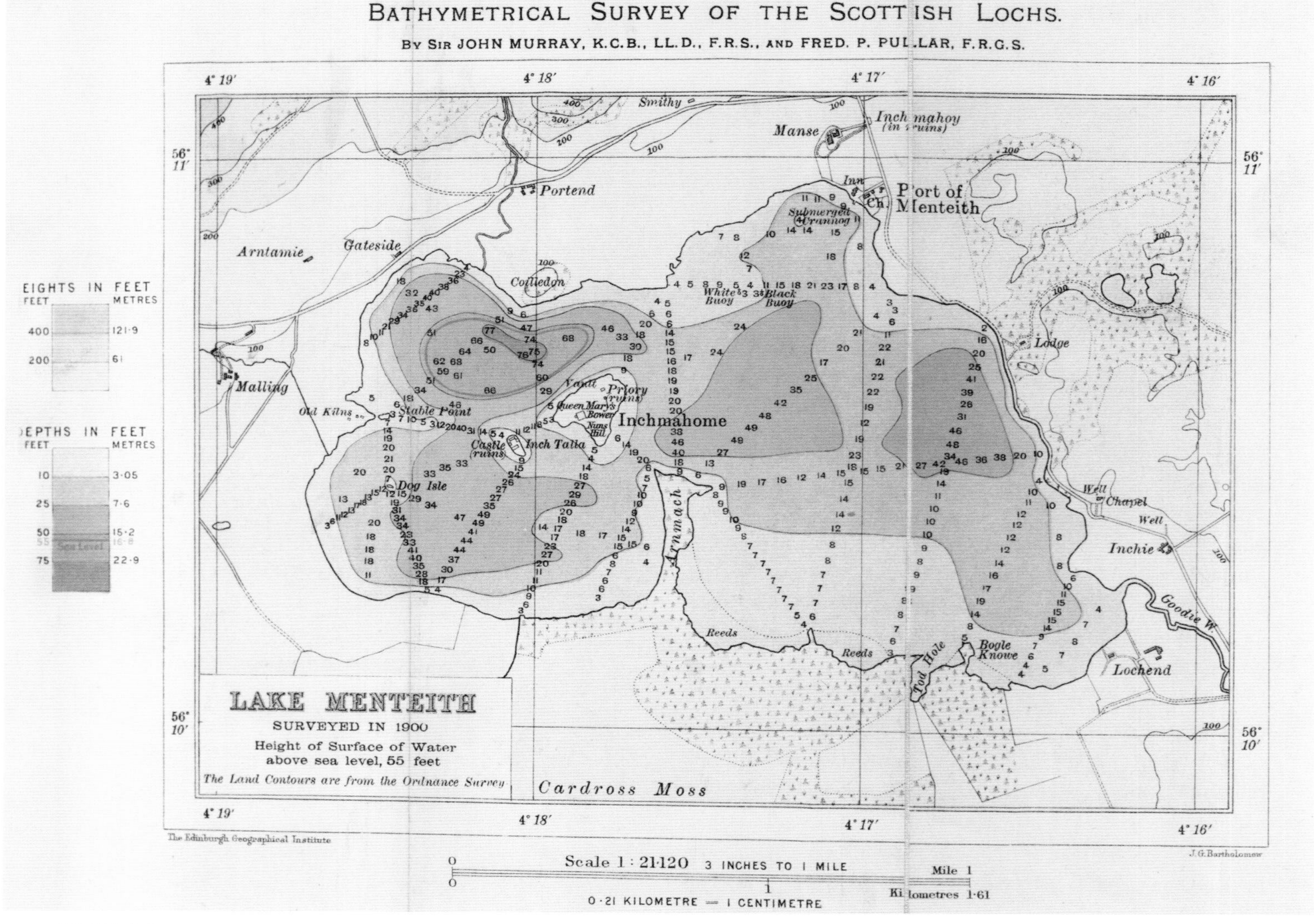

FIGURE 9.1

Inchmahome, a freshwater island situated in the Lake of Menteith three miles east of Aberfoyle, and, with its Augustinian Priory, a religious site and place of pilgrimage during medieval times, became a temporary place of retreat and refuge for Mary, Queen of Scots in 1547 and 1548. The threat of English invasion led those charged with protecting the queen to remove her to a place of relative safety – a freshwater island. This detailed bathymetric plan of the Lake of Menteith – produced some three hundred and fifty years after Mary's confinement there – demonstrates well the security of her refuge. The varying depths of the water are shown through the subtle use of colour shading and by including the lines and numerical values of soundings. That Inchmahome's ruined priory remained a place of escape is reflected in its selection in the twentieth century as the burial place of one of Scotland's most colourful writers, travellers, journalists and politicians, R. B. Cunninghame Graham. He was frequently referred to as 'Don Roberto' from his long association with and interest in South America. Cunninghame Graham was for a short time the first President of the Scottish Nationalist Party.

Source: Sir John Murray and Laurence Pullar, 'Lake Menteith', surveyed 1900, from *Bathymetrical Survey of the Fresh-water Lochs of Scotland* (1897–1909).

island engenders in some people a desire to escape, for others islands are places that people deliberately choose to retreat to or escape to. This chapter examines several of the themes associated with escaping to and from Scotland's islands – for different purposes – as illustrated through maps and charts.

Islands as places of retreat

Confinement need not be at the command of others. It can be self-imposed. And an island's lure and attraction as a place distant and isolated from the wider world, culturally if not geographically, may not always remain unchanged. Many of the remote islands off the Atlantic coasts of the British Isles, once deliberately selected by Irish monks of the early Celtic or Culdean Church for their isolation, their wind-swept solitude and their ideal conditions for a life of prayer and contemplation, have today become places of seasonal mass pilgrimage and thus of congregation and congestion (fig. 9.2). During the summer months, some islands off Scotland's coasts today demonstrate what might best be described as conflicting ideals regarding escape and retreat. Those persons wishing to retreat from the crush of urban life for a period of solitude or for a quiet holiday find themselves caught up with and unable to escape from what may seem like an almost constant stream of people wishing to view historic sites, once the epicentre of religious quiet and contemplation. Examples of such sites include Iona in the Inner Hebrides for its ecclesiastical importance and Orkney in the Northern Isles for the archaeological pre-eminence of Maeshowe, the Stones of Stenness and the Ring of Brodgar (fig. 2.2a and b).

Escape, particularly in its modern context of 'retreat' from the too-rapid rhythms of 'modern life', can therefore on islands sometimes be turned into its antithesis – the periodic assembly and amassing of transient people – with the concomitant requirements necessary to service such groups only exacerbating the situation. The same situation can arise, of course, for those booking package holidays to Pacific islands or other island retreats, where such islands, advertised as 'a place to

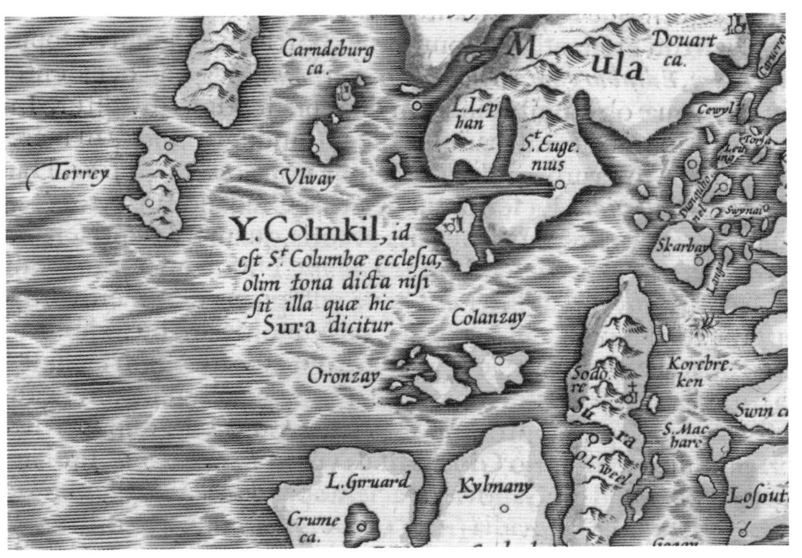

FIGURE 9.2
It is not uncommon for map makers to include additional text on their map, sometimes a brief anecdote or geographical fact. In this detail from a late-sixteenth-century map of Scotland by Gerard Mercator, the map maker has deliberately added a Latin text on St Columba's special association with Iona. Mercator was renowned for his calligraphic as well as his map-making skills. But despite Iona's geographical proximity to the island of Oronsay (shown immediately below the lettering relating to St Columba and Iona), which is by tradition also associated with St Columba and his disciples, Mercator either did not know of this tradition, or chose not to make reference to it on his map. By the mid-fourteenth century, Oronsay was also an important religious site in its own right, with an Augustinian Priory under the patronage of the Lord of the Isles. Today, it remains a place of quiet escape, a site of contemplation with far fewer visitors than Iona.
Source: Gerard Mercator, [*Scotiae regnum*], in two sheets – detail from northern sheet (1630).

escape to' in the travel literature, are often far from free of people. On summer days in Skye, Iona and elsewhere, later generations of North American and Antipodean Scots – anxious in their genealogical wanderings to visit and connect with the homes of their island forebears – must sometimes be puzzled about the 'clearances', given that such places now appear so crowded.

Sacred isles and sites for contemplation

At the north-western edge of the known world in Classical times and into the early medieval period, the islands and waters off the north and west coasts of Scotland were not only seen as remote but also often perceived as stepping stones or potential gateways to what might lie beyond, in a physical sense and in a spiritual sense. In the minds of people, questions arose about the nature of those small islands and isles to the north and west of what was regarded as the centre of the world, around the Mediterranean – were there further lands out there? Where did the world end? Where did the souls of the dead journey to? The Classical Greek concept of the Islands of the Blessed, of Elysium – where the souls of their heroes were said to go – or, within the Celtic tradition, of Tir na n-Óg, literally 'the Land of the Young', led to the view that these places must lie beyond the ends of the then known world – which, in this case, was taken to be somewhere beyond Scotland's Northern and Western Isles/Na h-Eileanan an Iar. Did Paradise indeed lie there? Could it be reached by boat?

Assays into these unknown waters include the legend of St Brendan, an Irish monk commonly known as Brendan the Navigator, who left his native Ireland in the fifth century AD and first sailed to parts of the west coast of mainland Scotland and to the island of Iona. In Brendan's case, his curiosity is said to have led him further, on an epic sea journey across the Atlantic, possibly reaching the Faroes, Iceland, Newfoundland and even the east coast of the North American continent. In his 1976–77 re-creation of St Brendan's voyage, the explorer Tim Severin was able to prove that such a journey would have been possible, by sailing from the Dingle Peninsula in Ireland via the Hebrides/Innse Gall and Iceland to Newfoundland in a boat modelled on one which St Brendan might have used.

Although the islands in these Celtic seaways were stepping stones for further exploration, they were also places chosen as ideal for seclusion and spiritual contemplation. Other Irish monks associated with the Celtic Church, together with their followers, penetrated the Scottish islands during this period (fig. 9.3). The most famous of these was St Columba, around AD 563. Evidence of the monks' domicile on some of these islands is still visible and was recorded by Ordnance Survey surveyors working in the mid to late nineteenth centuries (see fig. 2.3 for example). Some two hundred years after the Celtic monks established their cells and churches on these islands, Viking seafarers setting out on their journeys south-west from Scandinavia viewed these islands not as sites of retreat but as opportunities for plunder. Long-term retreat to an island for a life of prolonged spiritual contemplation is easier than hurried retreat from an island in the face of pressing secular despoliation. Although no maps of the Scottish islands survive from the Viking period – if indeed they were ever created – the existence of such islands has been known or conjectured about from Classical times. Later, tentative attempts to indicate their existence on maps of the known world that survived from the tenth to the twelfth centuries usually show what we know to have been inhabited sites as nebulous shapes only, features of the map makers' imagination rather than their inhabitants' sense of sacred place in the world.

Of the 750 individuals in the list of Scotland's recorded saints, Columba, Ninian and Magnus are perhaps best-known to the public today – they are certainly the best recorded on maps. Columba is particularly associated with Iona and with Inchcolm in the Firth of Forth (fig. 9.4). Magnus is associated with the Northern Isles, and Ninian with Whithorn in Galloway (fig. 9.5). Holy Island off the east coast of the larger island of Arran is another island location associated with the early Celtic Church; it has remained a spiritual site but with a new look in the present century (fig. 9.6). Scotland's islands encapsulate geographies of particular but enduring piety.

FIGURE 9.3
The earliest surviving maps to portray the islands off Scotland's western seaboard in any detail post-date the period of the Celtic Saints by some ten centuries. They therefore do not indicate the sites of their early refuges. Yet they do record several of those places which, into the early-modern period, either preserved the sites of, or were directly associated with, the early Church. This detail showing the Western Isles/Na h-Eileanan an Iar records the names of a few Irish-born saints such as Columba, Patrick and Clement. The map from which this detail is taken was published around 1573 by the Antwerp-based map and atlas maker and publisher Abraham Ortelius. Its content is based on that included on a larger-scale, multi-sheet map of the British Isles produced ten years earlier in Duisburg by his friend and rival, Gerard Mercator. In terms of Scottish content, there is some reason for attributing this to John Elder, a Caithness-born priest who is known to have produced a map of Scotland in the 1540s and to have travelled to Skye and to Lewis/Leòdhas, though his map has not survived. *Source*: Abraham Ortelius, *Scotiae tabula* (*c.*1573).

Evading capture

Islands have often been natural destinations for those wishing to escape from their pursuers. Perhaps one of the most famous island escapees was Charles Edward Stuart, better known as Bonnie Prince Charlie. After the disastrous Battle of Culloden (16 April 1746), which effectively marked the end of '45 Jacobite Rebellion, Charles was a wanted man: his pursuit over the Scottish Highlands and Islands was soon commemorated in broadsheets, books and maps (fig. 9.7). With British army forces hunting for him on land and sea, and clear dangers for those offering him shelter or support, Charles lived on the run for five months. His initial destination was the Western Isles/Na h-Eileanan an Iar but he found Stornoway/Steòrnabhagh hostile to him, and the Minch full of navy ships. He was lucky that MacDonald of Clanranald initially offered him shelter in South Uist/Uibhist a Deas for a few weeks, but he was forced to flee to Benbecula/Beinn na Faoghla, and, from there, to Skye, where Flora MacDonald famously helped disguise him as her female servant. But his time in Skye was precarious – he was fired upon at Waternish (his size made him an unlikely maidservant) – and he fled to Raasay by early July before quickly returning to Skye and then heading for Mallaig. The next two months were spent being pursued around parts of Lochaber, living in the open and hiding in caves. It must have been with huge relief that information reached him of two French ships, sent to rescue him from Loch nan Uamh. Charles sailed away in the early hours of 20 September, and after narrowly avoiding a British squadron off Britanny, he landed in Roscoff on 30 September. The prince over the water would never return.

FIGURE 9.4

While St Columba is known for his arrival and landing on the island of Iona in the Inner Hebrides in AD 563 and for the founding of the original abbey there, he is also associated with other holy sites. One of the most important of these is on the small island of Inchcolm (shown here as 'St Columb's I.') in the Firth of Forth. The remains of an early hermit's cell can still be seen on the island. The island continued as a holy site and from the twelfth century was the site of an Augustinian abbey. This detail from a chart of the Firth of Forth dating from around 1730 and based on an earlier survey and chart of the Forth by John Adair some thirty years earlier (see fig. 4.3), shows the position of Inchcolm. It is depicted as a place of retreat and tranquillity in what the chart's engraver, Richard Cooper, otherwise portrays as a busy Firth, highlighted by the addition of ships plying its waters and by the manner in which he has rendered the water. His criss-cross lines extending from the northern shores of the Forth give the idea of an unsafe coastal strip in terms of the rocky outcrops offshore and the dangerous stretches of shallow water with underlying sand even at high tide: these features are visible today when on the train journey between Inverkeithing and Kirkcaldy. The island of Inchcolm was not only a place for escape and monastic seclusion from the early Celtic period onwards but also, in later periods, somewhere off which ships suspected of carrying dangerous forms of pestilence were confined until the contagion had passed and they were permitted to make for their intended destination. *Source*: John Adair/Richard Cooper, *The River and Frith of Forth* (*c*.1730).

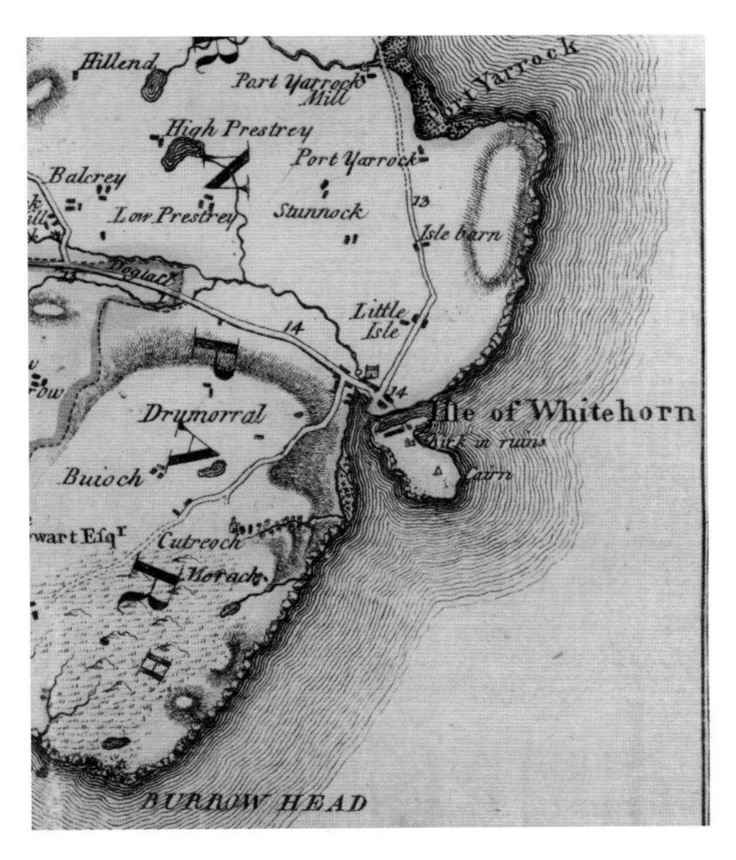

FIGURE 9.5
When does an island cease to be an island? The Isle of Whithorn is no longer an island in the true sense, being now permanently connected to the mainland of Wigtownshire by the man-made construction of Harbour Row. But in the past it was only connected to the mainland at low tide by a causeway. It is recorded as an island by the Jedburgh-born surveyor, map maker and publisher John Ainslie, in this enlarged detail from his map of Wigtownshire published in 1782. A century later, the village of Isle of Whithorn was described in Francis Groome's *Ordnance Gazetteer of Scotland* thus: 'The most southerly village in Scotland, it stands upon what was once a rocky islet, and conducts some commerce with Whitehaven and some other English ports having a well-sheltered harbour, with a pier erected about 1790.'
Source: John Ainslie, *A Map of the County of Wigton or the Shire of Galloway* (1782).

Literary retreats

Islands throughout the world and throughout history have always been also places to retreat to in the hope that the creative muse may inspire more readily by virtue of one's isolation from distraction. Painters, poets, writers and musicians of all nationalities, past and present, have sought heightened creativity by retreating to an island outwith their normal geographical milieu. Think, for instance, of the painter Paul Gauguin in Polynesia, or of the poet and writer Robert Louis Stevenson in the South Seas, and, in the twentieth century and nearer to home, the novelist George Orwell, who wrote most of his influential and ground-breaking novel *Nineteen Eighty-Four* at Barnhill, an isolated house located off an unmade track on the eastern side of the Isle of Jura, some three miles from the northern end of the island (fig. 9.8). On Orkney, one would perhaps think of the composer Peter Maxwell Davies, late Master of the Queen's Music, on Hoy in Orkney. In each of the examples cited, creation was spawned by isolation: many of their finest works result from the writerly or painterly retreat only easily afforded by islands.

Islands can symbolise not just a retreat but also a means to restitution – creation stimulated by virtue of managed separation. This is also true of islands as repositories of past geographies. The collection of folklore on and about the Scottish islands is one good expression of this. In the Gaelic culture of the Western Isles/Na h-Eileanan an Iar, much of what we take now to be folklore is history remembered in a certain form, passed down in an oral tradition and commemorated in poetry, tale and song. In the Hebrides/Innse Gall, John Francis Campbell of Islay – Iain Og Ile – and John Lorne Campbell to name two prominent figures in this respect, were

a. b.

FIGURE 9.6

Holy Island, an early religious site and retreat associated in the sixth and seventh centuries with the Irish monk St Molios, and, in the twelfth century, the site of a small monastery, has today reinvented itself as a religious site and retreat with a difference. Purchased in 1991 by the Scottish-based Buddhist Centre of Samye Ling in upper Eskdale, Dumfriesshire, the island is now a spiritual centre open to all.

This island retreat overlooks a great natural harbour, described in 1828 in *Lumsden & Sons' Steam Boat Companion* as 'the best anchoring place in Arran, and . . . the safest retreat, for vessels in a gale, of any port on the frith of Clyde'. Both the island and the waters of Lamlash Bay to its west side were then clearly viewed as retreats. Map extract (a) shows Ordnance Survey's representation of the relief features of the island, using – to dramatic effect – contour lines depicted in a burnt-sienna shade overlying hachured lines in a shade of dull brown. In contrast, the British Admiralty's mid-nineteenth-century chart (b) uses hachures only, to far less effect, though its first priority was of course to indicate the nature of the safe anchorage in Lamlash Bay. It does this well and does so by concentrating on providing numerous soundings and describing the Bay on the chart as a 'good anchorage' with 'scarcely any tide'.

Source: (a) Ordnance Survey of Scotland, One-Inch to the mile, *Island of Arran, Parts of Sheets 21.13* (1911). (b) Hydrographic Office, *Arran Island – Lamlash Harbour*. Admiralty Chart 1974 (1849).

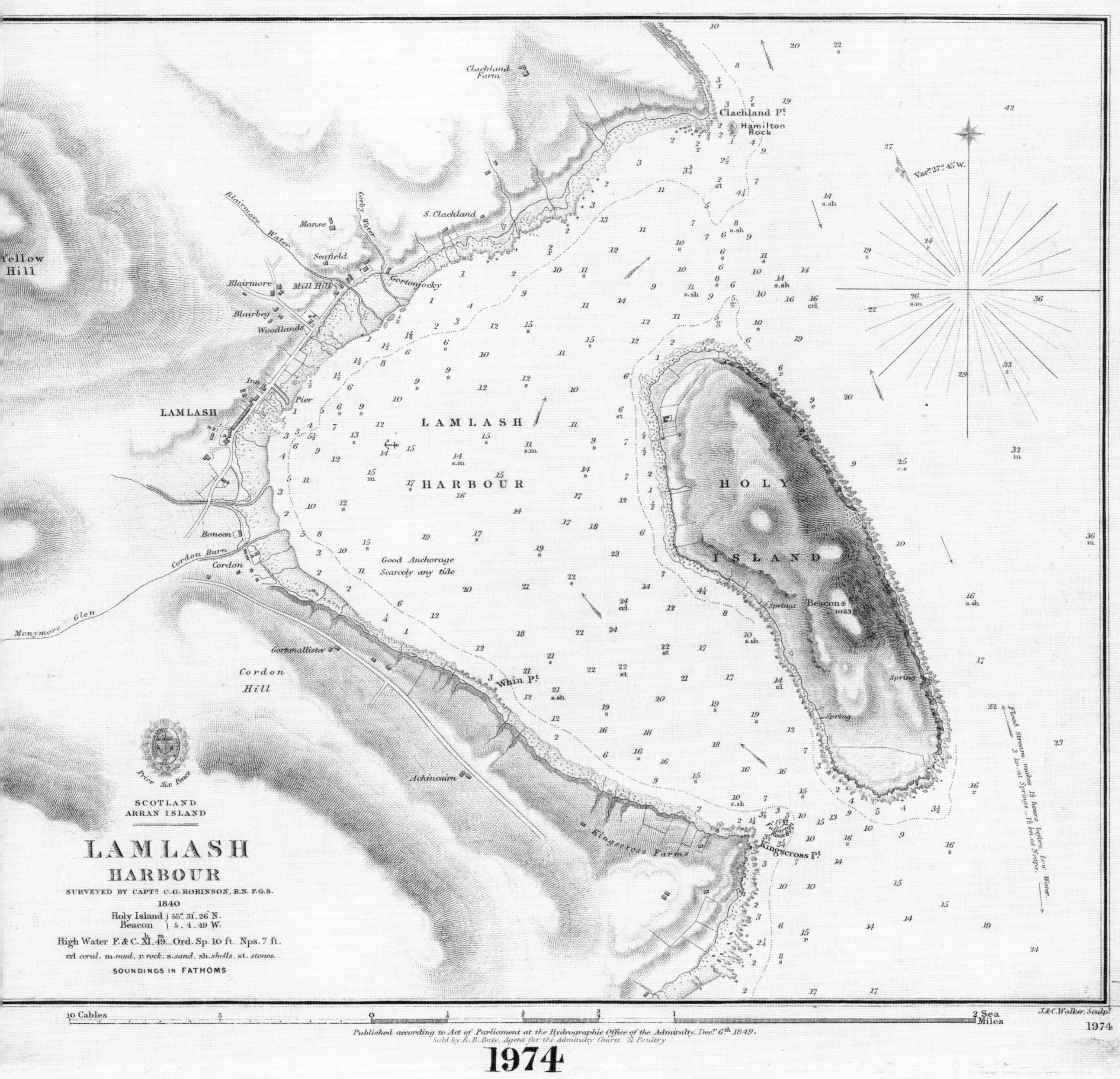

Yellow Hill
Clachland Farm
Clachland Pt.
Hamilton Rock
Varn. 27° 45' W.
Blairmore Water
Corby Water
Manse
S. Clachland
Seafield
Blairmore
Mill Hill
Cortonjocky
Blairbeg
Woodlands
Inn
Pier
LAMLASH
LAMLASH
HARBOUR
HOLY
ISLAND
Beacon
1023
Springs
Spring
Boneen
Cordon Burn
Cordon
Monymore Glen
Good Anchorage
Scarcely any tide
Cortonallister
Cordon Hill
Whin Pt.
Achincairn
Kingscross Farms
Kingscross Pt.
Flood Stream makes 1½ hours before Low Water.
3 kn. at Springs 1½ kn. at Neaps.
Price Six Pence
SCOTLAND
ARRAN ISLAND
LAMLASH
HARBOUR
SURVEYED BY CAPTN. C. G. ROBINSON, R.N. F.G.S.
1840
Holy Island Beacon 55° 31' 26" N. 5. 4. 49 W.
High Water F. & C. XI. 49. Ord. Sp. 10 ft. Nps. 7 ft.
crl coral, m. mud, r. rock, s. sand, sh. shells, st. stones.
SOUNDINGS IN FATHOMS
10 Cables
2 Sea Miles
J. & C. Walker, Sculpt.
1974
Published according to Act of Parliament at the Hydrographic Office of the Admiralty, Decr. 6th 1849.
Sold by R. B. Bate, Agent for the Admiralty Charts, 21 Poultry

1974

Stornway
L. Stornway
LEWIS
Cap Faro
I. Erifort
Strath naverr
I. Shifort
I Glas
HARRIS
Ross
Rona
Rarsa I.
I. Nord Wist
I. Isac
I. Benbicula
Rossiness
I. Whie
C. Ussiness
L. Skibbord
Corodel
I. S KIE
Portry
Sluppa
STRATH
SCLEAT
Kilmirg
Fort de Bernera
I. Sudvist
Kannay
Rum
Egg
INVERNES SHIRE
Ch. de Glingary
Ch. de Glin
L. Boisdel
Erisca I
I. Bara
Muc
Cap D'ard Namurchin
Boradel
ARISAG
Kenlochmoyedart
Daleilie
L. Shiel
Glinfinan
L. Arkek
Etablissement de l'Etendard Royal
Moy
Fassfern
Col I.
MUL
L. Swinard
LOCH
A BIR
Fort Guillaume
Turrif I.
le Prince allant en Ecose
le Prince venant en France
Dunstaffage.
F. d'Innersnau
LORN
Inverary
Assemblée des Campbels
LE

FIGURE 9.7

(*Opposite*) This detail of the Western Isles/Na h-Eileanan an Iar showing Bonnie Prince Charlie's escape from Scotland after Culloden is from one of nine sheets, which together form an impressive map, measuring 159 × 108 cm, covering all of Great Britain. It was drawn by James Alexander Grante, who described himself in the map's title as 'Colonel of the Artillery of the Prince in Scotland', and the map also includes a detailed written description of the main events of the '45. Grante was a Jacobite French officer of Scottish descent, who accompanied Bonnie Prince Charlie throughout most of the 1745 Rebellion. There are a number of variant states of this map, and Grante went on to produce another, smaller, map in 1749 showing the campaign on a single sheet, also attempting to rally support and favour for the Jacobite cause among French Jacobite supporters. On the map here, note Charles' open-topped rowing boat, fleeing back and forth across the Minch, with British naval ships all around. While the coastal outline and shapes of the major islands were intended only as a backdrop to the main itinerary, it is striking how dated and idiosyncratic they are. One of Grante's other maps credits Robert Morden (an English map maker who died in 1703) as his main cartographic source, and so, in many ways, the coastal outline here reflects the Blaeu outlines of a century earlier.
Source: James Grante, *Carte où sont tracées toutes les différentes routes, que S.A.R. Charles Edward Prince de Galles, a suivies dans la grande Bretagne* (*c*.1747).

FIGURE 9.8

(*Right*) The isolation of the writer George Orwell's retreat at the far north end of Jura is shown well on this detail from a contemporary Admiralty chart, though Barnhill, the house he lived in, is not marked. Orwell wrote *Nineteen Eighty-Four* there between 1947 and 1948. The house was situated almost immediately to the right of 'Cot Fank' on the chart. Orwell's stay there was not without incident. Together with some companions, Orwell set out for a sail in the Gulf of Corryvreckan with its notorious whirlpool, got into difficulties when their boat overturned and was lucky to be rescued from the uninhabited island of Eilean Mor. The 1974 edition of the Clyde Cruising Club's *Sailing Directions and Anchorages West Coast of Scotland* makes this comment about the waters of the Gulf of Corryvreckan: 'This gulf is considered the worst in the West Highlands, and strangers are warned against it . . . There is only ¼ hr. of slack water at Sp. and 1 hr. at Np. in the gulf, when it is possible to pass through.'
Source: Hydrographic Office, *Loch Killisport to Cuan Sound – Including the Sound of Jura*. Admiralty Chart 2326 (1944).

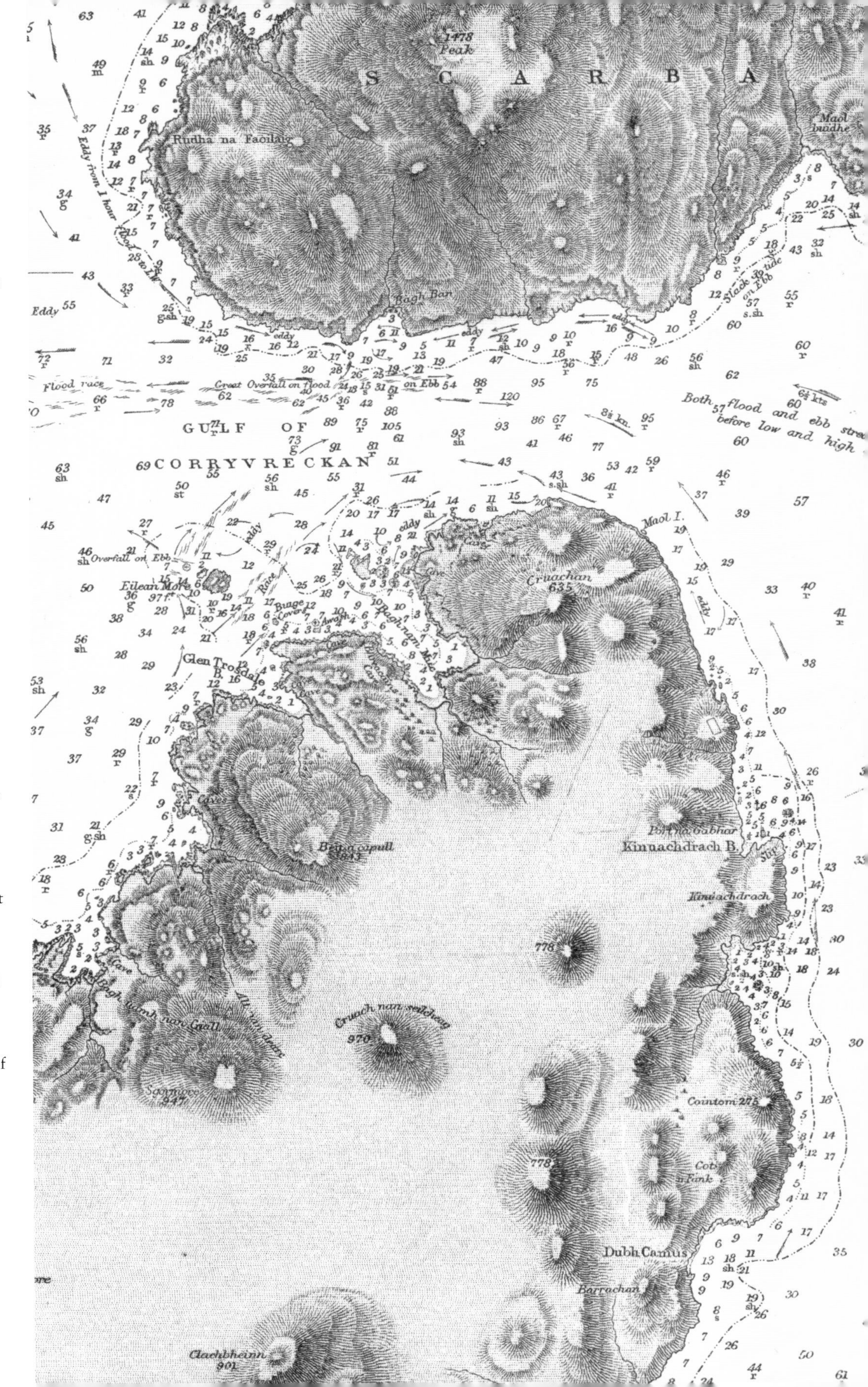

both fascinated by the islands' legends and folklore and travelled extensively in the Hebrides to collect, record and publish these. Both had island bases – Islay in the case of Iain Og, on his family's estate near Bridgend, and Canna for John Lorne Campbell – and these, in their respective ways, acted as literary retreats, sites of literary accumulation as well as of artistic reflection. Writers like Sir Compton Mackenzie, a self-confessed island lover, lived on a variety of islands including Jethou in the Channel Islands, and the Shiant Islands in the Minch off Lewis/Leòdhas, before finally settling on Barra/Barraigh where his creative talents produced perhaps his best-known work, *Whisky Galore*, immortalised in the Ealing Studios film of 1949.

Tourist retreats and summering isles

We have already touched on the paradox of escape for the purpose of deliberate retreat and the fact that such escape sometimes leads to congestion caused by there being too many other people intent on the same thing. Although the west coast of Scotland and the waters surrounding its islands were known and traversed by local peoples from time immemorial – and without the need for maps or charts – tours of the Scottish islands and Hebrides/Innse Gall by outsiders were relatively unusual until the later eighteenth century. This was due to a combination of factors: political and religious conditions between the sixteenth and mid eighteenth centuries, the difficulties of travel and accommodation, the relative isolation of the islands themselves and the fact that aesthetic categories such as 'the sublime' and 'the picturesque' which attracted visitors to the margins of Britain were not common cultural currency or applied to the islands until the beginning of the nineteenth century. Recorded tours by figures such as Thomas Pennant in 1772 (fig. 9.9), Boswell and Johnson in 1773 and by Sir Walter Scott in 1814 made the reading public increasingly aware of Scotland's islands, either as the locus of a presumed antiquated way of life teetering on the edge of extinction, as the venue for cultural expressions less and less

FIGURE 9.9
(*Right*) Thomas Pennant (1726–1798) was a leading writer on natural history in the eighteenth century and a noted antiquarian with a keen eye for topography and landscape. Some have also explained his zeal for travel – initially in Europe in the mid 1760s, latterly in Scotland in 1769 and 1772 – as a desire to escape from domestic troubles, particularly the death of his father, wife and son in close succession. Pennant's published account of his 1769 *Tour*, chiefly in the Highlands, was well-received, and encouraged his return visit in 1772, this time accompanied by the botanist, the Rev. John Lightfoot, the Rev. John Stuart (a Gaelic scholar) and his personal draughtsman Moses Griffith. The 1772 *Tour* involved a two-month voyage, sailing on a 90-ton cutter from Greenock. They visited many of the west coast islands including Bute, Arran, Gigha, Jura, Islay and Skye, before landing at Loch Broom, where their journey continued by land.

Pennant's map, which was included in editions of his *A Tour in Scotland MDCCLXIX* and *Tour in Scotland and Voyage to the Hebrides* after 1777, clearly shows this voyage, but it is interesting for other reasons too. This is, in fact, the earliest printed map of Scotland to correctly show the Great Glen as a straight line – a fact well known to William Roy and his assistants, but their 1747–55 Military Survey was not revealed to the public for the next half-century. It is possible that Pennant caught sight of the original Roy map through his friendship with Roy, but more likely, perhaps, that he saw a proof copy of Roy's 'Mappa Britanniae Septentrionalis', eventually published in 1793. His shared interest with Roy in antiquities also explains the Ptolemaic names given to various islands (such as Dumna for Skye, and the Five Ebudæ for the Western Isles/Na h-Eileanan an Iar), but many other topographical details were added through his extensive correspondence with local people. Pennant continued to add details to his draft map for at least two years before publication, and updated his original outlines based on Dorret with the new surveys of Murdoch Mackenzie (chapter 4). Pennant's tours were important in inspiring many later travellers to visit Scotland and the islands.
Source: Thomas Pennant/John Bayly, 'A Map of Scotland, the Hebrides and Part of England adapted to Mr Pennant's Tours' from *A Tour in Scotland MDCCLXIX* (London: Printed for Benj. White, [1777]).

Five EBUDÆ
of Ptol
Harris
Glass I.
L. Tarbet
Housaness
Slakanish B.
Rowranish
C. Ore
Kirkabol
Monach I.
NORTH UIST
L. Madic
L. Rona
L. Rond
The Minch
Maiden-Rock
Dunvegan Head
BENBECULA I.
C. Urich
Uisness
Childonan
L. Chusbay
L. Skipper
Gill
SOUTH UIST
L. Eynord
L. Boysdale
Humbla
Fuda I.
Borg
BARRA I.
Kismuil C.
Watersay I.
Dear I.
Bishops I.
COL I.
TIREY I.
Duntulm C.
Rona
Dundonald
Rownisdale
Kingsburgh
Dunvegan
DUMNA
I. of SKIE
Kesto
Brakadale
Kilmorney
Eynard L.
Dunan
Dunan P.
RAASAY I.
Scalpa I.
Mackinnons C.
Canay I.
Sand I.
Rum I.
Eyford L.
Ardslate
Egg I.
Muck I.
Aich I.
Ardnamurchan P.
Row Ru Pt.
L. Ewe
L. Broom
L. Gare
Torrisdale
Pluckhirt
Everdale
Smith T.
L. Mari
Sherriefs
Fern
Cullakill
Applecross B.
L. Dibea
Torridon
Applecross Kirk
Annale
Borridale
L. Toscaik
L. Kisserne
L. Carran
L. Duit
L. Russan
KINTAIL
L. Duich
Glendochart
Invershal
Glenshiel
Bernera Barrack
Glen Elg
Barrisdale
L. Urn
L. Nevish
Kinloch
L. Morrer
Oban-Morer
Arisaig P.
L. Na Nuna
Glenbeasdale
L. Ranach
L. Hallyort
Essa
Hill
C. Tyro
L. Moydart
L. Shiel
Gorlanforn
L. Sundart
Strontian
Tobirmoire
Killvore
Sound of Mull
Druman
Acharn C.
Aros
Staffa I.
Jona I.
Glenkanner
MALEOS MULL
Burg
Duart C.
Tirergan
Glenbair
Kerera I.
Easdale I.
Scarba
Du Hirtach
Colonsa I.
Oronsay
Ina Gyed
I. Eun
Loch Tarbat
ISLAND of JURA
Paps of Jura
JURA
SOUND OF JURA
Small Isles
L. Graynard
Finlagan Cas.
EPIDIUM ILAY
Dunvegan
L. in daal
Sound of Ilay
Giga I.
Cara
Killbery
Ardpatrick
Skipness
L. Tarbat
Duppan
Killian
Machrthanish Bay
EPIDII
Saddle C.
Campbeltown
Dunaverty C.
Kilkeran
Sanda I.
Mull of Cantire
EPIDIUM PROM.
RICNEA PLIN.
Rathry or Racline I.
Giants Causey
Broom
Ardmore
Logie
Craigsaven
Grunard
Dundonel
L. R. Strathnaskelly
Pracklaid
L. Kellan
L. Druim
Benaldiel
Dornyne
L. Damve
L. Crosk
L. Bran
Dingwall
Foulis
L. Lwychart
Cannon R.
L. Monar
L. Ouhoy
Fairry R.
Real
Combir
Tullich
C. Dunie
Kiltmore
Carrymony
Inverness
Beluachrack
Urquhart C.
St. Ninians
Benlogd
Loch Ness
Old Ft. George
Belvacract
Invermorison
Kinckie
L. Garry
Invergarry
Ft. Augustus
Aberlarf
L. Spey
Murlagan
L. Arkek
Achnacary
Achanellan
L. Lochy
Inverglue
Castle of Inverlochy
Old Ft. William
Tulloch
Ura
Laggan
Kiltros
Moye
Adrering
Inverlair
Galachy
L. Ericht
L. Rannach
Annon
Glenco
Kildurer
Appen
Ellenstalker C.
Linnhe L.
L. Lismore
Beregonium
L. Etive
Dunstaffnage C.
Stonfield
Glen Urchy
Invergaunn
Glen Larichenan
Inverurchy
L. Lyon
Killmore
Kilicharn
Scotstown
Port Sonachan
Soak
Lochy
Tiendrum
Glen Lochy
Finlayrig
Strathfillan
Glen dochart
Armady
Ardinaddie
L. Aw
Iverary
Dundera
Carndow
ARGYLL
Ard Kinglas
Kilenner
Kilmaklash
DUMBARTON
L. Lomond
Pass of Aber
Cardros
Kilmartin
L. Fine
Glenderawel
Otter
Kilfinan
Kilmore
L. Long
Kilmund L. Goyl
Buchannan
Dunbarton
Aird
L. Stryvan
Gurock
Greenock
Port Glasgow
Renfrew
Glasgow
Paisly
RENFREW
Halkek
Bute I.
Rothsay
Cumra I.
Kelly
Killburn
Kilbrial
Beith
Dalry
Saltcoats
Stewarton
Kilwinning
Irvine
Kilmarnock
Ramsay C.
GLOTA INS.
I. OF ARRAN
Goatfield Hills
Brodic
Lamlash
Kilmorie
Pladda
GLOTA ÆST. FIRTH OF CLYDE
Ladys I.
Prestwick
Irvine R.
Auchinluck
St. Quivox
Air
Ochiltrd
Stair
Cumnock
AIR
Heads of Air
Castlehill
New ark
Dalrymple
Glenhead
Doon R.
Kebryd
Bargenny
Cassillo
Maybole
Ailsa
Boghall
Kirkoswald
Turnbury
Girvane
Kilkerran
Carney
Arran

seen on the mainland or as the home of indigenous species of plant, bird and animal life. Such accounts prompted outsiders to consider visiting at least some of them, and led to the start of a small tourist industry, with Glasgow and the ports of the Firth of Clyde leading the way, as places from which to visit the Hebrides/Innse Gall. Aberdeen became the principal southernmost port of exit for visiting the Northern Isles, although trips across the Pentland Firth from Scrabster and John O' Groats to Orkney were available to those travelling by rail.

With the growth in shipbuilding on Clydeside during the nineteenth and twentieth centuries, such local sites developed into places from which both cargo and people could be transported to and from the west coast islands. Ports on the Clyde were used as emigration points for those evicted from their Highland homes and those who had a choice in seeking a new life overseas. Many of those who lived and work in the area saw islands not as a new home but as places for a temporary escape from the drudgeries of industrial labour. A tour 'Doon the Watter' or round the Kyles of Bute was once a common way of taking a breath of fresh air, or a longer holiday (fig. 9.10) and of course had a significant impact on local island towns such as Rothesay (fig. 9.11). From June 1875, the ss *Dunara Castle*, registered in Port Glasgow and built by the local firm of Blackwood & Gordon, was one of several ships which carried cargo, locals and tourists to and from some of the outlying islands on tours of the Hebrides, on occasion travelling to St Kilda. The ship was involved in the evacuation of the St Kildans in 1930 and continued in service until it was broken up in 1948.

Another more recent and popular recreational escape to Scottish islands has been on board yachts and boats, which of course require good charts, whether on paper or in digital form, and experience in using them. For navigation across wider channels and open sea, Admiralty charts continue to provide a vital function, and these can be supplemented by more detailed pilot guides for entry into harbours and ports, or charts by commercial map makers. However, leisure yachts often wish to navigate close to shore, to explore secluded inlets and bays away from harbours for anchorages, and, for this, the deficiencies of Admiralty charts have encouraged recent private surveying and mapping initiatives of west coast islands (fig. 9.12).

Retreat and seclusion do not require sea travel. Islands within freshwater lochs in mainland Scotland were often selected as places of refuge and escape. Many have been used as places for one's final rest: the small river island of Inchbuie within the River Dochart near Killin, for example (fig. 9.13), the burial place of Clan MacNab, or, as we have seen, Inchmahome in the Lake of Menteith (fig. 9.1).

Maps either escaped or lost

Maps also 'escape' and there are some good examples of this associated with the mapping of the Western Isles/Na h-Eileanan Iar. Where evidence exists in sources such as early newspaper advertisements or notes to published topographical accounts, remarks made on existing maps relating to their source survey, or from hearsay passed down the generations, can lead to questions over the survival of the maps. For the Western Isles/Na h-Eileanan an Iar, there is evidence (fig. 9.14a) that some detailed surveys were made of the islands towards the end of the eighteenth century and in the early nineteenth century, but that the maps themselves have not survived. Their location is now not known (fig 9.14b).

The idea of escaping thus has numerous connotations and implications when applied to maps and to islands. In one important sense, islands are only as clear in the mind as they are clear on the map. Details on the ground always escape the map maker's attention. Details on the map regularly escape the map reader's attention. Maps help facilitate an escape from the reality they purport to depict. We 'escape' – or try to at least – the confines of our normal life by retreating to islands: to write, to paint, to relax, to be ourselves. If, in many ways, islands are places not just of escape it is because the some*where* that they are allows us to be some*how* different, if only for a while.

FIGURE 9.10
This striking poster produced for the Caledonian Railway company illustrates rail and steamer routes in the Clyde environs in 1888. The adverts at the base provide further detail, promoting the 'Wemyss Bay Route – Quickest and Best for Innellan, Rothesay, Largs and Millport and the Ivanhoe trip to Arran' as well as the 'Greenock Route – convenient for Dunoon, Kirn, Holy Loch, Loch Long' and for many places further afield with 'Mr McBrayne's Royal Mail Steamers'. The Caledonian Railway's Gourock Railway and Pier was 'expected to be opened for the season of 1889', which it just managed – by 1 June 1889. Note the inconspicuous thin black lines of the rival North British Railway's routes. In the late nineteenth century, firms such as Bartholomew printed huge volumes of mapping for railway companies: route maps, timetable maps, tourist itineraries, posters and advertisements. Posters often have a low survival rate, and so we are lucky to still have this poster through the Bartholomew Archive Printing Record. *Source*: John Bartholomew & Co., *Caledonian Railway: Clyde Watering Places and the Western Highlands* (1888).

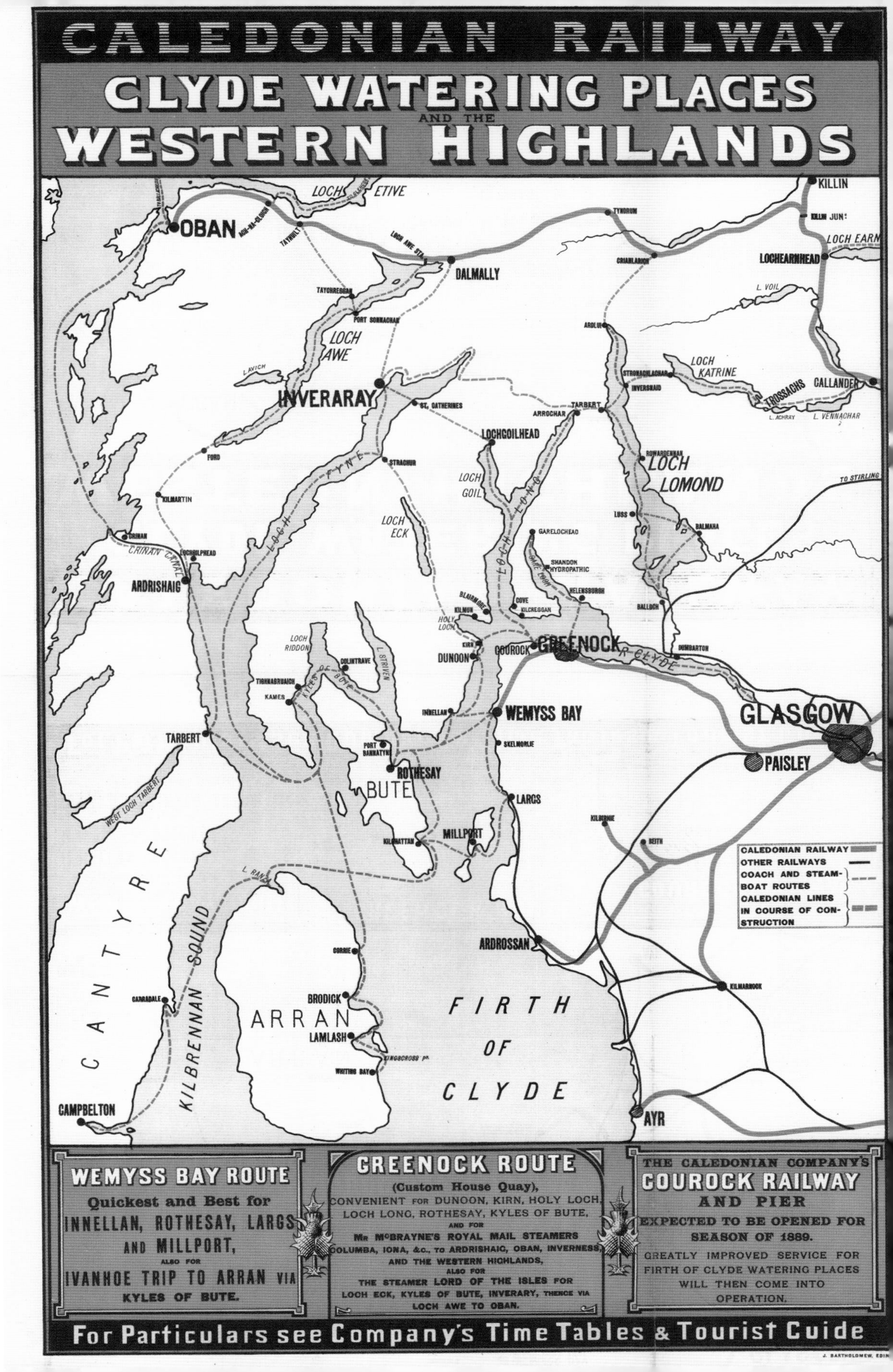

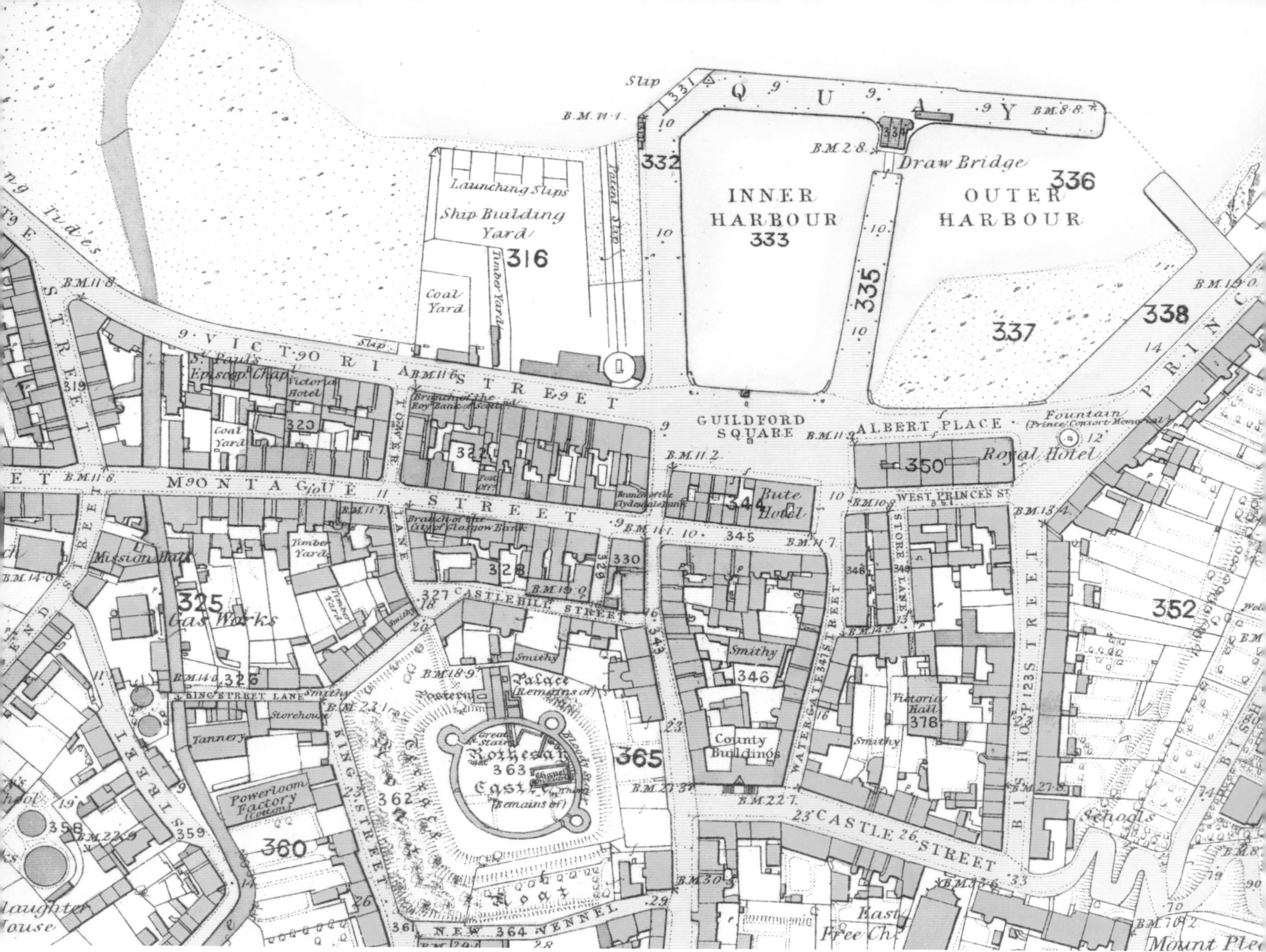

a.

FIGURE 9.11

These two detailed Ordnance Survey 25 inch to the mile maps of Rothesay, in (a) the 1860s and (b) the 1890s, illustrate the growing predominance of summer visitors, particularly those arriving by steamer from the Clyde environs. Large hotels, many of them named on the 1860s map, dominated Victoria Street on both editions, but with local philanthropy, major investments had been made during the 1870s, converting four acres of foreshore into ornamental gardens and an esplanade to the west of the harbour, with an octagonal bandstand. A new tramway had been laid out, taking visitors along the coast to Port Bannatyne to the north, and a Royal Aquarium was erected between 1875 and 1876 on the site of a former battery at the east end of the town. The traditional former industries of Rothesay, particularly textiles and fishing, were in rapid decline during

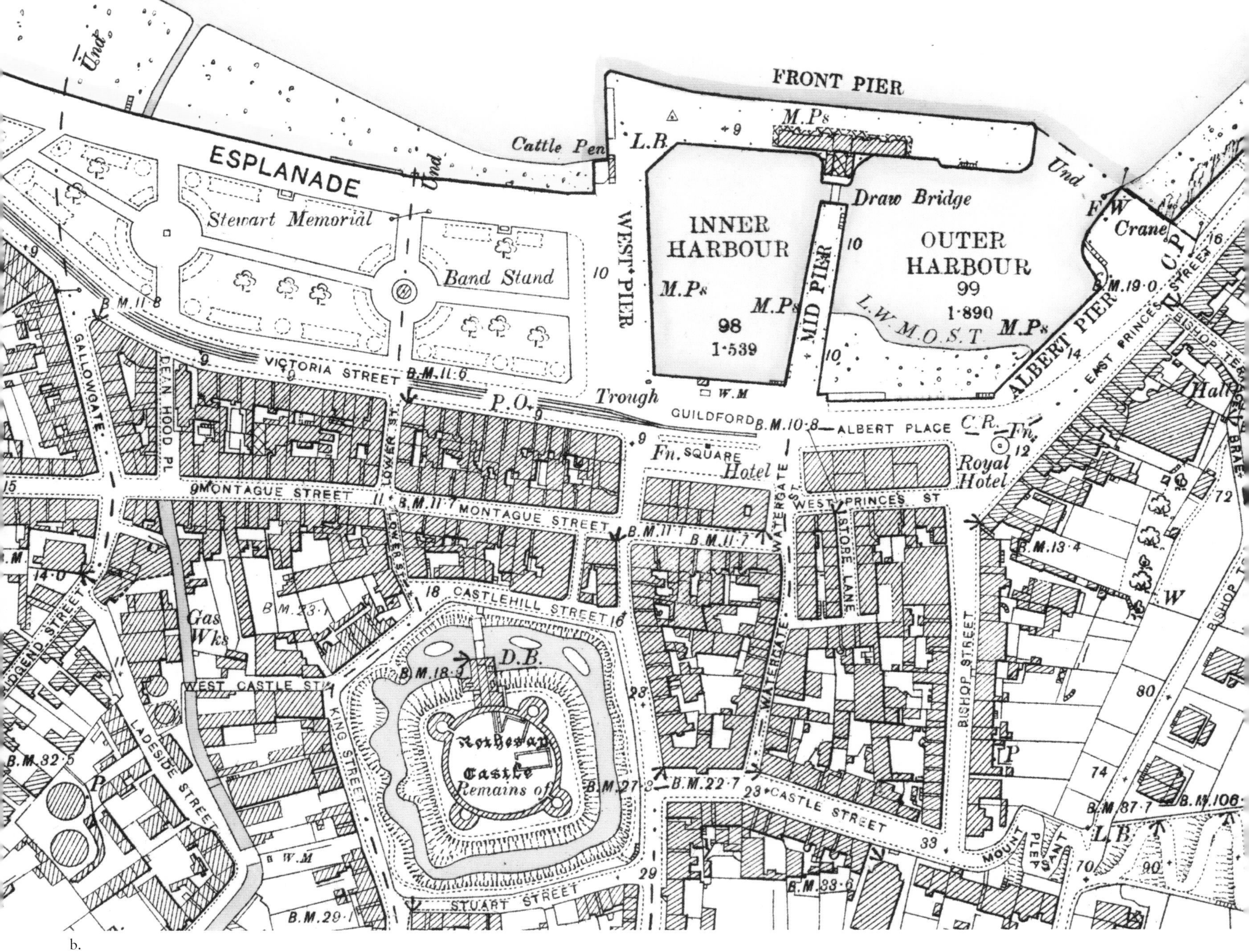

b.

the nineteenth century, and the town increasingly depended for its income on tourism. As noted by Francis Groome, 'In any weather and under any circumstance [Rothesay] would attract the eye, but it looks its best under a bright summer sun, with its blue waters dotted with skiffs and white-sailed yachts, and ploughed by the keels of gaily-crowded steamers.'
Source: (a) Ordnance Survey, 25 inch to the mile, *Argyll and Bute, Sheet CCIV.6 (North Bute)* (surveyed 1866, published 1869). (b) Ordnance Survey, 25 inch to the mile, *Argyll and Bute, Sheet CCIV.6* (revised 1896, published 1896).

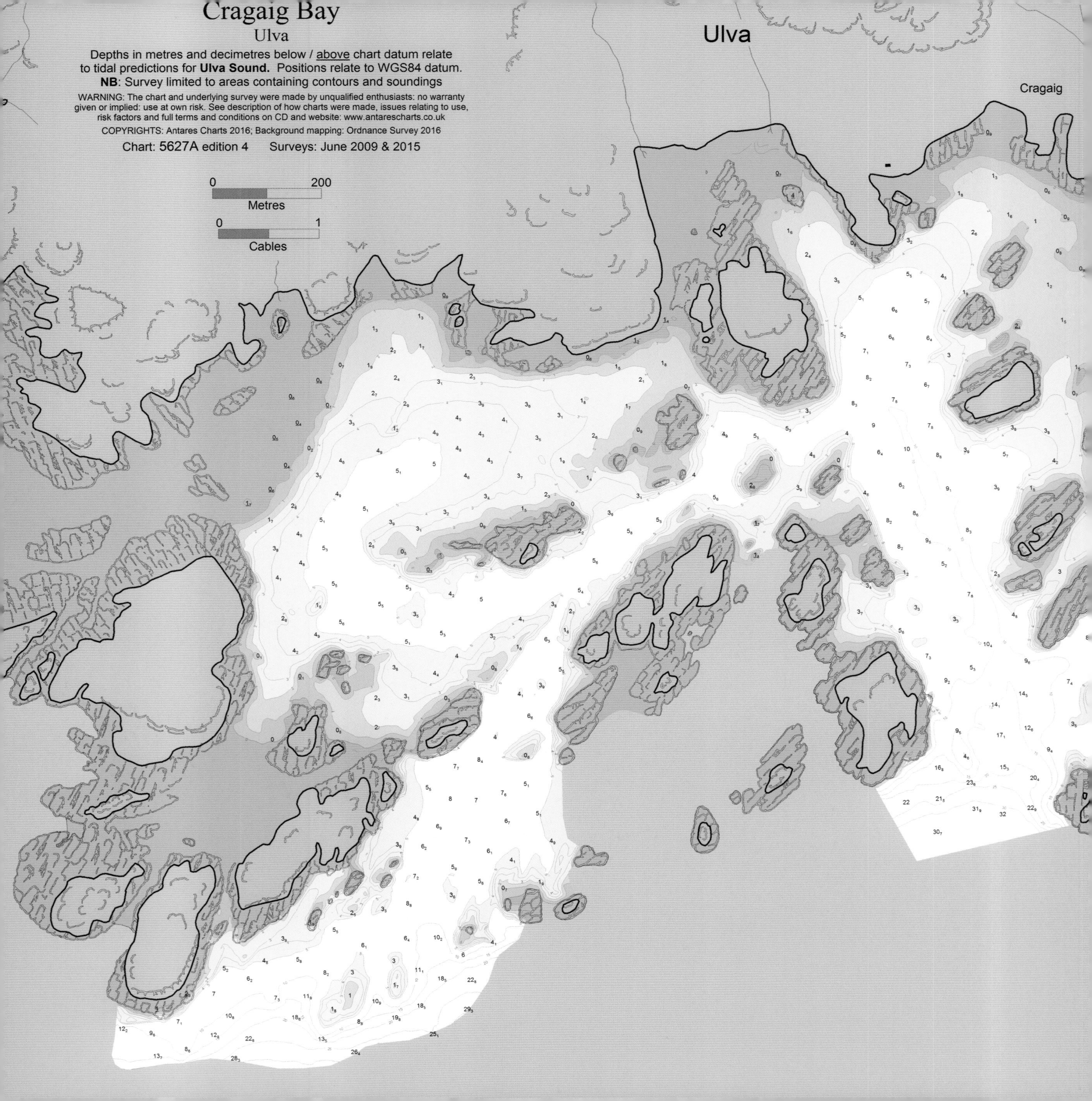

Cragaig Bay
Ulva
Depths in metres and decimetres below / above chart datum relate
to tidal predictions for Ulva Sound. Positions relate to WGS84 datum.
NB: Survey limited to areas containing contours and soundings
WARNING: The chart and underlying survey were made by unqualified enthusiasts: no warranty given or implied: use at own risk. See description of how charts were made, issues relating to use, risk factors and full terms and conditions on CD and website: www.antarescharts.co.uk
COPYRIGHTS: Antares Charts 2016; Background mapping: Ordnance Survey 2016
Chart: 5627A edition 4 Surveys: June 2009 & 2015
0 200
Metres
0 1
Cables
Ulva
Cragaig

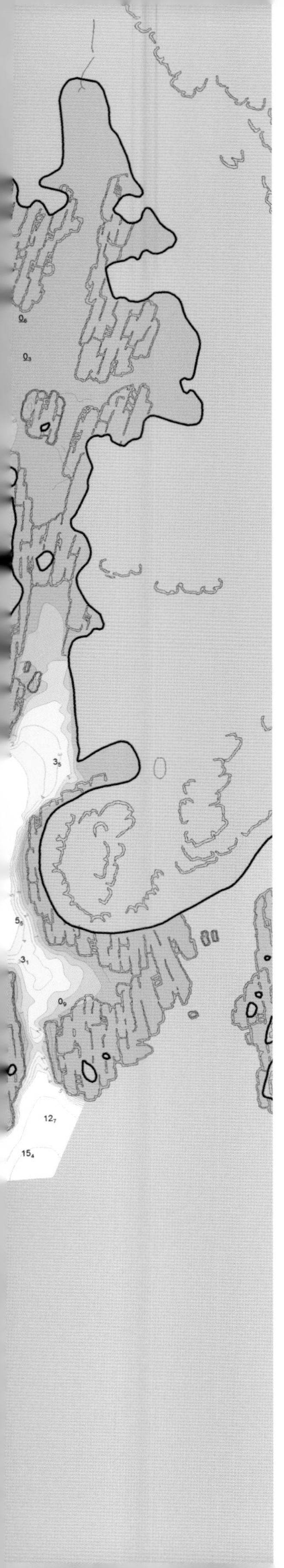

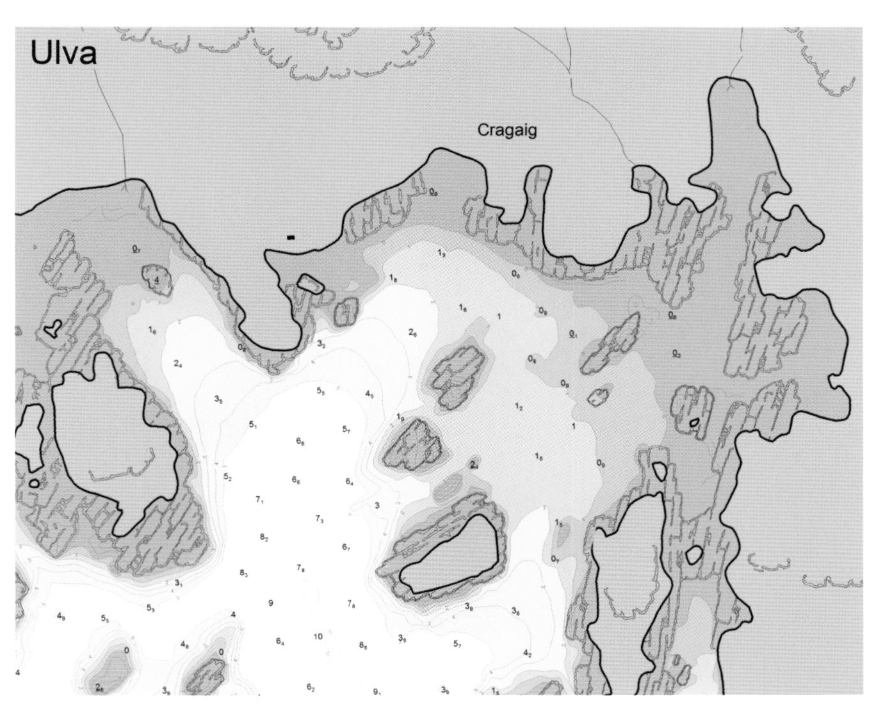

FIGURE 9.12
(*Opposite and above*) In recent decades, the Western Isles have become a very popular destination for yachtsmen, but the best recent Admiralty charts often do not provide sufficient detail for leisure yachting. Owing to the expense of detailed surveying, the usual practice of updating previous charts rather than redrawing afresh, and the priorities given to larger shipping, Admiralty charts may miss hazards close to shore. Bob Bradfield, a former engineer who founded Antares Charts in 2009, has recently addressed this problem as a retirement hobby. He surveys popular anchorages and bays off the west coast of Scotland in much greater detail than the Admiralty by criss-crossing specific anchorages and other popular locations in an inflatable boat, equipped with an accurate GPS and depth sounder. The raw data is brought together and double-checked with a sidescan sonar device and resurveyed as necessary, before being plotted into a readable, very large-scale chart with coloured depth shading. This Antares chart of Cragaig Bay, on the south of the island of Ulva, looks clear at scales of around 1:2,500, but the most detailed current Admiralty chart of this area (Admiralty Chart 2652 of Loch na Keal and Loch Tuath) is at a scale of 1:25,000. Although the Antares charts are 'unofficial', they follow similar colour and style conventions to the Admiralty charts, thereby helping their interpretation and use by their main target audience. By early 2016, Bob Bradfield had published over 300 Antares charts, and, with an already wide and enthusiastic user community, he plans to survey many more areas. *Source*: Antares Charts, *Cragaig Bay, Ulva,* Antares Chart 5627A, ed. 4 (surveyed June 2009 and 2015, published 2016). © Antares Charts. This chart should not be used for navigation; further information can be found at: http://www.antarescharts.co.uk/.

Killin
Union Bank of Scotland
Weighing Machine
Manse
Saw Mill
Sluice
Carding Mill
B.M. 449·9
Well
Monomore
Innis Bhuidhe
Mc Nab's Burial Ground (Disused)
Yellow Cottage
Craobh an Easain
Bridge of Dochart
Millmore (Corn)
B.M. 404·9
Nursery
Well
B.M. 416·2
Falls of Dochart
Clachaig House
Clachaig Check
T.P.
Garbh Innis
240
198
239
238
237
396
424
236
234
235
233
228
227
231
258
232
230
229
226
400
225
224
260
259
403
223
222
220
219
404
221
261
213
410
212

FIGURE 9.13

(*Opposite*) Islands have been used as places of escape for the living, and as burial places – peaceful retreats or final escapes – for the dead. As an example of the latter, the Clan MacNab, whose lands once stretched from south of Loch Tay in Perthshire up the length of Glen Dochart to the west, lost most of their clan lands over time, but were able more recently to re-acquire the ancestral burial ground on the small island of Inch Buie or Innis Bhuidhe in the River Dochart, near the village of Killin. This enlarged extract from a mid-nineteenth-century large-scale Ordnance Survey plan records the MacNab burial site (at the north-east tip of the island) as then disused.

Source: Ordnance Survey, 25 inch to the mile, *Perthshire*, *Sheet LXXX.2*, *Killin Parish* (surveyed 1861, published 1867).

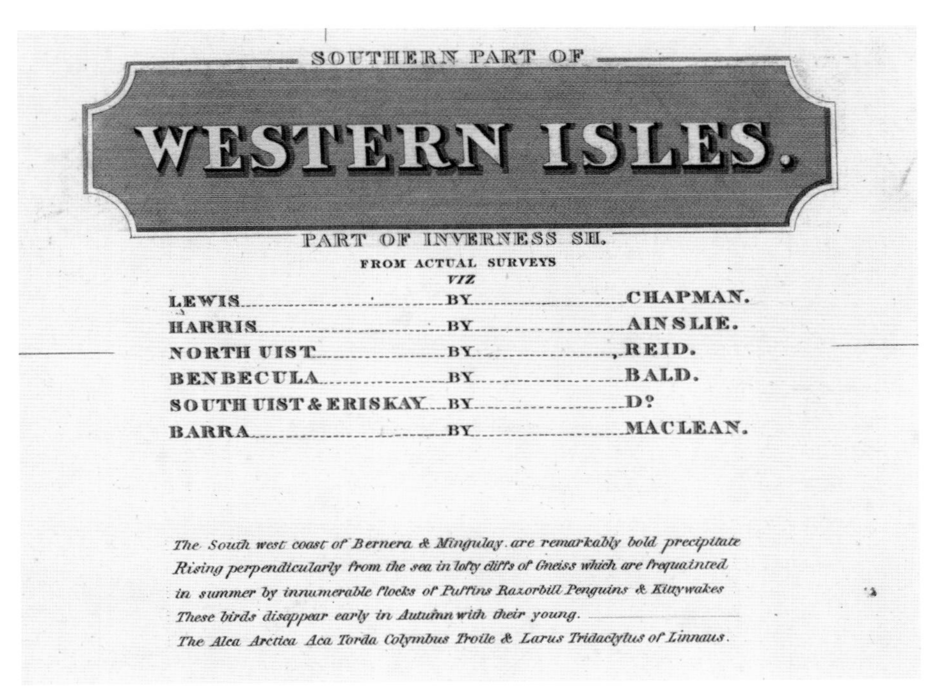

a.

FIGURE 9.14

(*Above and overleaf*) Image (a) is a detail from under the title of one sheet of a triptych of map sheets covering the Western Isles/Na h-Eileanan an Iar. It lists the 'actual surveys' made of these islands during the late eighteenth and early nineteenth centuries and by whom these were undertaken, with the map producer, John Thomson, an Edinburgh-based publisher, acknowledging his reliance on these earlier sources for his own maps. Of the surveys listed, three of the maps produced from these surveys are missing in their original manuscript map form. Robert Reid's original manuscript map of North Uist/Uibhist a Tuath and William Bald's original manuscript maps of Benbecula/Beinn na Faoghla (fig. 6.2), South Uist/Uibhist a Deas and Eriskay/Èirisgeigh survive, as do later, much-reduced and lithographically printed versions of Bald's map of Harris/Na Hearadh (fig. 7.5) and a manuscript copy at reduced scale of James Chapman's map of Lewis/Leòdhas, together with a later, reduced-scale lithographically printed version of the Chapman map (fig. 6.4).

Of the three missing manuscript maps known to have been produced, that of Barra/Barraigh by John Maclean is the only one for which there seems to be either no surviving original, or a later reduced and lithographically printed version. Thomson's map therefore provides us with the only clue as to what Maclean's original survey may have contained in terms of information. Map (b), which covers Barra/Barraigh and the associated islands from Thomson's map, gives us an idea of how much local knowledge must have filled the original lost map as it is crammed with place names, and even manages to squeeze in existing tracks on the island, most, if not all, of which remain on the ground in some form today. This detail, which we may presume to have been copied closely from Maclean's map, shows Maclean had a profound local knowledge of the island. It is yet possible, of course, that John Maclean's original map, or a version of it, will surface. It might lurk hidden in a lawyer's office or in an attic, castle, basement, outhouse or shed, unrecognised for what it is and for its potential significance to historians and geographers. Such reappearances are not unknown.

Source: William Johnson, 'Western Isles', from John Thomson's *Atlas of Scotland* (1832).

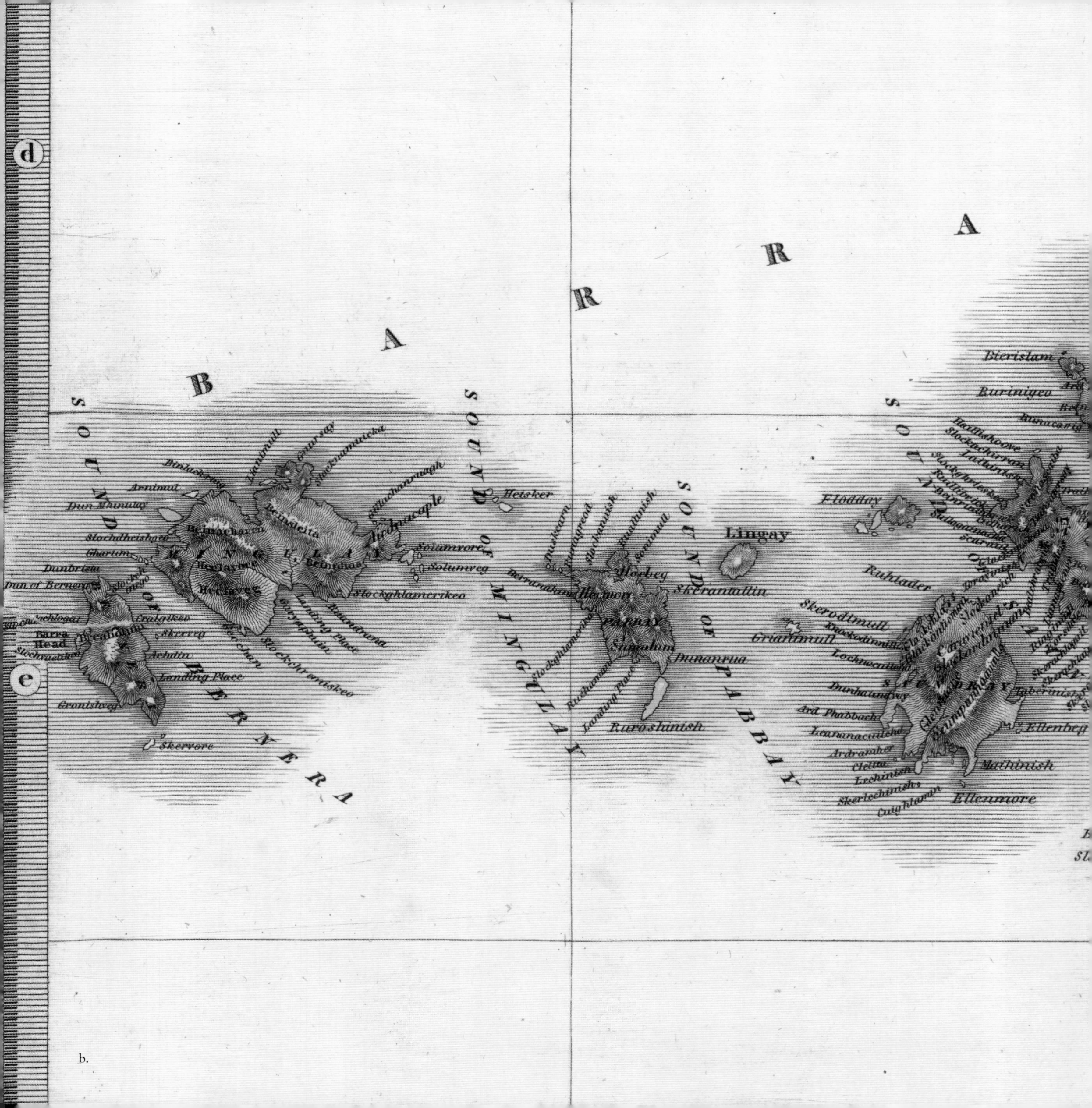
d
e
B A R R A
SOUND OF BERNERA
SOUND OF MINGULAY
SOUND OF PABBAY
Lingay
Flodday
Hetsker
Solumvore
Solumveg
Skerantallin
Ruhlader
Skerodimull
Grianmull
Dunanrua
Ruroshinish
Barra Head
Dun of Bernera
Dunbrista
Arnimul
Gronisheeg
Skervore
Landing Place
Slockhlamerikeo
Ellenmore
Ellenbeg
Maihinish
Bierislam
Buriniged
Ard Phabbach
Lechinish
Cuighlamin
b.

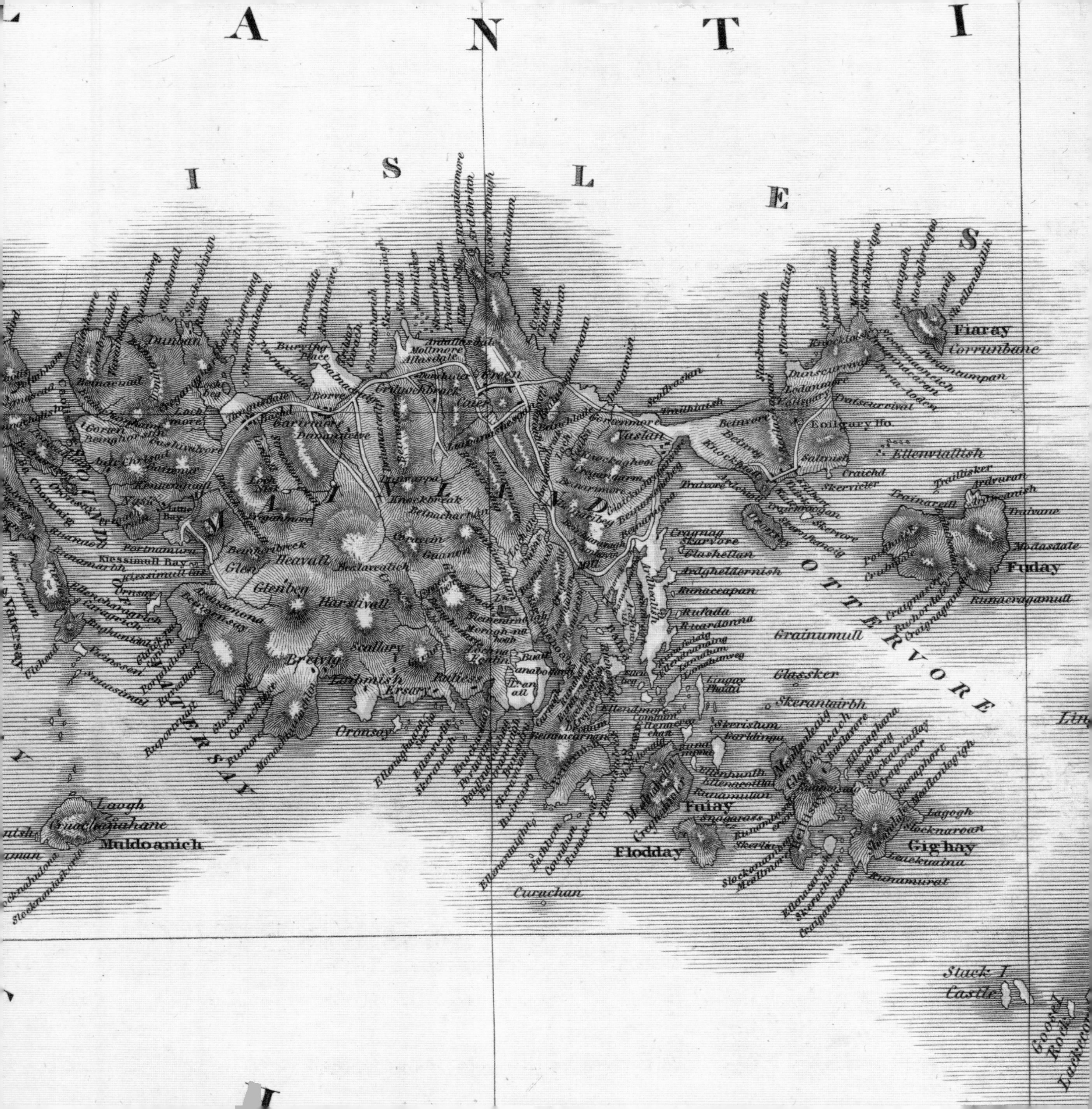
A N T I
I S L E S
MAINLAND
OTTERVORE
VATERSAY
Fiaray
Fuday
Fulay
Flodday
Gighay
Muldoanich
Oronsay
Heavall
Kiessimull Bay
Eoilgary Ho.
Grainumull
Glassker
Skerantairbh
Curachan
Stack I.
Castle

GUIDE TO SOURCES AND FURTHER READING

Chapter 1 Introduction

The quotes from David Greig are from his *Outlying Islands* (London: Faber and Faber, 2002), pages 9 and 1, respectively. There is an extensive literature on map history and the history of cartography. For the fullest up-to-date illustrated accounts, see the several volumes in the *History of Cartography* project by the University of Chicago Press. The idea of 'power' in maps is explored by J. B. Harley, *The New Nature of Maps: Essays in the History of Cartography* (Baltimore: The Johns Hopkins University Press, 2001). Among several essays which explore the complex history and historiography of map making, see J. B. Harley, 'The Map and the Development of the History of Cartography', in J. B. Harley and D. Woodward (eds), *The History of Cartography. Volume 1: Cartography in Prehistoric, Ancient and Medieval Europe and the Mediterranean* (Chicago, IL: University of Chicago Press, 1987), 1–42, and M. Edney, 'Cartography with "Progress": Reinterpreting the Nature and Historical Development of Mapmaking', *Cartographica* 30 (1995), 54–68. On the *isolarii*, see G. Tolias, '*Isolarii*, Fifteenth to Seventeenth Century', in D. Woodward (ed.), *The History of Cartography. Volume 3: Cartography in the European Renaissance* (Chicago, IL: University of Chicago Press, 2007), 263–84. On portolan charts, see T. Campbell, 'Portolan Charts from the Late Thirteenth Century to 1500', in Harley and Woodward, *The History of Cartography. Volume 1*, 371–463 and T. Campbell, 'A Critical Re-examination of Portolan Charts with a Reassessment of Their Replication and Seaboard Function', Map History/History of Cartography website. Available from: http://www.maphistory.info/portolan.html [accessed 18 April 2016].

The best single history of marine map making in Britain, although much of its argument has been extended by more recent research, is A. H. W. Robinson, *Marine Cartography in Britain: A History of the Sea Chart to 1855* (Oxford: Leicester University Press, 1962). The brief discussion and list of marine charts of Scotland produced before 1850 published in D. G. Moir (ed.), *The Early Maps of Scotland to 1850: Volume 2* (Edinburgh: Royal Scottish Geographical Society, 1983), 1–26, should be supplemented by the relevant online map listings of the National Library of Scotland (http://maps.nls.uk). This important national resource is searchable by type and date of map and by map maker. Other important online map repositories include Scot-landsplaces (http://www.scotlandsplaces.gov.uk/), which allows access to the records and map images held by Historic Environment Scotland, the National Records of Scotland and the National Library of Scotland. *OldMapsOnline* (www.oldmapsonline.org) provides a portal to several online map collections. E. Graham, *A Maritime History of Scotland, 1650–1790* (East Linton: Tuckwell Press, 2002) pays brief attention to mapping Scotland's islands.

The brief discussion of the possible origins and purpose of the early map of Lewis and Harris included here as fig. 1.4 is taken from G. A. Hayes-McCoy (ed.), *Ulster and Other Irish Maps, c.1600* (Dublin: Dublin Stationery Office for the Irish Manuscripts Commission, 1964), 34. The broader history and geography of Scotland's maps and mapping is addressed in C. Fleet, M. Wilkes and C. W. J. Withers, *Scotland: Mapping the Nation* (Edinburgh: Birlinn in association with the National Library of Scotland, 2011). C. Delano-Smith and R. J. P. Kain pay scant attention to the mapping of Scotland's islands in their *English Maps: A History* (Toronto: University of Toronto Press, 1999), 224–8, choosing to see the work of Murdoch Mackenzie in the 1770s and Ordnance Survey's work in Orkney and Shetland in the early nineteenth century as elements of Britain's map history. Numerous essays on the work of Ordnance Survey in Scotland appear in *Sheetlines*, the journal of the Charles Close Society for the Study of Ordnance Survey Maps. For an institutional history of Ordnance Survey, see R. R. Oliver, *The Ordnance Survey in the Nineteenth Century: Maps, Money and the Growth of Government* (London: Charles Close Society, 2014). The mapping work of the Department of Geography at the University of Glasgow is discussed in G. Petrie, 'Mapping for the Field Sciences', *Scottish Geographical Journal* 125 (2009), 321–8.

On nissology as the study and coherent theory of islands and islandness in general, see for example F. Péron, 'The Contemporary Lure of the Island', *Tijdschrift voor Economische en Sociale Geografie* 95 (2004), 326–39; G. Baldacchino, 'Islands – Objects of Representation', *Geografiska Annaler* 87B (2005), 247–51; P. Hay, 'A Phenomenology of Islands', *Island Studies Journal* 1 (2006), 19–42; S. A. Royle, *Islands: Nature and Culture* (London: Reaktion Books, 2014). Scotland's many islands have numerous books on their geography and history: see, for example, J. Dow, *Islands Galore: A Handbook of the Scottish Islands* (Edinburgh: Black and White, 2005), and H. Haswell-Smith, *The Scottish Islands: The Bestselling Guide to Every Scottish Island* (Edinburgh: Canongate, 2008 edition). On St Kilda, A. Gannon and G. Geddes, *St Kilda: The Last and Outmost Isle* (Edinburgh: Historic Environment Scotland, 2015) is particularly good on archaeological evidence as well as on former surveys and maps. On the mapping of St Kilda, see M. Harman, *An Isle called Hirte* (Waternish: Maclean Press, 1977), chapter 3.

Chapter 2 Peopling

On the genetic geography of what, later, becomes Scotland, see B. Sykes, *Blood of the Isles: Exploring the Genetic Roots of our Tribal History* (London: Bantam Press, 2006) and S. Oppenheimer, *The Origins of the British: A Genetic Detective Story* (London: Constable, 2006). Scotland's early archaeo-logy and history is the subject of extensive research in the field and in related academic publications: accessible accounts of the general themes and chronology include G. and A. Ritchie, *Scotland: Archaeology and Early History* (Edinburgh: Edinburgh University Press, 1981, with numerous later reprints and editions), I. Armit, *Scotland's Hidden History* (Stroud: Tempus Publishing, 1998); and K. Edwards and I. Ralston (eds), *Scotland After the Ice Age: Environment, Archaeology and History, 8000 BC–1000 AD* (New York and London: Wiley and Sons, 1997). The quote from Principal Gordon is cited in [Principal Gordon], 'Remarks Made in a Journey to the Orkney Islands', *Archaeologia Scotica* 1 (1792), 256–68, quotes from pages 264 and 267 respectively.

On crannogs in general, and Pont's depiction of them in particular, see N. Dixon, *The Crannogs of Scotland: An Underwater Archaeology* (Stroud: Tempus Publishing, 2004) and M. Shelley, 'Timothy Pont and the Freshwater Loch Settlements of Late Medieval and Early Modern Mainland Scotland', *Scottish Geographical Journal* 127 (2011), 108–16. On the recognition of Scotland's history in its geography, see S. Piggott, *Ruins in a Landscape: Essays in Antiquarianism* (Edinburgh: Edinburgh University Press, 1976). The remarks here on the archaeological work of Ordnance Survey and the pioneering endeavours of O. G. S. Crawford are based upon information in R. Hellyer, 'The Archaeological and Historical Maps of the Ordnance Survey', *Cartographic Journal* 26 (1989), 111–33, and K. Hauser, *Bloody Old Britain: O. G. S. Crawford and the Archaeology of Modern Life* (London: Granta, 2008).

The definitive account of Scottish population history remains M. Flinn (ed.), *Scottish Population History from the 17th Century to the 1930s* (Cambridge: Cambridge University Press, 1977): later publications have added to and refined, rather than replaced, its principal findings. On Alexander Webster's 1755 survey, see J. G. Kyd (ed.), *Scottish Population Statistics including Webster's Analysis of Population 1755* (Edinburgh: T. and A. Constable, 1952) and, more recently and critically, M. Anderson, 'Guesses, Estimates and Adjustments: Webster's 1755 "Census" of Scotland Revisited Again', *Journal of Scottish Historical Studies* 31 (2011), 26–45.

There is a voluminous literature on the social and economic history of the Northern and Western Isles, much of it as part of work on 'the Highlands and Islands of Scotland' which fails to differentiate island-by-island differences. For the period before *c.*1840, we have drawn from F. J. Shaw, *The Northern and Western Islands of Scotland: Their Economy and Society in the Seventeenth Century* (Edinburgh: John Donald, 1980); A. I. MacInnes, *Clanship, Commerce and the House of Stuart, 1603–1788* (East Linton: Tuckwell Press, 1996); and R. A. Dodgshon, *From Chiefs to Landlords: Social and Economic Change in the Western Highlands and Islands, c.1493–1820* (Edinburgh: Edinburgh University Press, 1998). Dodgshon is especially attentive to regional and inter-island (even intra-island) differences: much of his book is based on the detailed study of rentals for townships and estates on Islay, Mull, Skye, and Tiree. The statistics on the Small Isles are taken from J. Munby, *Lost Ancestors: Island Families in 1765 on Eigg, Muck, Rum & Canna – An Edition of Neill McNeill's Census of Small Isles Parish, Inner Hebrides, in 1764/5* (Oxford: privately printed, 2007). On famine, clearance and emigration from the islands from the early nineteenth century, see J. Hunter, *The Making of the Crofting Community* (Edinburgh: John Donald, 1976); T. M. Devine, *The Great Highland Famine: Hunger, Emigration and the Scottish Highlands in the Nineteenth Century* (Edinburgh: John Donald, 1988); D. Craig, *On the Crofters' Trail: In Search of the Clearance Highlanders* (London: Jonathan Cape, 1990); E. Cameron, *Land for the People? The British Government and the Scottish Highlands, c.1880–1925* (East Linton: Tuckwell Press, 1998); E. Richards, *A History of the Highland Clearances: Agrarian Transformation and the Evictions 1746–1886* (London: Croom Helm, 1982) and *A History of the Highland Clearances Volume 2: Emigration, Protest, Reasons* (London: Croom Helm, 1985). The quote on Tiree in 1848 is taken from the *Tenth Report by the Glasgow Section of the Central Board on the Fund for Relief of Destitution in the Highlands and Islands of Scotland* (Glasgow: William Eadie and Co. Ltd, 1848), on page 17. The quote on Mingulay, and the figures on the numbers receiving famine relief in the islands in 1849, are taken from *Report on the Outer Hebrides or Long Island, by A Deputation of the Glasgow Section of the Highland Relief Board (August 1849)* (Glasgow: William Eadie, 1849). On the peopling, and re-peopling, of St Kilda, see A. Gannon and G. Geddes, *St Kilda: The Last and Outmost Isle* (Edinburgh: Historic Environment Scotland, 2015).

Chapter 3 Naming

The story of the naming of Rothesay as 'Penis Island' was widely reported in the Scottish media in late August 2015: see, for example, *The Scotsman*, 25 August 2015. On maps, offensive names and renaming (with especial reference to the United States of America), see Mark Monmonier, *From Squaw Tit to Whorehouse Meadow: How Maps Name, Claim, and Inflame* (Chicago, IL: University of Chicago Press, 2006). The extract from Blaeu's typology of island types with reference to Orkney is taken from the 'New Chorographic Description of the Orkneys' in *The Blaeu Atlas of Scotland* (Birlinn, in association with the National Library of Scotland: 2006), page 114.

For a fuller discussion of Speed's names on the maps of the Uists in the Hebrides, see Roger Auger, *An Interpretation of Early Uibhist Maps*, Uist Summer Wine website, available from: http://uistsummerwine.weebly.com/ uist-mapping.html [accessed 8 March 2016]. There are numerous works, in print and online, which detail the origins and meanings of islands' place names: here, we have drawn from William J. Watson, *The History of the Celtic Place-Names of Scotland* (Edinburgh and London: William Blackwood & Sons Ltd, 1926) and James B. Johnston, *Place-Names of Scotland* (Wakefield: S. R. Publishers Ltd, 1976) [first published 1892].

The work of Ordnance Survey in naming places is explored in several publications. The evolution of Survey policy in Wales is discussed in J. B. Harley and G. Walters, 'Welsh Orthography and Ordnance Survey Mapping, 1820–1905', *Archaeologia Cambrensis* 131 (1982), 98–135. Of Ireland, see the detailed work of John Andrews, chiefly in his *A Paper Landscape: The Ordnance Survey in Nineteenth-Century Ireland* (Oxford: Clarendon Press, 1975). Of Gaelic Scotland, see M. Robson, 'The Living Voice' in F. MacLeod (ed.), *Togail Tìr: Marking Time, The Map of the Western Isles* (Stornoway: Acair, 1989), 97–104, and C. W. J. Withers, 'Authorizing Landscape: "Authority", Naming and the Ordnance Survey's Mapping of the Scottish Highlands in the Nineteenth Century', *Journal of Historical Geography* 26 (2000), 532–54.

Sorley MacLean's 'Hallaig' appears in the collected edition of his poems *O Choille go Bearradh* (*From Wood to Ridge: Collected Poems in Gaelic and English*) (Edinburgh: Carcanet Press, 1999) and is also available online, in its original Gaelic, and in translation by Seamus Heaney, the late Irish poet and Nobel Laureate. Community mapping in Lewis is discussed in R. F. Freeman, 'Place Identity in Lower Shader, Isle of Lewis: The Evidence of Mapping, Narrative and Communal Activities', unpublished M.Phil. thesis (University of Edinburgh, 1997). Finlay

MacLeod includes a memory map of Shader on Lewis in his *Togail Tìr* (1989). The Ness Historical Society on Lewis and other community groups are doing detailed important work on local names on the island. This and other work is discussed by Robert Macfarlane in his *Landmarks* (London: Penguin, 2015) as part of his rediscovery of Britain's 'lost' or remembered toponomy. We acknowledge the kind assistance of Finlay Macleod in this chapter. The clash between poetic and remembered landscape and the abstract language of profit over the mountain of Roineabhal on Harris is a theme explored in A. F. D. Mackenzie, '"The Cheviot, The Stag . . . and the White, White Rock?": Community, Identity, and Environmental Threat on the Isles of Harris', *Environment and Planning D: Society and Space* 18 (1998), 509–32.

Chapter 4 Navigating

For a summary of the history of maps in navigating, see J. R. Akerman, 'Finding our Way', in J. R. Akerman and R. W. Karrow Jr. (eds), *Maps: Finding Our Place in the World* (Chicago, IL: University of Chicago Press, 2007), 19–64. The phrase about the commonest use of the commonest types of maps is taken from C. Delano-Smith, 'Milieus of Mobility: Itineraries, Route Maps, and Road Maps', in J. R. Akerman (ed.), *Cartographies of Travel and Navigation* (Chicago, IL: University of Chicago Press, 2006), 16–68 (quote from page 16). Sibbald's geographical work and his relationship with John Adair is discussed in C. W. J. Withers, *Geography, Science and National Identity: Scotland since 1520* (Cambridge: Cambridge University Press, 2001) from which work (page 74) the extract from Sibbald's 1683 *Account of the Scotish Atlas* is taken. Adair's charting work is examined in J. N. Moore, 'John Adair's Contribution to the Charting of the Scottish Coasts: A Re-assessment', *Imago Mundi* 52 (2000), 43–65. Graham's *Maritime History of Scotland, 1650–1790* is unjustly critical of Adair's work. Mackenzie is rightly lauded in Robinson's *Marine Cartography in Britain* (pages 61–2), misrepresented by Graham, *Maritime History of Scotland, 1650–1790* (page 301) and, of his Orkney mapping especially, studied fully and fulsomely by D. C. F. Smith, 'The Progress of the *Orcades* Survey, with Biographical Notes on Murdoch Mackenzie Senior (1712–1797)', *Annals of Science* 44 (1987), 277–88.

There is a large literature on Scotland's lighthouses though only a few works pay attention to their cartographic representation. For good overall accounts, see B. Bathurst, *The Lighthouse Stevensons* (London: Harper, 1999), and A. Morrison-Low, *Northern Lights: The Age of Scottish Lighthouses* (Edinburgh: National Museum of Scotland, 2010). The quote from the Parliamentary Commission over the Bell Rock is from the *Memorial of the Commissioners appointed by Act of Parliament for erecting Lighthouses in the Northern Parts of Great Britain, relative to the Erection of a Lighthouse upon the Cape or Bell Rock* (Edinburgh: Printed for D. Willison, Craig's Close, 1806), page 11.

The commentary on Dalrymple's method of sea triangulation and his tenure as Hydrographer to the Admiralty is from Robinson, *Marine Cartography in Britain*, 128–30. G. S. Ritchie, *The Admiralty Chart: British Naval Hydrography in the Nineteenth Century* (Edinburgh: The Pentland Press, 1995 edition) gives useful detail on Dalrymple and the Hydrographic Office. The French metre-based mapping scheme, which involved the Orkneys and Shetland from 1817, is discussed at length by K. Alder, *The Measure of All Things: The Seven-Year Odyssey and Hidden Error That Transformed the World* (London: Little, Brown, 2002). The French–British mapping disputes in the Northern Isles are discussed by Rachel Hewitt, *Map of a Nation: A Biography of the Ordnance Survey* (London: Granta, 2010), 226–33. This episode, and the variably paced triangulation of Scotland in the first half of the nineteenth century, is the subject of articles by D. Walker: 'The Initial Triangulation of Scotland from 1809 until 1822', *Sheetlines* 98 (2013), 5–15; 'Balta Sound and the Figure of the Earth', *Sheetlines* 99 (2014), 5–17; 'The Troubled Progress of the Scottish Triangulation 1823–1858', *Sheetlines* 100 (2015), 5–18.

For the work on islands as bird observatories, see W. Eagle Clarke, 'Bird Migration in the British Isles: Its Geographical and Meteorological Aspects', *Scottish Geographical Magazine* 12 (1896), 616–26; W. Eagle Clarke, *Studies in Bird Migration*, two volumes (London and Edinburgh: Gurney and Jackson and Oliver and Boyd, 1912); and J. A. Love, *A Natural History of Lighthouses* (Dunbeath: Whittles Publishing, 2015), chapter 11. Chapter 14 of C. R. Perkins and R. B. Parry, *Mapping the UK* (London: Bowker Saur, 1996) looks at aeronautical charts from a British context. For a brief summary of aero-nautical charts in international context, see 'Aeronautical Chart' in M. Monmonier (ed.), *The History of Cartography Volume 6: Part 1* (Chicago, IL: University of Chicago Press, 2015), 22–30.

Chapter 5 Defending

A general overview of Scottish military mapping can be found in Chapter 4 'Scotland Occupied and Defended' of C. Fleet, M. Wilkes and C. W. J. Withers, *Scotland: Mapping the Nation* (Edinburgh: Birlinn in association with the National Library of Scotland, 2011). For Pont's depiction of crannogs, see M. Shelley, 'Timothy Pont and the Freshwater Loch Settlements of Late Medieval and Early Modern Mainland Scotland', *Scottish Geographical Journal* 127 (2011), 108–16. For more details on the crannogs of Loch Tay, see N. Dixon, *The Crannogs of Loch Tay* (Edinburgh: The Scottish Trust for Underwater Archaeology, 2000). Although mapping is only one source of evidence for brochs in Scotland, I. Armit, *Towers in the North: The Brochs of Scotland* (Stroud: Tempus, 2003) and D. W. Harding, *The Iron Age in Northern Britain* (London: Routledge, 2004) provide good summaries of recent research. On the work of Captain Fred Thomas on brochs and other early buildings in the Northern and Western Isles, see F. W. L. Thomas, 'On the Primitive Dwellings and Hypogea of the Outer Hebrides', *Proceedings of the Society of Antiquaries of Scotland* 7 (1866–88), 153–95. The quotation from the Blaeu Atlas of Scotland (1654) was translated by Ian Cunningham, *The Blaeu Atlas of Scotland* (Birlinn, in association with the National Library of Scotland, 2006).

On the work of Lewis Petit, see C. Fleet, 'Lewis Petit and his Plans of Scottish Fortifications and Towns, 1714–16', *Cartographic Journal* 44 (2007), 329–41. On the Board of Ordnance in Scotland, see C. J. Anderson, 'State Imperatives: Military Mapping in Scotland, 1689–1770', *Scottish Geographical Journal* 125 (2009), 4–24, and C. J. Anderson, 'Cartography and Conflict: The Board of Ordnance and the Construction of the Military Landscape of Scotland, 1689–1815', in B. Lenman (ed.), *Military Engineers and the Development of the Early-Modern European State* (Dundee: Dundee University Press, 2013), 131–52. The quote is from B. Lenman, *The Jacobite Clans of the Great Glen 1650–1784* (London: Methuen, 1984). Roy's 'Great Map' is available online at http://maps.nls.uk/roy/ and as a facsimile edition: [W. Roy], *The Great Map: The Military Survey of Scotland, 1747–1755* [with introductory essays by Y. Hodson, C. Tabraham and C. W. J. Withers] (Edinburgh: Birlinn in association with the National Library of Scotland, 2007). R. Hewitt, 'A Family Affair: The Dundas Family of Arniston and the Military Survey of Scotland', *Imago Mundi* 64 (2012), 60–77 includes research on the family links and aristocratic support for the Military Survey.

A detailed account of the sinking of HMS *Royal Oak* is provided by D. Turner, *Last Dawn: the HMS Royal Oak Tragedy at Scapa Flow* (Argyll: Argyll Publishing, 2008). For accounts of the use of the Hebrides for military rocket and aircraft testing, see F. MacDonald, 'The Last Outpost of Empire: Rockall and the Cold War', *Journal of Historical Geography* 32 (2006), 627–47; F. MacDonald, 'Geopolitics and "The Vision Thing": Regarding Britain and America's First Nuclear

Missile', *Transactions of the Institute of British Geographers* 30 (2006), 53–71.

The Russian military mapping of other countries is discussed by J. Davies, 'Uncle Joe Knew Where You Lived – Part 1', *Sheetlines* 72 (April 2005); Part 2 in *Sheetlines* 73 (August 2005). D. Watt, 'Soviet Military Mapping', *Sheetlines* 74 (December 2005) describes the history of the Russian Military Topographic Directorate (VTU) from 1812 to the present day. All these and other resources are available at https://www.sovietmaps.com/.

Chapter 6 Improving

A general overview of Scottish estate mapping and rural improvement can be found in chapter 6 of C. Fleet, M. Wilkes and C. W. J. Withers, *Scotland: Mapping the Nation* (Edinburgh: Birlinn in association with the National Library of Scotland, 2011). R. Gibson, *The Scottish Countryside: Its Changing Face, 1700–2000* (Edinburgh: John Donald in association with the National Archives of Scotland, 2007) is especially good on the unpublished maps held by the National Records of Scotland, and the primary purposes behind their creation, with specific examples relating to division of commonty, enclosure, clearances, land settlement, planned villages and forestry. A valuable source for identifying who was at work as a land surveyor before the mid nineteenth century is S. Bendall (ed.), *Dictionary of Land Surveyors and Local Map-Makers of Great Britain and Ireland, 1530–1850* (London: British Library, 1997 edition). On estate maps in local history, see D. Smith, *Maps and Plans for the Local Historian and Collector* (London: Batsford, 1988), chapter 4.

The quote from Martin Martin is from his *A Description of the Western Islands of Scotland* (London: printed for Andrew Bell, 1703). For an informed description of this account and on other improvers, see T. C. Smout, 'A New Look at the Scottish Improvers', *Scottish Historical Review* 91 (2012), 125–49. On estate surveyors in the Western Isles, see J. B. Caird, 'The Contribution of Historical Geography to the Understanding of Land Use Patterns and Population Distribution in the Outer Hebrides, Scotland', *Contributi Geografici* (1978), 61–78; 'Land Use in the Uists since 1800', *Proceedings of the Royal Society of Edinburgh* 77B (1979), 505–26; and 'Early 19th Century Estate Plans', in F. Macleod (ed.), *Togail Tir Marking Time: The Map of the Western Isles* (Stornoway: Acair, 1989). On William Bald, see Margaret C. Storrie, 'William Bald, F.R.S.E., *c.*1789–1857: Surveyor, Cartographer and Civil Engineer', *Transactions of the Institute of British Geographers* 47 (1969), 205–31. The development of agriculture in Orkney is discussed in D. Omand (ed.), *The Orkney Book* (Edinburgh: Birlinn, 2003). The quote on the Fife Adventurers is cited in T. C. Smout and M. Stewart, *The Firth of Forth: An Environmental History* (Edinburgh: Birlinn, 2012). B. Lawson, *Lewis in History and Legend: the West Coast* (Edinburgh: Birlinn, 2008) is useful on the Port of Ness. For the development of Bowmore, see M. Storrie, *Islay: Biography of an Island* (Isle of Islay: The Oa Press, 2011 edition). For recent research on John Wood, and information about his work in Stornoway, see B. Robson, 'John Wood 1: The Undervalued Cartographer', and 'John Wood 2: Planning and Paying for his Town Plans', *Cartographic Journal* 51 (2014), 257–86.

Chapter 7 Exploiting

On the history of sea fishing in Scotland, see J. R. Coull, *The Sea Fisheries of Scotland: A Historical Geography* (Edinburgh: John Donald, 1996). D. Flinn, *Travellers in a Bygone Shetland: An Anthology* (Edinburgh: Scottish Academic Press, 1989), discusses visitors to Shetland who made maps: chapter 5 on the Dutch fisheries, chapter 7 on Arctic whaling and chapter 9 on Samuel Hibbert. Biographical information, as well as details on the charts of Colom and Blaeu, and the disputes between them, are discussed in C. Koeman (ed.), *Atlantes Neerlandici: Bibliography of Terrestrial, Maritime and Celestial Atlases and Pilot Books, Published in the Netherlands up to 1880* (Amsterdam: Theatrum Orbis Terrarum, 1967–85). The quotation from the Blaeu Atlas of Scotland (1654) was translated by Ian Cunningham, *The Blaeu Atlas of Scotland* (Birlinn, in association with the National Library of Scotland, 2006).

For details of the Sula Sgeir egg fisheries and Niseach fishermen, see B. Lawson, *Lewis in History and Legend: the East Coast* (Edinburgh: Birlinn, 2011). L. K. Schei and G. Moberg, *The Shetland Isles* (Grantown-on-Spey: Wendy Price, 2006) provide information on the fisheries of Balta Sound. See chapter 6 further reading (above) for sources on William Bald and his work in the Western Isles. Descriptions of many islands, including Harris, and their potential for future fisheries, appear in J. Knox, *A Tour through the Highlands of Scotland, and the Hebride isles, in MDCCLXXXVI* (London: Printed for J. Walter, Charing-Cross; R. Faulder, New Bond-Street; W. Richardson, Royal Exchange; Edinburgh: W. Gordon and C. Elliot; Glasgow: Dunlop and Wilson, 1787).

H. Haswell-Smith, *The Scottish Islands: The Bestselling Guide to Every Scottish Island* (Edinburgh: Canongate, 2008 edition) describes the geology of each island and the more significant quarries. On geological mapping by the Highland and Agricultural Society, see R. C. Boud, 'Agriculture and Geology: The Cartographic Activities of the Highland and Agricultural Society of Scotland, 1832–1875', *Bulletin of the Society of University Cartographers* 20 (1986), 7–15. On Hibbert's work in Shetland, see R. C. Boud, 'Samuel Hibbert and the Early Geological Mapping of the Shetland Islands', *Cartographic Journal* 14 (1977), 81–8. R. C. Boud, 'The Early Geological Maps of the Isle of Arran, 1807–1858', *Canadian Cartographer* 12 (1975), 179–93 describes the pioneers of geological mapping in an island often referred to as a 'geologists' paradise' because it shows in miniature much of the geology of Scotland. The mining at Easdale is described in the Statistical Accounts and in F. C. Groome's *Ordnance Gazetteer of Scotland: A Survey of Scottish Topography, Statistical, Biographical, and Historical* (Edinburgh: T. C. Jack, 1883–85).

Two detailed descriptions of the Harris superquarry, both written by opponents of it, are M. Scott and S. Johnston, *The Battle for Roineabhal: Reflections on the Successful Campaign to Prevent a Superquary at Lingerbay, Isle of Harris, and Lessons for the Scottish Planning System* (Perth: Scottish Environment LINK, 2006), available from: http://www.scotlink.org/pdf/ Lingerabay_hi-res.pdf [accessed 21 March 2016], and A. McIntosh, *Lafarge Aggregates Ltd (ex Redland Aggregates) Isle of Harris Superquarry Briefing*, available from: http://www.alastairmcintosh.com/general/quarry_briefing.htm [accessed 21 March 2016].

Link Quarry Group, *The Case against the Harris Superquarry* (Scotland: Link Quarry Group, 1996) states the original planning objections to the quarry.

The protest that followed the conversion of crofting land into shooting estates and deer forests, and earlier the loss of arable land and crofting smallholdings to sheep farms, is discussed in J. Hunter, *The Making of the Crofting Community* (Edinburgh: John Donald, 1976) and, for the twentieth century in Pairc, Lewis and throughout the islands, in I. J. M. Robertson, *Landscapes of Protest in the Scottish Highlands after 1914* (Ashgate: Farnham, 2013). A. Wightman, *The Poor Had No Lawyers: Who Owns Scotland (And How They Got It)* (Edinburgh: Birlinn, 2010) analyses the background to the rise in sporting estates in Scotland. The map in fig. 7.10a was originally printed for A. Geddes, 'Lewis', *Scottish Geographical Magazine* 52 (1936), 217–31 and subsequently included in his book *The Isle of Lewis and Harris: A Study in British Community* (Edinburgh: Edinburgh University Press, 1955). For an informative account of Leverhulme and Lewis, see R. Hutchinson, *The Soap Man: Lewis, Harris and Lord Leverhulme* (Edinburgh: Birlinn, 2003).

Chapter 8 Picturing

For fuller exposition of the connections here made between art making, map making and the geography of art, see Thomas DaCosta Kaufman, *Toward a Geography of Art* (Chicago, IL: University of Chicago Press, 2004). The place of the cartouche as a feature of maps in the early modern period especially is the subject of James A. Welu, 'The Sources and Development of Cartographic Ornamentation in the Netherlands' in David Woodward (ed.), *Art and Cartography: Six Historical Essays* (Chicago, IL: University of Chicago Press, 1987), 147–73. On the connections between Dutch mapping and art, see Svetlana Alpers in the Woodward volume (pages 51–96), and for a history of use of colour in cartographic depiction, see also the essay by Ulla Ehrensvärd, 'Color in Cartography: A Historical Survey', also in the Woodward volume above (pages 123–46). On the ways maps were used as pictures and to picture space in different ways, see Charles W. J. Withers, 'Art, Science, Cartography and the Eye of the Beholder', *Journal of Interdisciplinary History* 42 (2012), 429–37. For more information on the Van Keulen family and firm and their contributions to navigation and cartography, see Part I of Dirk de Vries et al. (eds), *The Van Keulen Cartography, Amsterdam 1680–1885* (Alphen aan den Rijn: Uitgeverij Canaletto/Repro-Holland, 2005). The connections between map making and painting in the work of Paul Sandby are discussed in Nicholas Alfrey and Stephen Daniels (eds), *Mapping the Landscape: Essays on Art and Cartography* (Nottingham: University Art Gallery/Castle Museum, 1990), and more recently in John Bonehill and Stephen Daniels (eds), *Paul Sandby – Picturing Britain* (London: Royal Academy of Arts, 2009). For other recent discussions of the connections between art and cartography, see Edward. S Casey, *Earth Mapping: Artists Reshaping Landscape* (Minneapolis, MN: University of Minnesota Press, 2005), and Katharine Harmon, *The Map as Art: Contemporary Artists Explore Cartography* (New York: Princeton Architectural Press, 2009). A detailed history of the genesis and development of Bartholomew's 'half-inch to the mile' series can be found in T. Nicholson, 'Bartholomew and the Half-Inch Layer Coloured Map 1883–1903', *Cartographic Journal* 37, (2000), 123–45. The extracts from the poetry of Norman MacCaig are taken, respectively, from Norman MacCaig, 'Centre of Centres' (February 1971), and from Norman MacCaig, 'Old Maps and New' (November 1970), both of which are in E. MacCaig (ed.), *The Poems of Norman MacCaig* (Edinburgh: Polygon, 2005).

Chapter 9 Escaping

Alan MacQuarrie's *The Saints of Scotland: Essays in Scottish Church History, AD 450–1093* (Edinburgh : John Donald, 1997) examines, through a collection of essays, some of the major Scottish saints within a broader context of religious development during this period. The 'Saints in Scottish Place-Names' website http://saintsplaces.gla.ac.uk/index.ph [accessed 19 May 2016] presents a fully searchable database and map of Scottish hagiotoponyms, including 5,000 places, 13,000 place names and some 750 saints potentially commemorated in these names. R. Oliver and R. Hellyer's *A Guide to the Ordnance Survey One-inch Third Edition Maps, in Colour: England and Wales, Scotland, Ireland* (London: Charles Close Society, 2004) provides a fully researched background essay and cartobibliography of this series of OS maps, illustrated in fig. 9.6a. For further details on James Grante's mapping of Bonnie Prince Charlie's routes, see Rodney W. Shirley, *Printed Maps of the British Isles 1650–1750* (Tring and London: Map Collector Publications and the British Library, 1988), 61–4. George Orwell's narrow escape from the Corryvreckan whirlpool is described and discussed in Bernard Crick's *George Orwell: A Life* (London: Secker & Warburg, 1980). G. Walters, 'Thomas Pennant's Map of Scotland, 1777: A Study in Sources, and an Introduction to George Paton's Role in the History of Scottish Cartography', *Imago Mundi* 28 (1976), 121–8, provides a full account of the map illustrating Pennant's 1772 Hebridean voyage, as well as details of his correspondence, which illuminates the story behind the map. For a recent, detailed and well-illustrated account of railway mapping in Scotland see David Spaven, *The Railway Atlas of Scotland: Two Hundred Years of History in Maps* (Edinburgh: Birlinn in association with the National Library of Scotland, 2015). Further details of Antares Charts can be found on their website at: http://www.antares charts.co.uk/ [accessed 18 May 2016]. More background to 'missing' maps can be read in Margaret Wilkes' chapter 'Missing, Presumed Lost' in F. MacLeod (ed.), *Togail Tìr Marking Time: The Map of the Western Isles* (Stornoway: Acair, 1989), 43–8. John Thomson's *Atlas of Scotland* 1832 is available online at http://maps.nls.uk/atlas/ thomson/index.html and in facsimile form: [John Thomson], *The Atlas of Scotland: Containing Maps of Each County* (Edinburgh: Birlinn in association with the National Library of Scotland, 2008). The introductory essays in this volume by C. W. J. Withers, C. Fleet and P. Williams provide both a broader context to the *Atlas*, as well as more detail on the sources of the maps and the production of the *Atlas* itself.

INDEX

Items in **bold** indicate figures or the caption text relating to figures.